Examining the Environmental Impacts of Materials and Buildings

Blaine Erickson Brownell
University of Minnesota, USA

A volume in the Practice, Progress, and Proficiency in Sustainability (PPPS) Book Series

IGI Global
DISSEMINATOR OF KNOWLEDGE

Published in the United States of America by
IGI Global
Engineering Science Reference (an imprint of IGI Global)
701 E. Chocolate Avenue
Hershey PA, USA 17033
Tel: 717-533-8845
Fax: 717-533-8661
E-mail: cust@igi-global.com
Web site: http://www.igi-global.com

Library of Congress Cataloging-in-Publication Data

Names: Brownell, Blaine Erickson, 1970- editor.
Title: Examining the environmental impacts of materials and buildings / edited by Blaine Erickson Brownell.
Description: Hershey, PA : Engineering Science Reference, [2020] | Includes bibliographical references and index. | Summary: "This book explores the environmental impact of building design, construction, maintenance, demolition, and related activities"-- Provided by publisher.
Identifiers: LCCN 2019039278 | ISBN 9781799824268 (hardcover) | ISBN 9781799824275 (paperback) | ISBN 9781799824282 (ebook)
Subjects: LCSH: Building materials--Environmental aspects. | Construction and demolition debris--Environmental aspects. | Architecture--Environmental aspects. | Sustainable construction.
Classification: LCC TD196.B85 E93 2020 | DDC 324.1/20286--dc23
LC record available at https://lccn.loc.gov/2019039278

This book is published in the IGI Global book series Practice, Progress, and Proficiency in Sustainability (PPPS) (ISSN: 2330-3271; eISSN: 2330-328X)

British Cataloguing in Publication Data
A Cataloguing in Publication record for this book is available from the British Library.

Practice, Progress, and Proficiency in Sustainability (PPPS) Book Series

ISSN:2330-3271
EISSN:2330-328X

Editor-in-Chief: Ayman Batisha, International Sustainability Institute, Egypt

MISSION

In a world where traditional business practices are reconsidered and economic activity is performed in a global context, new areas of economic developments are recognized as the key enablers of wealth and income production. This knowledge of information technologies provides infrastructures, systems, and services towards sustainable development.

The **Practices, Progress, and Proficiency in Sustainability (PPPS) Book Series** focuses on the local and global challenges, business opportunities, and societal needs surrounding international collaboration and sustainable development of technology. This series brings together academics, researchers, entrepreneurs, policy makers and government officers aiming to contribute to the progress and proficiency in sustainability.

COVERAGE

- Green Technology
- Sustainable Development
- Technological learning
- Socio-Economic
- Outsourcing
- Innovation Networks
- E-Development
- Eco-Innovation
- Strategic Management of IT
- Environmental informatics

IGI Global is currently accepting manuscripts for publication within this series. To submit a proposal for a volume in this series, please contact our Acquisition Editors at Acquisitions@igi-global.com or visit: http://www.igi-global.com/publish/.

The Practice, Progress, and Proficiency in Sustainability (PPPS) Book Series (ISSN 2330-3271) is published by IGI Global, 701 E. Chocolate Avenue, Hershey, PA 17033-1240, USA, www.igi-global.com. This series is composed of titles available for purchase individually; each title is edited to be contextually exclusive from any other title within the series. For pricing and ordering information please visit http://www.igi-global.com/book-series/practice-progress-proficiency-sustainability/73810. Postmaster: Send all address changes to above address.

Titles in this Series

For a list of additional titles in this series, please visit:
https://www.igi-global.com/book-series/practice-progress-proficiency-sustainability/73810

Cross-Border Cooperation (CBC) Strategies for Sustainable Development
Rui Alexandre Castanho (University of Dąbrowa Górnicza, Portugal)
Engineering Science Reference • © 2020 • 348pp • H/C (ISBN: 9781799825135) • US $225.00

Membrane Technology for Water and Wastewater Treatment in Rural Regions
Rosalam Sarbatly (Universiti Malaysia Sabah, Malaysia)
Engineering Science Reference • © 2020 • 354pp • H/C (ISBN: 9781799826453) • US $205.00

Smart Materials for Environmental and Industrial Applications
Khaled Salah Abou-El-Sherbini (National Research Centre, Egypt) Nasrallah Mohamed Deraz (National Research Centre, Egypt) and Alia Hosni Salama (National Research Center, Egypt)
Engineering Science Reference • © 2020 • 300pp • H/C (ISBN: 9781799829317) • US $215.00

Handbook of Research on Health and Environmental Benefits of Camel Products
Omar Amin Alhaj (University of Petra, Jordan) Bernard Faye (Independent Researcher, France) and Rajendra Prasad Agrawal (Independent Researcher, India)
Medical Information Science Reference • © 2020 • 480pp • H/C (ISBN: 9781799816041) • US $285.00

Interdisciplinary Approaches to Public Policy and Sustainability
Rituparna Das (Adamas University, India) and Nivedita Mandal (Adamas University, India)
Information Science Reference • © 2020 • 320pp • H/C (ISBN: 9781799803157) • US $195.00

African Perspectives on Reshaping Rural Development
Mavhungu Abel Mafukata (Department of Development Studies, University of South Africa, South Africa) and Khathutshelo Alfred Tshikolomo (Limpopo Department of Agriculture and Rural Development, South Africa)
Information Science Reference • © 2020 • 403pp • H/C (ISBN: 9781799823063) • US $195.00

IGI Global
DISSEMINATOR OF KNOWLEDGE

701 East Chocolate Avenue, Hershey, PA 17033, USA
Tel: 717-533-8845 x100 • Fax: 717-533-8661
E-Mail: cust@igi-global.com • www.igi-global.com

This book is dedicated to the memory of John Carmody, a pioneer in sustainable building research.

Editorial Advisory Board

Table of Contents

Detailed Table of Contents

It is only in the case of fire that materials are considered by the American International Building Codes across an aggregation of scales (i.e., a building, a block, a district) leaving many other essential factors of material performance neglected. Mostly ignored are environmental and social parameters that also present forms of risk. This chapter uses the cities of New York and Chicago and three performance characteristics as case studies to examine additional material impacts at the city-scale. Case studies analyze material maintenance requirements against urban disinvestment, moisture absorption capacity against mold rates within flood-prone communities, and embodied carbon against material lifespan averages across cities. Findings reveal connections between material performance and economic, health, and energy implications across the city, suggesting the need for more broadly defined urban material performance standards.

Replacing older homes with new ones constructed to higher efficiency standards is one way to raise the operating efficiency of building stocks. However, new buildings require large amounts of embodied energy to construct, and it can take years before more efficient operations offset carbon emissions associated with new construction. This chapter looks at the carbon dioxide emission payback period of newly constructed, efficient single-family homes in Vancouver, British Columbia, where the authors find that it takes over 150 years for the operation to equal the embodied carbon associated with the of a typical high-efficiency new home. The findings suggest that current policies aimed at reducing emissions by replacing older homes with new high-efficiency buildings should be reconsidered.

Section 2
Material Topics

This section consists of in-depth studies of various material topics. The chapters offer insights regarding the environmental performance of plastics, cementitious materials, and water in building construction. Approaches include linking disparate fields of inquiry, proposing a new assessment methodology, and developing and testing a new composite. These works demonstrate that in-depth explorations of specific materials can contribute to the broader knowledge base of building construction as a whole.

The goal of sustainable design and development is threefold, including economic, environmental, and social sustainability. While there are well-established methods for assessing the economic and environmental performance of products and buildings, the determination of social performance is less clear. This chapter explores the emerging field of social life cycle assessment (S-LCA), particularly as it relates to building materials and construction. This chapter includes 1) an introduction to and overview of S-LCA, summarized case studies of S-LCA; 2) a discussion of the relevance of S-LCA in sustainable design practice and education; 3) an examination of the role of environmental life cycle assessment (E-LCA) in building performance standards and certifications as a model for the incorporation of S-LCA; and 4) a reflection on areas for future research, including the addition of social science theory and practice for methodology, criteria, and metric development.

This chapter proposes an approach to thermal comfort that increases occupant pleasure and reduces energy use by connecting architecture's material and environmental dimensions. Today's dominant thermal comfort model, the predicted mean vote (PMV), calls for steady-state temperatures that are largely unrelated to building design decisions. A more recent alternative approach, the adaptive thermal comfort (ATC) model, ties comfort to outdoor conditions and individual experience. Yet reliance on HVAC technology to provide building comfort hampers how such ideas are integrated into building design. This chapter outlines the historical background of the PMV and ACT models to understand the current status of thermal comfort research and practice. It then uses four recent buildings to outline how the insights of adaptive comfort research can be translated to bespoke comforts through spatial, material, formal, and other design strategies.

Preface

Climate change, resource depletion, and widespread pollution are fundamental global challenges today. Because nearly half of all resource use is associated directly with the building environment, the architecture/engineering/construction (AEC) industry is closely connected to, and in many ways responsible for, these problems. Many architectural scholars, design professionals, and product manufacturers have demonstrated the motivation to make changes that are beneficial to the environment. However, despite advances in determining the environmental impact of buildings and construction materials, there remains a dearth of tools, processes, and benchmarks that can reliably indicate ecological performance relative to holistic goals. Unanswered questions include: How do building-scale material decisions correlate to global CO_2 emissions targets? How can architects design buildings that consume no more than one Earth's share of resources? How can a building product industry optimally establish a net positive industrial ecology in the context of complex chains of custody? For the AEC industry to achieve measurable progress, the science of environmental impact assessment must be developed further.

GAPS AND OPPORTUNITIES

Current knowledge limitations are based on three primary knowledge gaps: the lack of understanding of trans-scalar impacts (part to whole), the paucity of environmental characterization data for materials (depth), and the absence of normalized relationships across different measurement categories that would enable comprehensive environmental assessments (breadth).

The trans-scalar disconnect is apparent in the different ways ecological impacts are measured at different scales. For example, materials and buildings are evaluated via methods such as life cycle assessment and environmental checklists. Meanwhile, planetary health is estimated via approaches like atmospheric CO_2 measurements and ecological footprint (EF) accounting. Yet there are no formulas that reliably link the two scales, establishing the precise EF of a particular material unit, for example, or the unit's compounding effect on worldwide carbon targets (Brownell, 2019). Furthermore, connections between other scales, such as a city or a region, are also poorly understood. Without these translated relationships, material decisions are simply wild guesses with no reliable reference to broader effects.

Materials are extensively characterized in terms of engineering attributes (e.g., compressive strength) but not environmental characteristics (Ashby, 2012). One problem is the proprietary nature of manufacturer processes and the hesitancy to divulge material ingredients and methods. In addition to protecting trade secrets, there is little incentive for manufacturers to admit any use of chemicals of concern or energy-intensive techniques. Another hurdle concerns widely varying contextual impacts, such as different fuel sources, when attempting to determine general material information—a point that relates to the trans-scalar disconnect above. Yet another difficulty concerns establishing consensus on materials' effects on human and ecological health. Without adequate environmental characterization of materials, impact measurements will remain a matter of loose conjecture.

The ability to establish data relationships across different categories is key to advancing sustainable building research. This objective is similar to resolving the trans-scalar disconnect, but in this case, it refers to knowledge areas rather than scales of operation. A common struggle concerns establishing meaningful connections between various life cycle impact categories. For example, how do we relate the impacts of acidification and eutrophication? Are there ways to normalize the multiple factors that influence human health and performance, such as material toxicity, ground-level ozone, and CO_2 emissions? How do we prioritize environmental impacts as a result? The sustainability movement has spawned thousands of disparate measurement systems, definitions, standards, ecolabels, and guidelines—a veritable "Tower of Babel" of information that hinders the aim for consistent, reliable, and standardized metrics. As pockets of deep expertise continue to advance, a fundamental next phase of work concerns the development of translation mechanisms between these areas.

AIMS AND TARGET AUDIENCE

The objective of this book is to disseminate the latest scholarship regarding the environmental performance measurement of buildings and materials. This book is intended to fill the gap in ecological accounting by providing a platform for the development of new approaches for documenting and communicating the environmental impact of building design, construction, maintenance, demolition, and related activities. It contains research related to the three gap areas outlined above and addresses the utilization of various tools such as life cycle assessment (LCA), material flows analysis (MFA), ecological footprint (EF), environmental product declaration (EPD), ecolabeling, and other measures of material-focused ecological performance. Some chapters feature innovative methods of visualizing environmental impacts—a welcome inclusion given that AEC disciplines rely heavily on visual means of thinking and communication. Others include novel material applications and building design methods. Effective techniques that advance both the science and art of environmental impact assessment for building materials and construction will not only assist AEC personnel in making better design choices but also facilitate the decision-making process for a broader set of stakeholders, including the general public. It is hoped that such approaches will also offer a significant toolkit for the growing population of students and educators studying humanity's impact on the environment.

This book is intended to serve the disciplines of architecture, landscape architecture, engineering, planning, construction management, building product manufacturing, and related fields in understanding and communicating the environmental impacts of the built environment. A secondary audience consists of those studying the environmental sciences and related areas such as ecology, environmental policy, and public health. Additionally, the visual nature of some of the research is appropriate for the disciplines of graphic design, informatics, business, and other fields that translate quantitative data into visual information.

ORGANIZATION OF THE BOOK

The book is organized into 10 chapters organized into three sections. A brief description of each section and a summary of each chapter follows:

Section 1 acknowledges the significant allocation of resources in the existing built environment. Authors explore trans-scalar inquiries related to embodied energy methodology, the influence of building codes on the material composition of cities, and the economic ramifications of replacing existing buildings with more operationally efficient yet energy-intensive versions. The overarching message is that embodied resources deserve more considerable attention in building environmental performance assessments.

Chapter 1, "The Embodied Impact of Existing Building Stock," assesses the environmental effects of the embodied resources within the existing built environment. The embodied energy of construction materials comprises a significant portion of all energy consumption, yet this quantity is often ignored in energy metrics that emphasize the operational or use phase of buildings. This chapter evaluates current multidisciplinary research activities in this area, summarizing them according to five primary parameters and three major components. Based on these results, the author determines that there can be discrepancies between embodied energy and embodied environmental impact and that there is a need for more consistent assessment methodologies. The chapter concludes with a discussion of embodied impact assessment barriers and potential solutions.

Chapter 2, "Building Codes Don't Measure Up: A Case for Urban Material Performance Standards," analyzes the under-appreciated influence of building codes on environmental and socioeconomic dimensions of urban materiality. Based on a specific set of materials and their anticipated performance criteria, building codes have dramatically shaped the built environment, prescribing a limited set of physical systems for various types of occupancy. Yet, as this chapter demonstrates, this restrictive approach has emphasized product-scale over city-scale performance, resulting in a set of unforeseen risks and missed opportunities. The author focuses on case studies within New York and Chicago, quantifying various building assemblies by construction type and visualizing these using material mapping techniques. She reveals unexpected connections between these material-focused restrictions and adverse human and environmental health effects, arguing for a more holistic consideration of these influences in future code development.

Chapter 3, "Teardown Index: Emissions of Single-Family Homes in Vancouver," highlights the discrepancies between the assumed benefits of new sustainable construction versus maintaining more poorly performing existing buildings. A study of single-family residential development in Vancouver, Canada, demonstrates that the payback period of embodied resources in new buildings is sufficiently long that it is more advantageous to preserve existing structures, despite their inherent inefficiencies. Based on an established formula used to predict the demolition of a building upon its sale, the authors present a new statistical model called the "teardown index" with similar aims, based on an expanded set of physical factors. The authors also estimate the total emissions of single-family housing stock in Vancouver by 2050 and include a variety of scenarios of real estate appreciation. Although confined to a single city and building type, the study offers a compelling method for conducting informed environmental impact assessments that is broadly applicable.

Section 2 consists of in-depth studies of various material topics. The chapters offer insights regarding the environmental performance of plastics, cementitious materials, and water in building construction. Approaches include linking disparate fields of inquiry, proposing a new assessment methodology, and developing and testing a new composite. These works demonstrate that in-depth explorations of specific materials can contribute to the broader knowledge base of building construction as a whole.

Chapter 4, "Toxicity in Architectural Plastics: Life-Cycle Index of Human Health in Building (LCI-HHB)," addresses the pervasive, adverse health effects of plastics used in building construction and proposes a more significant inclusion of human health considerations in traditional life cycle assessment. The author evaluates the various polymers used in buildings and analyzes their carbon impacts.

Special attention is paid to polyvinyl chloride (PVC), the most commonly used plastic in building construction. The author introduces the model of a Life-Cycle Index of Human Health in Building (LCI-HHB) as a more appropriate indicator of the potential impacts of material decisions on human health. The remainder of the chapter assesses policy priorities related to existing tools, professional education, and material disclosures.

Chapter 5, "Green Charcoal: Developing Biodegradable Construction Materials for a Circular Economy," proposes a new building material to fulfill the aspirations of cradle-to-cradle material flows within India's construction industry. The authors evaluated a variety of widely available materials, from agricultural waste to different formulations of concrete, and devised a novel composite composed of charcoal, luffa fiber, and cement. The resulting building blocks are lightweight, biodegradable, and encourage plant growth and insect habitation. The biocompatible units also perform favorably in compressive strength and water absorption tests. Part functional product development, part theoretical investigation—the green charcoal experiment raises thought-provoking questions about the functions of decay and ecosystem services in buildings.

Chapter 6, "Fluid Matters: The Water Footprint of Building Materials," probes the water consumption associated with major categories of building products. Unlike energy use or carbon emissions, embodied water is less frequently studied, yet the growing scarcity of global freshwater makes it a critical topic for more in-depth scrutiny. The chapter connects established Water Footprint Assessment (WFA) methodology with materials specific to building construction for which WFA has not yet been fully translated. The author assesses water uses by quality-based "color" (e.g., blue water, greywater), types, and material processes. She also analyzes building integration measures and methods for calculating water use impacts in building life cycles. The chapter further considers regional variations, historical analysis, and emerging technologies. The author concludes with an argument for the prioritization of water impact accounting, recognizing that human survival depends more on water than on modern forms of energy production that dominate environmental impact studies.

Section 3 navigates territories that exist outside of conventional material performance methods, ranging from incremental transformations to disruptive paradigms. Authors consider the future of life cycle assessment, the need for increased emphasis on social performance, the adaptation of interior climates, and "post-normal" approaches to building design and construction. Common themes concern the potential for new contributions in environmental accounting to emerge by questioning limitations, exploring interdisciplinary connections, and subverting entrenched models.

Chapter 7, "Life-Cycle Assessment-Based Environmental Performance Targets for Buildings: What Is Next?" addresses the current limitations and potential opportunities for this fundamental approach to determining environmental performance. Despite advances in LCA, the process remains inconsistent regarding holistic system boundaries, performance targets, and appropriate incentives. The chapter identifies pathways to more meaningful benchmarks, referring to The Natural Step and Planetary Boundaries (PB) models as examples. The author explores opportunities for LCA to influence the early design stage of buildings considering industry standards, guides, and tools that support this integration. A critical next step is for LCA to relate measures of material-scale performance to considerations of planetary limits.

Chapter 8, "Social Life-Cycle Assessment for Building Materials," explores the emerging topic of Social Life Cycle Assessment (S-LCA) in connection with building materials. Compared with economic and environmental performance measures, which are more established, social performance metrics are less well known or utilized—yet they are arguably fundamental. The chapter provides an overview of S-LCA methodology and compares it to its more recognized E-LCA (environmental life cycle assessment) counterpart. The author offers a series of case studies to demonstrate the inherent similarities and differences between these models. A list of recommendations for incorporating S-LCA methodology into existing environmental performance appraisals follows. The chapter also identifies future research areas in application development and visual frameworks.

Chapter 9, "Improving the Weather: On Architectural Comforts and Climates," focuses on the mechanical side of building performance related to indoor environmental quality. The author argues that our growing planetary climate crisis motivates the reevaluation of the interior climate as well as traditional notions of physical comfort. The chapter features a comparison between the conventionally applied Predicted Mean Vote (PMV) standard, a low-tolerance approach to maintaining specific temperature and humidity levels at a significant energy cost, with the Adaptive Thermal Comfort (ATC) model, which establishes an acceptable range of indoor climate criteria to reduce buildings' environmental footprint. The author surveys a history of architectural comfort standards, followed by a summary of new perspectives on thermal comfort. He then argues that comfort, which is generally viewed as an unremarkable criterion, is an underdeveloped characteristic of architecture. An assessment of five contemporary case studies provides a defense of this claim, suggesting that indoor climate should be treated as a fundamental architectural design opportunity.

Chapter 10, "Post-Normal Material Practice: 'Building' as Inquiry," offers a thought-provoking counterargument to the trend of quantifying building environmental performance according to narrowly defined criteria. The author proposes the application of Post-Normal Science (PNS) theory, developed in the late

20th century as a critique of what was perceived to be overly constrained "normal" scientific inquiry, to considerations of contemporary building construction. He makes a case for employing abductive reasoning and "operationalizing ignorance" in the generation of novel solutions to so-called wicked problems. This recommendation is based on the notion that such challenges are sufficiently complex and uncertain that they defy traditional problem-solving techniques. The author summarizes the design and construction processes of several recent built projects as a demonstration of this approach in practice. He concludes with an argument that the goal of building science should not merely be to answer known questions but to ask new ones.

CONCLUSION

Recognition of the built environment's significant global impact has led to its emphasis as a "key sector" in recent reports by the Intergovernmental Panel on Climate Change (IPCC, 2014). When Paris hosted the United Nations' 21st Conference of the Parties (COP21) in 2015, the forum notably included its first-ever Buildings Day. Cities worldwide utilize approximately three-quarters of global energy and generate a similar proportion of emissions, in large part to construct and operate buildings (McDade, 2017). It therefore may be expected that widespread improvements in building environmental performance, even in an incremental fashion, will have measurable planetary effects. In short, because buildings are a fundamental part of the problem, they must be part of the solution. Now that buildings' influence has been appropriately recognized, the environmental science must advance sufficiently that informed decisions may be made at multiple scales, regarding myriad building materials, and with an understanding of broadly translatable effects. It is hoped that this book, as a concise collection of focused scholarship on the environmental impacts of buildings and materials, can support this multidimensional aim.

Blaine Brownell
University of Minnesota, USA

REFERENCES

Ashby, M. F. (2012). *Materials and the Environment: Eco-Informed Material Choice* (2nd ed.). Amsterdam: Elsevier.

Brownell, B. (2019). Determining Architecture's Footprint: Preliminary Methods for Measuring the True Environmental Impact of Buildings. In G. Koç & B. Christiansen (Eds.), Reusable and Sustainable Building Materials in Modern Architecture. Hershey, PA: IGI Global.

IPCC. (2014). AR5 Synthesis Report: *Climatic Change*, 2014. IPCC.

McDade, E. (2017). Beyond Zero: The Time Value of Carbon. In B. King (Ed.), The New Carbon Architecture: Building to Cool the Planet. Gabriola Island, Canada: New Society Publishers.

Acknowledgment

I would like to extend my gratitude to all the individuals involved in this endeavor and, more specifically, to the contributors and reviewers who participated in the blind peer-review process. Without their tireless efforts, this project would not have come to fruition.

I wish to acknowledge all of the authors for their chapter contributions. I sincerely appreciate the time and expertise they devoted to this book.

I would like to thank each of the reviewers for their valuable contributions to the quality, content, and coherence of the book chapters. Most of the authors also participated as referees, and I would like to acknowledge this double-duty.

I appreciate the participation and assistance of the Editorial Advisory Board members in helping shape the aims of the book, recruiting potential authors, and serving as peer-reviewers.

I would also like to thank Parisa Mokhtari for her invaluable assistance as my research assistant in surveying and recruiting potential authors for this project, and the University of Minnesota School of Architecture for supporting this work.

Blaine Brownell
University of Minnesota, USA

Section 1
Embodied Impact

These chapters acknowledge the significant allocation of resources in the existing built environment. Authors explore trans-scalar inquiries related to embodied energy methodology, the influence of building codes on the material composition of cities, and the economic ramifications of replacing existing buildings with more operationally efficient yet energy-intensive versions. The overarching message is that embodied resources deserve greater attention in building environmental performance assessments.

Chapter 1
The Embodied Impact of Existing Building Stock

Ming Hu
https://orcid.org/0000-0003-2583-1161
University of Maryland, USA

ABSTRACT

This chapter provides the reader with a better understanding of the life cycle environmental impacts, with a focus on the embodied impact of existing building stock. A systematic literature review is conducted to paint a clear picture of the current research activities and findings. The major components of embodied impact and parameters influencing the embodied impact are outlined and explained. Lastly, this chapter discusses the major barriers for the embodied impact assessment, and a potential analysis framework is proposed at the end.

BACKGROUND AND MOTIVATION

Definition

Embodied energy includes the energy consumed through the life cycle of a building as well as the energy expended for raw material extraction, the manufacturing of materials, and transportation to the construction site; the building construction, maintenance, repair, and replacement of building components during operation; and the demolition, transportation of materials, and their end-of-life management (Chastas et al., 2016). Embodied energy excludes the operational energy consumed within the building when it is in use; for example, the heating, cooling, and lighting in buildings (Moncaster & Symons, 2013). It is measured in kWh, including direct

DOI: 10.4018/978-1-7998-2426-8.ch001

and indirect energy use, according to the CEN/TC350 standard "sustainability of construction works" (ECFS, 2019). In this chapter, embodied impact refers to the environmental impact embodied in the building, with carbon emissions representing an important measurement. Embodied emissions (EI) is the sum of fuel-related and process-related embodied emissions. Process-related embodied emissions are non-fuel related and can arise from a chemical reaction due to the building materials and assembly production (Moncaster & Symons, 2013).

Significance

The significance of embodied energy and embodied impact is reflected in three areas. First, the building and construction industry is a big energy consumer and carbon emitter; for instance, in the United States, building sectors contributed 39% of the country's carbon emissions (USGBC, 2019a) in 2018. In the United Kingdom, the construction materials sector alone accounted for 5–6% of total UK emissions (Rawlinson, 2007). Although emissions from the global building sector have leveled off since 2015, overall, building construction and operations still accounted for 36% of global final energy use and nearly 40% of energy-related carbon emissions in 2017 (2018 Global Status Report, 2018) as new construction and existing building renovation projects are energy-intensive (United Nations Environmental Programme-Sustainable Buildings and Climate Initiative, 2008; H.M. Government, Innovation and Growth team, 2010).

Second, operational energy saving could potentially be lost through sub-optimal management or an accelerated refurbishment cycle (USGBC, 2019b), meaning a shorter building use life span leads to quicker retrofit cycles, resulting in increased total embodied energy and embodied emissions. Therefore, more attention should be focused on embodied emissions embedded in renovation materials and components.

Lastly, savings in embodied emissions can effectively reduce the overall carbon emissions; however, the emissions reduction goal cannot be achieved through increasing operational emission savings alone (USGBC, 2019b; Dixit et al., 2010; Acuqaye, 2010). These three areas suggest that building designers can greatly influence the carbon footprint of buildings (Brummer et al., 2008), as designers and architects have the most control over the embodied energy used in buildings. Any initiative that is focused on reducing such embodied energy and embodied impact will significantly contribute to meeting energy reduction targets (Dowden, 2008).

Research Motivation

An increase in embodied impact has unintended consequences, such as shifting the environmental impact hot spots from one life cycle stage to another (e.g., the operational stage to the product stage) (IEA). Several recent studies have proven that embodied impact could contribute up to 50–70% of the life cycle's environmental impact of a low-energy or nearly net zero energy building (IEA, Birgisdottir et al., 2017), 32–38% for a passive designed building, and 9–22% for a conventional building (Chastas et al., 2018). Since operational impact decreases as energy efficiency improves, embodied impact has grown in significance, and its integration into legislation should be considered as a parallel focus. However, the majority of current legislation excludes the measurement and reduction of embodied energy and embodied environmental impact over the building's life cycle (Birgisdottir et al., 2017). Research tying the environmental impacts with embodied energy remains a largely unexplored area, partially due to a lack of clearly understood embodied impact attributes and inconsistencies in analysis methodologies (Pomponi & Moncaster, 2016).

RESEARCH METHOD, MATERIALS, AND TOOLS (LITERATURE REVIEW)

The literature review entailed three steps. In the first step, the key search terms "embodied energy" and "embodied impact," in combination with the keywords "building" and buildings," were used to scan the Web of Science databases. The search included only peer-reviewed publications: articles, conference proceedings, books, and book chapters. There were 195 publications found from a variety of disciplines in the research domain, from the period of 1970 through 2019. The different research fields included in the search were architecture, construction, engineering, environmental technology, and science.

In the second step, citation analysis and text data mining were carried out with VOSviewer software to determine major research clusters/areas and research synergies. In this research project, a combination of Citation analysis (CA) and Text data mining (TM) method / technique is used. CA is a commonly used bibliometric method for the quantitative evaluation of scientific and academic literature. It is used to assess the quality of a published paper, journal or the impact of authors, journals, organizations, or research projects. TM is a technique used for content analysis. Content analysis for text draws on techniques developed in natural language processing, information retrieval, and text mining. Over the past several years TM has boomed in the social and humanities fields (O'Connor et al. 2011). It is often used to investigate the text-based documents to identify the trends and patterns from a large number of articles (Hu & Pavao-Zuckerman 2019). In a VOS-constructed map, different cluster maps represent different research areas; the bigger the size of the nodes, the more influential and relevant of the items, the items could represent the research keywords, authors, journals, or published articles. The distance between the nodes indicates the intellectual connections of the research topics and authors/ researchers (Hu & Pavao-Zuckerman 2019).

In the third step, after creating an overview of research activities based on VOSviewer maps and identifying the major active research clusters, an in-depth qualitative review of all studies was conducted to reduce the articles to the ten most influential studies (with the most citations) in each cluster. Altogether, 40 papers were thoroughly reviewed to examine the findings and create synergies and conclusions. Through the focus review, from the 40 papers, **five primary parameters** (section 3.3) influencing embodied impact and **three major components** (section 3.2) that contribute to embodied impact were extracted and explained. The following section highlights and discusses the main results of the literature review.

FINDINGS: RESEARCH CLUSTERS, COMPONENTS, PARAMETERS

Current Research Activities and Trends

The research of embodied energy and related impacts in the building industry began to emerge in the 1990s based on existing records on Web of Science. In 2015, research peaked with a total of 33 publications, with 2017 ranking second with 18 research publications (refer to figure 1).

The disciplines that are interested in this topic are very diverse, with the top three fields leading the research efforts being engineering, construction building technology, and energy fuel (refer to figure 2). The most influential countries and regions with the highest number of publications are the United States, United Kingdom and China.

When diving into the research projects, four main research clusters (areas) emerged. Furthermore, these four areas were interdisciplinary and correlated. For example, the calculations of embodied energy (cluster one) and embodied emissions of buildings (cluster three) are related to the parameters included in the calculation (cluster two), and the design process is often intertwined with the selection of building materials. Table 1 indicates the data size of those clusters and keywords, and figure 3 illustrates the relationships and connections of those clusters and keywords within each cluster.

Certain clusters are more intertwined with others. In the following sections, each cluster is explained and its relationship to others is discussed.

Figure 1. Number of research publications over the years (data and diagram were extracted from web of science)

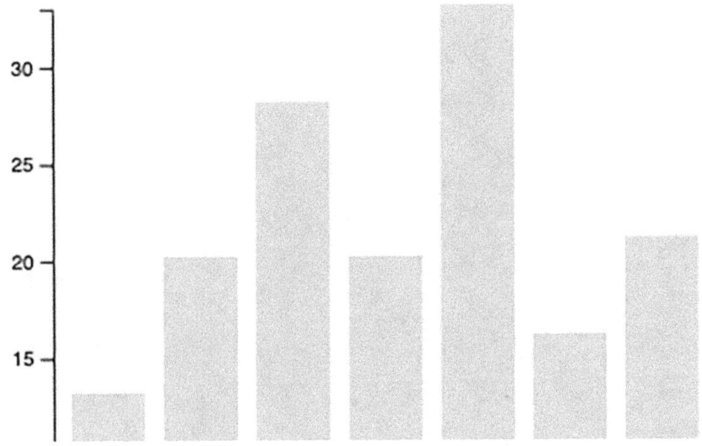

Figure 2. Research disciplines (data and diagram were extracted from web of science)

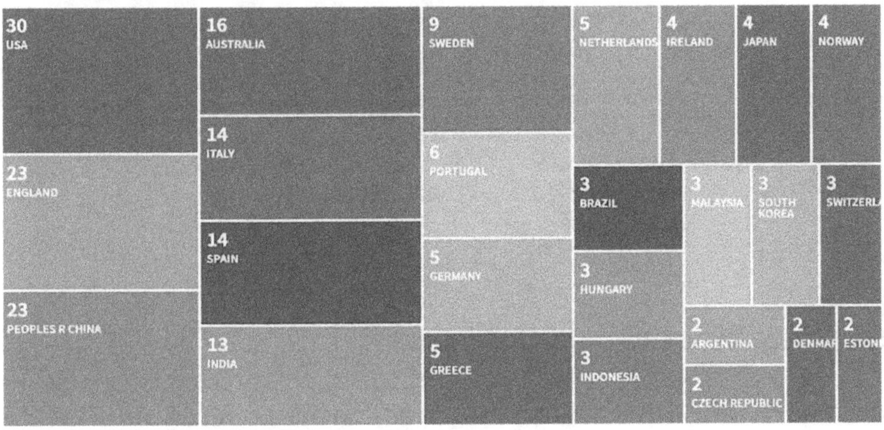

Table 1. Characteristics of clusters (size, keywords, color)

Cluster ID	Color	Size (keyword counts)	Representative keywords
One	Yellow	17	Structure, transportation, steel, environmental performance, cement, concrete
Two	Green	16	Calculation, LCA, GHG, framework, methodology
Three	Blue	15	Reduction, importance, CO_2 emission, operational energy, residential building, increase, relationship
Four	Red	13	Construction material, model, parameter, review, design problems

Figure 3. Research clusters in embodied energy (generated from VOSviewers)

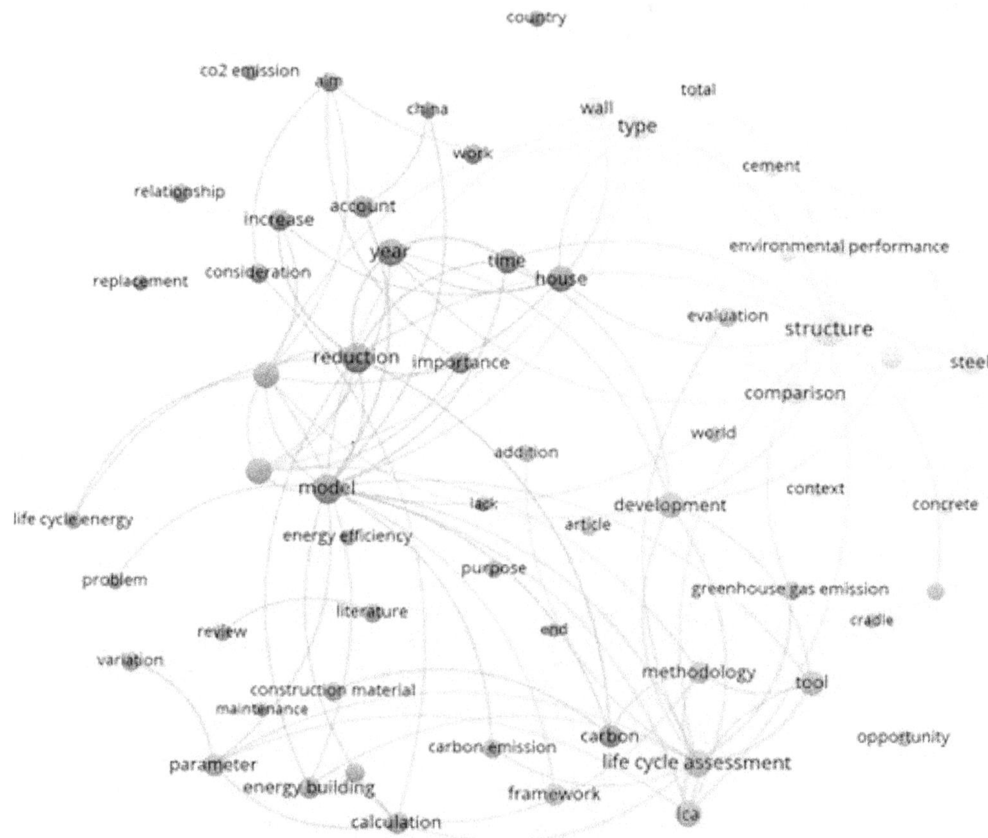

Cluster One: Structure and Materials

Cluster one is organized around the major materials used in a building's superstructure—such as steel, concrete (cement), and wood—and the environmental impact generated through their whole life cycle. Many studies have focused on understanding the impact of different structural systems and how to optimize the design by using green materials to reduce the embodied energy and embodied emissions (Zeng & Chini, 2017). For example, Goggins et al. (2010) compared the embodied energy of a typical reinforced concrete structure in Ireland using two concrete mix designs. The large embodied energy saving through using industrial slag instead of cement demonstrates the significant environmental benefits that could be achieved through the use of recycled building materials (Goggins et al., 2010). Yeo and Gabbai focused on optimizing reinforced concrete structures through embodied energy optimization. The techniques the team used were traditionally employed to minimize the total cost or weight of a building's structure. However, their results indicated a reduction of approximately 10% in embodied energy with a cost increase of only 5% (Yeo & Gabbai, 2011). Vukotic et al. (2010) used a simple single-story structure as a case study to perform comparative analyses between two structural design alternatives: glue-laminated timber panels and a steel frame with infill concrete blockwork. They found that material selection, material sourcing, and waste handling at the end of a building's life were the most important phases of the building's life cycle (Vukotic et al., 2010). They also suggested that extending the building life, improving construction methods, and developing benchmarks for structural designers could be an effective way to reduce the building's embodied energy and impact (Vukotic et al., 2010).

A study of residential buildings in China found that the embodied energy consumption of steel members, concrete, and cement accounts for more than 60% of all building components' contributions. Furthermore, steel buildings only consume approximately 73% of energy during the construction phase—compared to conventional concrete buildings—due to their construction simplicity. Also, the increase of embodied energy in steel buildings has a greater association with larger heights, rather than larger floor plates (Su & Zhang, 2016). Miller et al. (2015) demonstrated that steel is a critical building structure material for both embodied energy and impact reduction because, on average, steel accounts for less than 4% of a building's total weight but up to 59.6% of its embodied energy (Miller et al., 2015), due to the energy-intensive manufacturing process of steel. In Singapore, reinforced concrete, compared to steel, was found to produce lower carbon dioxide emissions and incur less embodied energy. These results could be altered by using more sustainable primary steel manufacturing technologies (Kua & Maghimai, 2017). A similar conclusion was obtained in the United States; Griffin et al. (2013) compared three existing parking structures with one-way spans. The three types researched were precast concrete, post-tensioned concrete, and cellular steel. Griffin concluded that when steel with a high recycled content was used, there was little difference in the embodied energy of the structural system used for parking garages (Griffin et al., 2013).

Besides the most commonly used structural materials, building structure components and special building types were studied as well. Sedláková et al. researched and compared three variations of concrete foundation. Their study indicated that a higher embodied energy consumption produced higher CO_2 and SO_2 emissions. Those emissions are mostly related to insulation materials (i.e., extruded polystyrene, concrete, and polystyrene concrete) (Sedláková et al., 2014). Tall buildings—a special building type (20 to 70 stories)—were studied, with results proving that the embodied energy depends mainly on the structural flooring system (Foraboschi et al., 2014) and not the building height, which is contradictory to regular mid- to low-rise buildings. Timber was also studied as an alternative structural material for high-rise building (Skullestad et al., 2016), and the results indicated that timber structures could potentially save 34–84% of embodied energy use compared to conventional reinforced concrete structures (Skullestad et al., 2016), which presents a promising direction for building structure technology in high-rise buildings.

Within cluster one, researchers also considered building blocks and infrastructure in addition to individual building studies. A close look at infrastructure projects provides first-hand data regarding how embodied energy considerations could be expanded to the larger built environment. Krantz et al. (2015) studied the embodied energy of a prefabricated bridge, integrating Building Information Modeling (BIM) and Discrete Event Simulation (DES) based on the life cycle assessment method. The consensus built within this cluster was that building materials have the promising potential to significantly reduce energy use in the construction industry as embodied energy gains importance among researchers, professionals, builders, and material manufacturers (Dixit et al., 2010).

Cluster Two: Calculations, Framework, and Method

Within cluster two, many studies focused on the validation, verification, and comparison of existing methods. The literature suggests that there are no usable standards or guidelines for embodied energy analysis (Dixit et al., 2012). The methods used for calculating embodied energy, emissions, and impact include different *carbon calculators, statistical/mathematic analysis* (Abanda et al., 2013), and *life cycle assessment*. Each method has advantages and disadvantages.

Carbon calculators were introduced to design professionals before the 1990s due to their ease of use. For instance, the carbon calculator, created by researchers at Portland State University, was used to calculate the impact of structural systems (Portland State University, 2019). A UK-based company, BSRIA, published the embodied energy calculations design tool to help designers make optimized decisions (BSRIA, 2019). While a good first step, carbon calculators have been criticized for their oversimplicity and inaccuracy, which has limited their application and reliability.

A statistical model determines the totality of logical connections, formalized dependencies, and formulas, which enables the studying of real-world objects without experimental analysis (Gertsev & Gertseva, 2004). Statistical/mathematic models have been used for energy simulations (Clarke, 2007), analyses of buildings' environmental impact (Jaffal et al., 2012), predictions of environmental impact during a building's life cycle, and design decisions about a building or product (Abanda et al., 2013). The most commonly accepted and used assessment of embodied energy is life cycle assessment (LCA). Embodied energy analysis is a subpart of LCA that appears in the life cycle energy analysis stage (Atkinson et al., 1996). There are tremendous interests in developing more reliable and accurate life cycle assessment methods to evaluate the embodied energy and embodied impact. Currently, there are three major LCA methods used in calculating embodied energy and impact: process-based analysis (P-LCA), input-output based analysis (IO-LCA), and the hybrid method (Azari & Abbasabadi, 2018).

Yohanis and North (2002) calculated the embodied energy of a generic single-story office building in the United Kingdom using the P-LCA method. Chang et al. (2010) used the IO-LCA method to study the embodied energy and environmental emissions of construction projects in China, based on 2002 Chinese economic and environmental data. The results indicated that the embodied energy from the construction industry in China accounted for one-sixth of the total economy's energy consumption in 2007 and increased to one-fifth by 2015 (Chang et al., 2010). Applying the hybrid method, Chang et al. studied a high-rise education building in China. They found that embodied emissions mainly derived from the electricity sector, gas, and water production due to intensive coal consumption (Chang et al., 2012).

Different LCA methods could generate noticeably different outcomes. Lenzen and Treloar adopted the case project published by Borjesson et al. (2000). Instead of using the P-LCA method, they used the hybrid method, combining the P-LCA and IO-LCA methods and using Australian data instead of Swedish data. They found that Borjesson's study underestimated the embodied energy and led to errors in the net life cycle emissions of CO_2 and CH_4, with the magnitude of discrepancies being a factor of 2 (Lenzen & Treloar, 2002).

After understanding the advantages and disadvantages of the existing methods, a variety of revised methods were proposed to include the parameters affecting embodied energy that had been largely ignored. Dixit et al. (2015) presented a revised method based on the IO-LCA hybrid method, which would avoid errors and incorporate human and capital energy in calculations (Dixit et al., 2015). Multiple research teams proposed a hybrid method, which combines data quality indicators and the statistical method through a prescreening process based on the Monte Carlo rank-order correlation sensitivity analysis (Ozoemena et al., 2014; Wang & Shen, 2013). The Monte Carlo method has been integrated as a complementary method to the LCA methods to calculate embodied carbon (CO_2) emissions (Robati et al., 2019; Acquaye et al., 2011), the carbon footprint of multiple regions (Lenzen et al., 2010), building materials' life cycle embodied impact (Lee et al., 2011), and renewable energy generation (Kabir et al., 2012). In the last decades, digital design technologies have been introduced to embodied energy assessments as well. For instance, Krantz et al. (2015) integrated Building Information Modelling (BIM) and discrete event simulation (DES) to simulate carbon emissions and energy use (Krantz et al., 2015), demonstrating a promising direction.

Overall, methodological challenges remain, affecting the reliability of embodied energy analysis results and leading to less effective communication. While a variety of relatively robust quantitative methods exist for estimation, the lack of standardized quantification protocols remains a significant problem (Chastas et al., 2016) that decreases the reliability of the embodied energy and impact estimations.

Cluster Three: Carbon Emissions Reduction and Embodied Energy

Cluster three centers around how a reduction of embodied energy leads to a decrease in carbon emissions; this cluster also has the most conflicting research results. Residential buildings have been the focus for the past decades; however, recently, research on commercial buildings and other building types has become more prevalent. The urgency of reducing carbon emissions through minimizing embodied energy has not attracted the same level of attention as the focus on reducing operational energy. According to the United Nations Environment Programme's (UNEP) most recent report in 2017, the manufacturing of building materials contributes to 11% of the total global greenhouse gas emissions. This is a small percentage compared to the impact of operational energy from the existing building stock. However, for new construction, embodied carbon holds the same level of significance as energy efficiency and renewables. Moving forward, the embodied emissions that are produced up to 2050 will determine whether the goal of the Paris Agreement is met (Building Green Inc, 2019).

Carbon reduction strategies are not only applicable to new construction but are also feasible for existing buildings through embodied energy minimization. Brown et al. (2014) evaluated the importance of the embodied global warming reduction potential of Swedish existing residential building stocks through renovation and refurbishment. They investigated different renovation strategies and techniques and found that replacing existing windows lowered the global warming potential (GWP) compared to other techniques, such as additional insulation to the exterior walls or roof (Brown et al., 2014). The research revealed that improving the existing building operational efficiency could contribute to an overall carbon emissions reduction. Li and Colombier (2009) studied several policies and instruments, with results indicating that an improvement in existing buildings' energy efficiency can generate a considerable carbon emissions reduction. Almeida et al. (2018) studied six case projects representing different regions in Europe and suggested a potential carbon emissions reduction range of 2% to 32% with renovation strategies, such as replacing mechanical equipment and adding insulation to the exterior, among others (Almeida et al., 2018). Meanwhile, other studies have produced opposing results. For example, Seo et al. (2018) analyzed the embodied impact of different dwelling (built before 2005) stock retrofit programs using a combination of the top-down and bottom-up approaches. Their findings revealed that embodied impact is expected to increase from existing renovations, even with a net-zero energy goal, due to the additional embodied energy added from advanced building systems.

The overall consensus asserts a correlation between embodied energy and carbon emissions, although with variations in and occasional opposition to the detailed results, due to the diverse methodologies and databases used. Ngo et al. (2009) estimated that emissions could be reduced by 60% through different construction assemblies, such as a curtain wall façade over a traditional masonry building. Sazedi et al. (2014) found masonry buildings to have lower embodied energy and carbon emissions compared to conventional glass buildings, excluding the use stage. Based on the findings, different mitigation strategies have been identified to meet the future goal. The use of materials with lower embodied energy and carbon, application of better design, and reuse of materials with intense embodied energy could help transition the current built environment to a lower emissions environment.

Cluster Four: Building Materials Selection and Design

Cluster four represents the current focus on building design and building materials selection. The majority of research focuses on how to optimize the design and minimize the embodied impact, through lighter building systems, optimized building geometry, or strategically chosen building materials. The parameters contributing to the embodied energy and impact are investigated as well, and will be examined in further detail in section 4.0.

A variety of construction materials (besides major ones) have been evaluated for their embodied energy and related impact. Reddy and Jagadish (2003) compared a load-bearing brick building to a soil-cement block building (alternative material) and concluded that using the alternative material could reduce the embodied energy by 50%. Oztas and Ipekci investigated the initial embodied energy of cement, marble, glass, and aluminum in Turkey. Surahman et al. (2017) studied 544 buildings in Jakarta and Bandung and concluded that if the maximum recycled material reuse could be applied, then the embodied energy and carbon emissions could be tremendously reduced. Other than reusing and recycling existing materials, alternative building materials are studied for their contribution to the reduction of embodied carbon. Clay-alginate composite aerogels were found to be promising and could have a significant impact on carbon emission reductions (Dove et al., 2019). Green roofs, as an alternative to traditional building materials, have been studied extensively for their carbon sequestration potential (Getter et al., 2009) and embodied energy reduction effect (Saadatian et al., 2013).

Doh et al. (2014) investigated six parameters for their influence on embodied carbon intensity: technological changes, energy tariffs, primary energy factors, disaggregation constants, emission factors, and material price fluctuations. Technological changes, energy tariffs, and material prices were discovered to cause significant variations. Other parameters were investigated as well, such as building shape. Lotteau et al. (2017) studied the influence of building shape/geometrical characteristics on embodied energy. Using a proposed analysis method, they found the building compactness had a correlation to its embodied energy and embodied impact level. Building compactness is defined as the ratio between the overall building surface area and the gross square footage of the building.

Synergy of the Research Clusters

While all four clusters have some general overlapping, in particular, clusters three and four are closely related. The benefits of reducing embodied energy and carbon emissions are typically studied together, and researchers and practitioners have been trying to understand the parameters in order to define a reliable assessment framework. Thus far, **the research results indicate that *embodied energy is not an absolute equivalent to the embodied environmental impact***. Findings from case studies reveal that a reduction in embodied energy use is not always proportionally associated with a reduction in environmental impact. For example, certain low-embodied energy materials, such as cast-in-place concrete, could lead to a higher global warming potential (refer to figure 4). Conversely, materials such as metal and steel have a much higher embodied energy contribution but generate less smog formation potential than concrete due to the fact that more than 90% of steel and metal manufactured in the United States is from recycled steel (refer to figure 5). In order to reduce the embodied environmental impact of building materials, other factors, besides using low-embodied energy materials, need to be taken into consideration, such as construction methods, local manufacturers' capabilities, and access to recycled materials. In section 3.2 below, the parameters are explored and explained in greater detail. In summary, finding a balance between reducing both the embodied energy and embodied environmental impact presents a complex issue. Further studies would be beneficial, both to the research community and industry.

Major Components of Embodied Impact

Initial Embodied Impact (IEI)

Initial embodied impact is the impact generated from the raw material extraction, manufacturing, and transportation of products and components, and building construction onsite (Azari & Abbasabadi, 2018). Therefore, IEI represents all the impacts that occur before the building is occupied.

Figure 4. Embodied energy per material and global warming potential (by author)

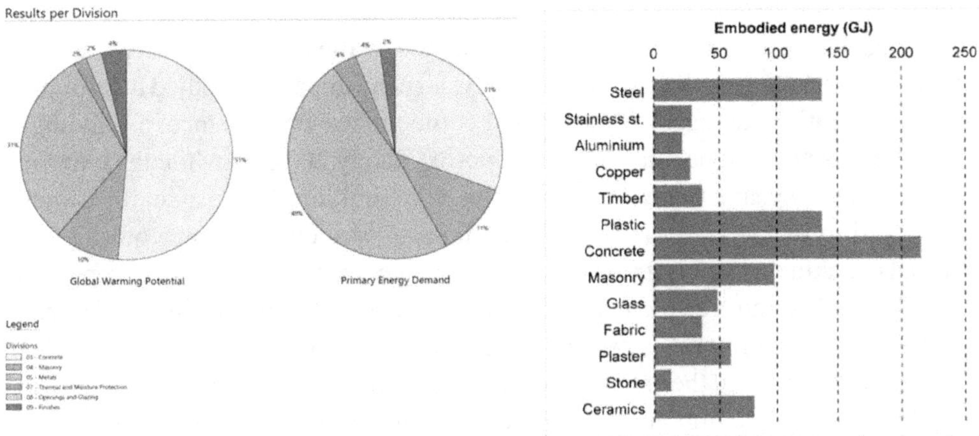

Figure 5. Embodied environmental impact (by author)

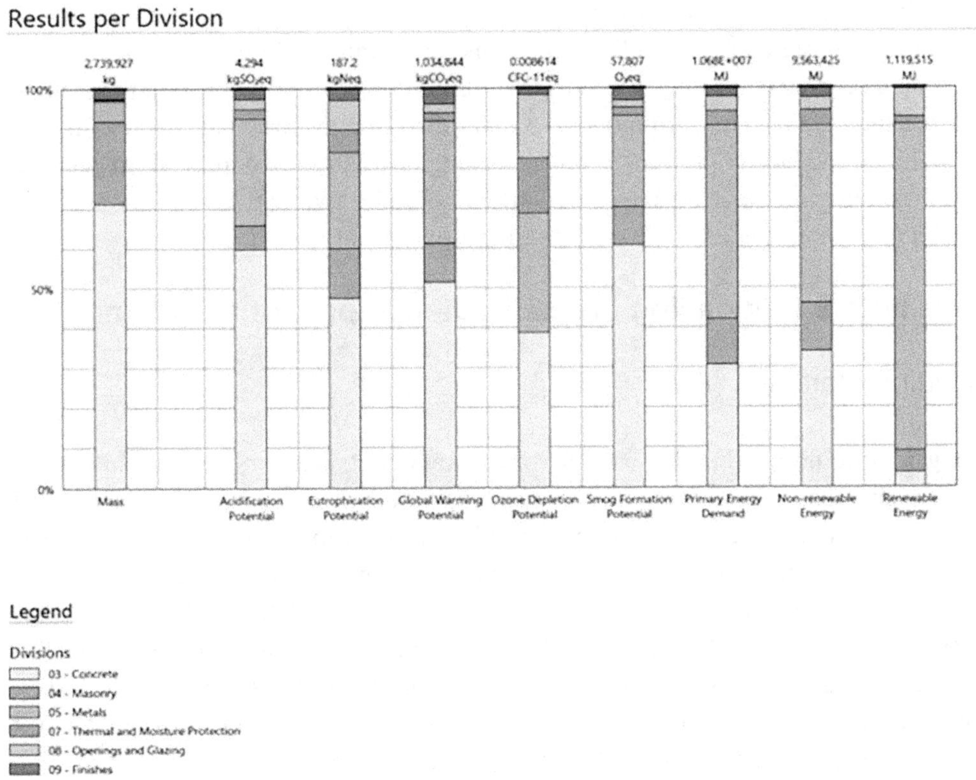

Recurring Embodied Impact (REI)

Recurring embodied impact is the impact related to the embodied energy used to maintain the building during its operation. It is also the impact embodied in the repair or replacement of damaged materials and components. For instance, a building's roof materials and components only last approximately 20 years while the building itself lasts for 60 years; therefore, during the entire building's life span, the owners might need to replace the roof twice. The impacts associated with acquiring raw materials, making products, transporting the products to the site, and then replacing the roof are all counted as recurring embodied impacts. The terms used to describe this energy use and impact is *recurring embodied energy* and *recurring embodied emissions* and represent the emissions incurred to maintain, repair, restore, refurbish, or replace materials, components, or systems during the effective life of the building (Chen et al., 2001).

Demolition Embodied Impact (DEI)

Demolition embodied impact is the impact caused by dissembling the building at the end of its life cycle, recycling and re-using certain building components, and disposing of others through landfills or incineration. DEI is very difficult to quantify and is a largely uncertain component of embodied impact due to limitations in data availability and lack of consensus on an acceptable quantitative method. In general, DEI has a negligible share in the overall embodied impact (Azari & Abbasabadi, 2018); therefore, it is often disregarded in research and practice.

Parameters Influencing the Embodied Impact of a Building

Building Materials (BM)

Long before embodied energy and life cycle assessments were known to architectural researchers, Cole and Keran (1996) compared wood, steel, and concrete structural systems from an office building and found that the embodied energy of the steel frame was 1.13 times greater than the wood frame, and the embodied energy of the concrete frame was 1.05 times greater than that of the wood frame. In a later study, Eaton and Amato (1998) researched the embodied carbon emissions between steel, composite, and concrete office buildings, which revealed different results. They found there was no significant difference between steel frames and concrete frames in terms of embodied energy and embodied carbon emissions.

Besides the materials used for structural systems, the material selection for the building envelope also provides a major contribution to embodied energy. The building envelope accounts for approximately 48–50% of the overall embodied energy from a typical house (Mithraratne & Vale, 2004), and components affecting the embodied energy from the building envelope are window systems (glazing types, frame material, etc.), insulation materials, and walls. Different window framing materials were studied and compared: PVC, aluminum, and wood. The researchers concluded that the highest and lowest embodied energy values were associated with aluminum double glazed windows with no recycled content (Azari & Abbasabadi, 2018).

Once the researchers dived deeper into the relationship between embodied energy and embodied impact, the conditions became more complicated. As previously discussed, the durability and longevity of building materials is of great importance when calculating the embodied energy and embodied carbon for the majority of existing buildings (Menzies, 2011). Embodied energy is dominated by the manufacturing of building materials, which represents 80–90% of embodied energy in a building's life cycle. Transportation embodied energy is approximately 3–5%, and construction embodied energy is around 6% (Chang et al., 2012; Nässén et al., 2007). For existing building renovations, the selection of building materials is the key issue, as it has the largest effect on the overall embodied impact. Higher embodied energy building materials, such as brick, may have a lower overall impact than a lower embodied energy product, such as wood cladding, because brick masonry lasts longer than wood. It has a longer life span and requires fewer repairs and maintenance needs. Additionally, the life cycle embodied energy is not based on initial embodied energy alone. For instance, a solid timber floor system can last 100 years or longer while a fiberboard floor may need replacement in 20 years, which adds additional/recurring embodied energy to the building.

A potential indicator associated with building materials is the building age. Based on a commercial building energy consumption survey, 67% of commercial buildings were built before 1990. Among the more than 5,000 buildings included in the survey, almost 50% of them were made with traditional masonry materials—like brick, stone, and stucco (refer to figure 6). Those types of traditional masonry materials generally have low embodied energy compared to modern building materials, such as aluminum or concrete.

Building Function (BF)

The data from the literature between 1994 and 2016 indicates that the majority of embodied energy assessments were focused on residential buildings, with more than 100 buildings in over 25 countries studied. In general, commercial office buildings have a higher embodied energy (9.5GJ/m^2) when compared to a residential building (1.5GJ/m^2–8.2GJ/m^2). This is because residential buildings are usually

Figure 6. United States commercial building age survey: number of buildings with different construction types (data extracted from CBECS 2015)

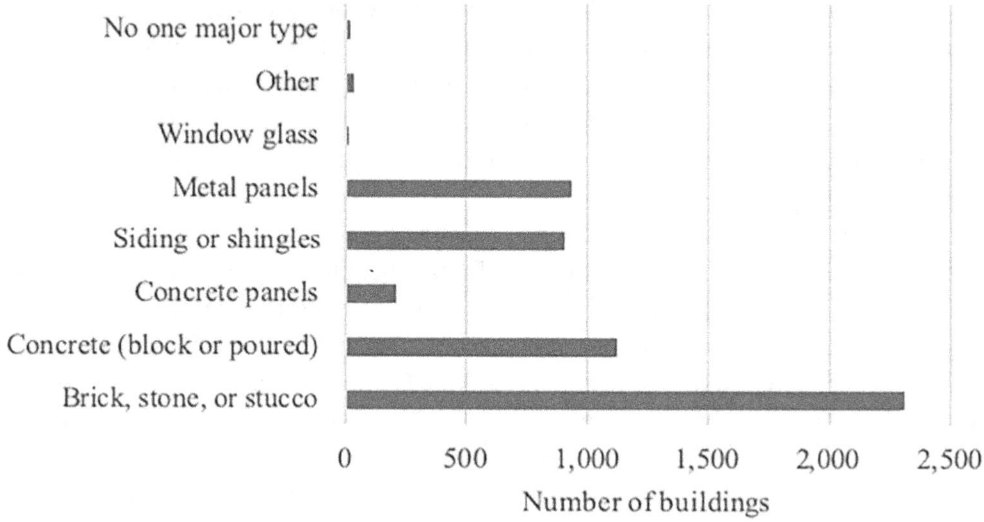

constructed with a wood frame, while commercial buildings are built principally using concrete or steel. Educational buildings typically have higher embodied energy (6.3GJ/m^2–11.7GJ/m^2) than commercial buildings since the former requires more specialized equipment and interior partitions. However, a large research gap exists between the relationship that different building types have with embodied energy and what design strategies and techniques could be applied to reduce the embodied energy and related impact.

Building Construction Type (BC)

In addition to building materials, the embodied energy of a building varies greatly with different construction types as well. For the purpose of discussion, we have categorized construction types as traditional buildings and non-traditional buildings. Traditional buildings can be defined as buildings that are constructed using techniques that were commonly used before the 1900s. This includes non-permeable materials—such as masonry, wood, and stone—and excludes membranes that act as damp-proof layers to promote the dissipation of moisture from the building fabric (Menzies, 2011). Non-traditional buildings are buildings with moisture and thermal barriers that are separated from the primary building construction materials. Certain modern renovations and upgrade solutions are suitable for non-traditional construction but not for traditional buildings. For example, traditional masonry buildings are made with solid masonry block, brick, and stone. Unlike the embodied impact directly related to building materials, the life cycle embodied energy and embodied impact are far more complicated as the embodied impact includes IEI, REI, and DEI (refer to section 3.2). The construction method and techniques using traditional building materials with a lower embodied impact do not necessarily correlate with a lower life cycle embodied energy and impact. For instance, infilling traditional walls with additional insulation is not applicable because there are no cavities in traditional walls. In order to augment thermal properties, traditional stone walls may need to be sealed off from the surface, although this can result in moisture accumulation within the building and cause mold growth and material decay, reducing the building's life span (Menzies, 2011) and increasing embodied energy through maintenance.

Within each of the above primary categories, there are sub-categories of building construction types. Iddon and Firth (2013) compared four different dwelling construction types: traditional masonry, heavyweight (concrete block), timber construction, and structural insulated panels. They found that the heavyweight construction type had the least embodied energy, which appears counterintuitive to conventional perspectives of the subject, since concrete block has a higher embodied content than other materials. However, unlike the other construction types, the walls are not finished with plasterboard, which results in a lower embodied carbon content in the inner wall and external walls. Also, because of the high acoustic value of concrete blocks, additional acoustic insulation is not required, which further reduces the embodied energy. Certain building systems and components with higher embodied energies can be justified due to their contribution to conserving operational energy. For example, insulation materials are high in embodied energy but could significantly reduce heating and cooling needs and lead to a reduction in operational energy. This parameter is less understood because the majority of research focuses on building materials instead of building systems or components. As building construction types continue to advance, the influence of this parameter will become more sophisticated.

Building Location (BL)

Geographical location also affects the embodied energy impact through unique manufacturing practices, construction methods, sources of primary energy, local energy infrastructures, modes of transportation, distances from building sites to manufacturers, supplier availability, and other economic factors that would be unique to the different building locations (Azari & Abbasabadi, 2018). Countries and regions differ from one another, not only in geographic and climatic characteristics but also in raw material quality, production processes, economic data, processes of delivered energy generation, transport distances, energy use (fuel) in transport, and labor (Dixit et al., 2010). The processes of industrial and economic sectors differ greatly and thus inñuence the calculated embodied energy values (Buchanan & Honey, 1994).

In different locations, varying production technologies could create a large variation when calculating the embodied energy (Tettey et al., 2014). For example, the metal studs that are used to manufacture partition walls for buildings in the United States may follow a different process than for a building in Europe or Asia. The difference in location translates to other discrepancies, such as transportation distance and mode (truck or ship), among others. Those all contribute to embodied energy from building assemblies or building materials. International standard ISO 14040 explicitly states the need for technological representativeness of data when assessing a building's life cycle embodied impact. Technological representativeness is an important quality of data that should be considered in order to fully understand the differences in embodied impact and energy in the same building materials and components (Dixit et al., 2010) and should include the technology for raw material extraction, processing, and manufacturing. Regarding economic factors, varying energy tariffs paid by material suppliers and manufacturers in different countries could lead to a 2–2.6% change in the embodied energy calculation (Pullen, 1996). A more complex issue is the export and import of building materials and components, particularly in trade between developed and developing countries.

Building Maintenance and Repair Status (BMR)

A wide range of materials is used during a building's construction. Several of these materials and components have a life span shorter than that of the building. More precisely, a typical building can last 40 to 60 years, whereas roofing systems usually last 20–30 years and doors and windows approximately 30 years, but other components might have shorter life spans of around ten years. As a result, they will be replaced before the building reaches its end of service life. Meanwhile, all buildings require regular maintenance, and the energy and emissions incurred for such repairs and replacements should be accounted for in the entire life cycle of the embodied energy of the building (Ramesh et al., 2010) (refer to section 3.2).

Buildings of varying construction types with differing life spans have different breakdowns of embodied energy throughout the life cycle (including operational energy). Yohanis and Norton (2001) estimated that the recurring embodied energies associated with a typical replacement and repair for a wood-structured building were 59%, 148%, and 339% (of overall life cycle energy) for a life span of 25 years, 50 years, and 100 years, respectively. The recurrent embodied energy of a building product may change according to the technological advancement and raw material availability. For instance, the roof assemblies used 20 years ago may no longer be available, and the construction method used may have improved due to the efficiency of machinery. All these changes could contribute to a reduction of embodied energy. Furthermore, if a building material needs to be imported from a far distance, then the embodied impact of the building product could increase. In general, the uncertainty and complexity of incurring embodied energy assessments make it difficult to integrate them into the decision-making process.

Knowledge Gaps

The primary problems in current research on influential factors that affect embodied impact are the isolation and fragmentation of the parameters and lack of an agreed-upon method. Much of the existing research on EI is subject to inconsistent and insufficiently reported assessment methodologies, poor data quality, and a lack of technological and geographical representativeness (Azari & Abbasabadi, 2018). A principal challenge is the assessment methodology. There are a variety of quantitative methodologies for EI in the engineering field; however, the absence of standardized quantification protocols still remains a major problem (Chastas et al., 2016). There is also a need to develop methods to integrate uncertainties in the EI assessment, such as the inconsistent environmental inventory databases, nontransparent data collection methods, and lack of data in certain geographic areas. The second knowledge gap is a deficiency in the benchmarks of EI performance metrics. In the past decades, researchers and practitioners have developed robust benchmarks to evaluate buildings' operational energy use and related environmental impact. However, no benchmarks exist for EI yet; therefore, it is difficult to evaluate buildings' performance based on their EI. Developing EI performance metrics and benchmarks would help designers to compare the EI results of different design options or compare one building's EI with others that have a similar function, size, and location, in order to make improvements. Related to the second knowledge gap, the third gap represents the disproportion between EI research and practice. More clear and specific measurement guidelines need to be developed for designers to demonstrate and communicate the impact of building materials (BM), building function (BF), building construction type (BCT), and other influencing parameters. Existing building renovation guidelines must also offer more reliable data on the relationship between OE and EE performance. Lastly, the industry needs to work closely with the academic community to translate research findings into practice.

DISCUSSION

Barriers for the Embodied Impact Assessment

For years, the concept of embodied energy has been part of the sustainability debate. However, despite all the advantages of its inclusion in the life cycle energy analysis of buildings, there is currently little incentive to integrate the calculation of embodied emissions into the construction decision-making process (Hamilton-MacLaren et al., 2009). The increase in embodied energy and environmental impact (carbon emissions) suggests that, in the future, building developers may need to place more importance on their choice of building materials and other factors, leading to higher embodied energy (Ibn-Mohammed et al., 2013). As energy efficiency continues to improve for building operations, the percentage of embodied energy will become high in a building's life cycle.

It is essential to understand why progress in the adoption of embodied energy and impact has been slow in order to effectively target the actions and strategies that will lead to wider uptake. There are three main challenges limiting the adoption of embodied energy and impact assessment: technological, methodological, and perceptual challenges. Among these challenges, the most severe are methodological challenges. There is currently no generally accepted method available to compute embodied impact accurately and consistently. Consequently, wide variations in measurements are inevitable (Ibn-Mohammed et al., 2013).

Technological Challenges

Quantifying embodied energy and impact requires expertise and technical know-how in order to deliver reliable results. In one of the most comprehensive literature reviews of embodied energy, Chatas et al. (2016) examined 90 residential projects, with samples of available case studies from Europe, the United States, Australia, Canada, New Zealand, Lebanon, Turkey, India, and Brazil. The results revealed that limitations of current assessments of embodied impact from buildings could be characterized by a lack of a standardized procedure (Chatas et al., 2016). Due to the inconsistent datasets and complexity of analyses, there is currently no generally accepted method available to calculate embodied energy accurately and reliably (Acquaye, 2010). Additionally, the required lengthy and demanding data collection process complicates the calculation of embodied CO_2 since tracking material sources from all their origins requires reliable data on established manufacturing processes and supply chains (Hamilton-MacLaren et al., 2009).

Perceptual Challenges

Including embodied energy in development has not been embraced as business as usual. Assumptions spread that calculations are technically difficult and therefore not a solid financial investment for any given project. Ambiguity and uncertainty have persisted as to whether including embodied energy calculations is useful or necessary, what methods and datasets should be deployed, and how the diverse parameters influence embodied energy. The reasons are partly due to methodological challenges, a focus on regulations regarding in-use energy and carbon, a lack of appropriate legislation, and a lack of interest in the impacts of embodied energy by public and industry stakeholders (Hamilton-MacLaren et al., 2009).

Potential analysis framework

A variety of methodological challenges affect the reliability of embodied impact assessment, although there are several existing robust quantitative methods that have been applied in embodied energy estimation. It is beneficial to develop a consensus-based quantiðcation framework and protocol based on ISO prescriptions to tackle this challenge (Azari & Abbasabadi, 2018). Most recently, an integrated life cycle assessment (LCA) and Multi-Criteria Decision Analysis (MCDA) were studied and tested, attracting researchers' and practitioners' attention as a practical framework to assess embodied energy and impact.

Life Cycle Assessment (LCA) use in the Building Industry

A range of life cycle assessment methods is used for the quantification of embodied energy. There are three primary methods used by LCA modelers, researchers, and practitioners: the process-based analysis (P-LCA), input-output analysis (IO-LCA), and hybrid method (Chastas et al., 2016; Zhang & Wang, 2016).

The process-based analysis estimates embodied energy based on physical quantities. It is the oldest and most familiar method, although it has many limitations and is characterized by its incompleteness and poor definition of the system boundaries within buildings (Treloar, 1998; Lenzen, 2000), which leads to an underestimation of embodied energy and impact (Lenzen, 2000; Crawford, 2008).

The input-output analysis is based on financial quantities and considers the price of the building materials linked to the economic sector. Extending the system's boundary into an entire economy leads to more comprehensive and representative results (Azar et al., 2018). However, the translation between the economic sectors and their energy intensity is typically done in a "black box" fashion that introduces inaccuracy and uncertainties (Zhang & Wang, 2016; Chastas et al., 2016).

In order to address the disadvantages of both methods, a hybrid analysis method was proposed by different researchers in the 1990s. For instance, Treloar proposed a hybrid method based on a disaggregation of the IO-LCA model into direct energy paths (Building Green Inc, 2019). Four variants of the hybrid LCA exist: *tiered, input-output (I-O) based, integrated,* and *augmented process-based* (Azar et al., 2018). The four methods vary in their models and data: tiered and I-O based methods rely on I-O data, whereas integrated and augmented process-based methods depend on a process-based framework and data. The analysis results based on different assessment methods cannot be ignored. Crawford revealed an average difference of 64% between P-LCA and IO-LCA (Crawford, 2008; Lenzen et al. 2000). The differences in the embodied energy of a passive house when applying three of the methods are 13.1%, 28.5%, and 56.9%, respectively, for the P-LCA, IO-PLA, and hybrid IO-LCA methods (Crawford & Stephan, 2013).

Multi-Criteria Decision Analysis (MCDA) use in the Building Industry

Multi-criteria decision analysis (MCDA) is defined as a set of tools and approaches that provide a mathematical methodology, which incorporates the values of decision-makers and stakeholders (Myllyviita et al., 2016) as well as a variety of attributes that have a reliable and accurate assessment of embodied impact. MCDA has been suggested to have the greatest potential in future use to support sustainability assessments based on quantitative measurements (Ramanujan et al., 2014). Arroyo et al. (2012) used MCDA methods to decide which design alternative was most sustainable, while Rowley et al. (2012) used MCDA to aggregate the assessment results to create a comprehensive environmental profile of buildings. MCDA was also used when assessing building components, such as a sustainable roofing system (Collier et al., 2013), building materials (Akadiri et al., 2013), and the combined heat and power system (Alanne & Saari, 2004). The main motivation for applying MCDA to sustainable design and assessment was to account for multiple attributes and indicate the trade-offs and interdependencies from a variety of attributes and criteria.

CONCLUSION

Construction contributes largely to our climate change problem as it represents the single largest contributor to global greenhouse gases arising from primary energy demand (Ibn-Mohammed et al., 2013). Energy is a key issue, and energy impact represents the primary measurement to evaluate buildings' performance across their life cycle and is particularly useful for evaluating existing buildings. This chapter discussed the growing signiðcance of embodied energy and embodied impact; as new construction becomes more energy-efficient, the importance of embodied energy has grown. Four research clusters were identified and explored to better understand the trend of research activities surrounding embodied energy. The overall goal is for both new construction projects and existing buildings to be constructed or renovated with minimum embodied energy. Embodied energy is also a genuine indicator and, therefore, a legitimate assessor of environmental impact (greenhouse gas emissions). However, the current state of research is plagued by a lack of accurate and consistent data as well as a standardized methodology. A robust methodological approach, supported by a mathematical analysis that integrates multiple attributes of embodied impact into a module that weights and ranks them, embodies the ideal decision-making tool, as is outlined at the end of the chapter. This combined LCA and MCDA framework has been tested in recent years and represents a promising method to assess the embodied energy of buildings in a more consistent way. As such, it is important to continue refining the framework and validating the process.

ACKNOWLEDGMENT

This research received no specific grant from any funding agency in the public, commercial, or not-for-profit sectors.

REFERENCES

Abanda, F. H., Tah, J. H. M., & Cheung, F. K. T. (2013). Mathematical modelling of embodied energy, greenhouse gases, waste, time–cost parameters of building projects: A review. *Building and Environment*, *59*, 23–37. doi:10.1016/j.buildenv.2012.07.014

Acquaye, A. (2010). *A stochastic hybrid embodied energy and CO2_eq intensity analysis of building and construction processes in Ireland*. Academic Press.

Acquaye, A. A., Duffy, A. P., & Basu, B. (2011). Stochastic hybrid embodied CO2-eq analysis: An application to the Irish apartment building sector. *Energy and Building*, *43*(6), 1295–1303. doi:10.1016/j.enbuild.2011.01.006

Acuqaye, A. (2010). *A Stochastic Hybrid Embodied Energy and CO2-eq Intensity Anal-ysis of Building and Construction Processes in Ireland* (PhD thesis). Dublin Institute of Technology.

Adalberth, K. (1997). Energy use during the life cycle of single-unit dwellings: Examples. *Building and Environment, 32*(4), 321–329. doi:10.1016/S0360-1323(96)00069-8

Adalberth, K., Almgren, A., & Petersen, E. H. (2001). Life cycle assessment of four multi-family buildings. *International Journal of Low Energy and Sustainable Buildings, 2*(1), 1–21.

Akadiri, P. O., Olomolaiye, P. O., & Chinyio, E. A. (2013). Multi-criteria evaluation model for the selection of sustainable materials for building projects. *Automation in Construction, 30*, 113–125. doi:10.1016/j.autcon.2012.10.004

Alanne, K., & Saari, A. (2004). Sustainable small-scale CHP technologies for buildings: The basis for multi-perspective decision-making. *Renewable & Sustainable Energy Reviews, 8*(5), 401–431. doi:10.1016/j.rser.2003.12.005

Almeida, M., Ferreira, M., & Barbosa, R. (2018). Relevance of Embodied Energy and Carbon Emissions on Assessing Cost Effectiveness in Building Renovation—Contribution from the Analysis of Case Studies in Six European Countries. *Buildings, 8*(8), 103. doi:10.3390/buildings8080103

Arroyo, P., Tommelein, I., & Ballard, G. (2012). Deciding a sustainable alternative by 'choosing by advantages' in the AEC industry. In *Proc. 20th Conf. of the International Group for Lean Construction (IGLC), San Diego, CA* (pp. 41-50). Academic Press.

Atkinson, C., Hobbs, S., West, J., & Edwards, S. (1996). Life cycle embodied energy and carbon dioxide emissions in buildings. *Industry and Environment, 19*(2), 29–31.

Azari, R., & Abbasabadi, N. (2018). Embodied energy of buildings: A review of data, methods, challenges, and research trends. *Energy and Building, 168*, 225–235. doi:10.1016/j.enbuild.2018.03.003

Birgisdottir, H., Moncaster, A., Wiberg, A. H., Chae, C., Yokoyama, K., Balouktsi, M., ... Malmqvist, T. (2017). IEA EBC annex 57 'evaluation of embodied energy and CO2eq for building construction'. *Energy and Building, 154*, 72–80. doi:10.1016/j.enbuild.2017.08.030

Blengini, G. A. (2009). Life cycle of buildings, demolition and recycling potential: A case study in Turin, Italy. *Building and Environment, 44*(2), 319–330. doi:10.1016/j.buildenv.2008.03.007

Börjesson, P., & Gustavsson, L. (2000). Greenhouse gas balances in building construction: Wood versus concrete from life-cycle and forest land-use perspectives. *Energy Policy, 28*(9), 575–588. doi:10.1016/S0301-4215(00)00049-5

Brown, N. W., Olsson, S., & Malmqvist, T. (2014). Embodied greenhouse gas emissions from refurbishment of residential building stock to achieve a 50% operational energy reduction. *Building and Environment, 79*, 46–56. doi:10.1016/j.buildenv.2014.04.018

BSRIA. (n.d.). *Embodied energy calculations design tool*. Accessed June 17, 2019. https://www.bsria.co.uk/news/article/embodied-calculations/

Buchanan, A. H., & Honey, B. G. (1994). Energy and carbon dioxide implications of building construction. *Energy and Building, 20*(3), 205–217. doi:10.1016/0378-7788(94)90024-8

Building Green Inc. (n.d.). *The urgency of embodied carbon and what you can do about it*. Accessed June 19, 2019. https://www.buildinggreen.com/feature/urgency-embodied-carbon-and-what-you-can-do-about-it

Cellura, M., Guarino, F., Longo, S., & Mistretta, M. (2014). Energy life-cycle approach in Net zero energy buildings balance: Operation and embodied energy of an Italian case study. *Energy and Building, 72*, 371–381. doi:10.1016/j.enbuild.2013.12.046

Chang, Y., Ries, R. J., & Lei, S. (2012). The embodied energy and emissions of a high-rise education building: A quantification using process-based hybrid life cycle inventory model. *Energy and Building, 55*, 790–798. doi:10.1016/j.enbuild.2012.10.019

Chang, Y., Ries, R. J., & Wang, Y. (2010). The embodied energy and environmental emissions of construction projects in China: An economic input–output LCA model. *Energy Policy, 38*(11), 6597–6603. doi:10.1016/j.enpol.2010.06.030

Chastas, P., Theodosiou, T., & Bikas, D. (2016). Embodied energy in residential buildings-towards the nearly zero energy building: A literature review. *Building and Environment, 105*, 267–282. doi:10.1016/j.buildenv.2016.05.040

Chen, T. Y., Burnett, J., & Chau, C. K. (2001). Analysis of embodied energy use in the residential building of Hong Kong. *Energy, 26*(4), 323–340. doi:10.1016/S0360-5442(01)00006-8

Cinelli, M., Coles, S. R., Nadagouda, M. N., Błaszczyński, J., Słowiński, R., Varma, R. S., & Kirwan, K. (2015). A green chemistry-based classification model for the synthesis of silver nanoparticles. *Green Chemistry, 17*(5), 2825–2839. doi:10.1039/C4GC02088J

Clarke, J. (2007). *Energy simulation in building design*. Routledge. doi:10.4324/9780080505640

Cole, R. J., & Kernan, P. C. (1996). Life-cycle energy use in office buildings. *Building and Environment, 31*(4), 307–317. doi:10.1016/0360-1323(96)00017-0

Collier, Z. A., Wang, D., Vogel, J. T., Tatham, E. K., & Linkov, I. (2013). Sustainable roofing technology under multiple constraints: A decision-analytical approach. *Environment Systems & Decisions, 33*(2), 261–271. doi:10.100710669-013-9446-5

Crawford, R. H. (2008). Validation of a hybrid life-cycle inventory analysis method. *Journal of Environmental Management, 88*(3), 496–506. doi:10.1016/j.jenvman.2007.03.024 PMID:17485159

Crawford, R. H., & Stephan, A. (2013, January). The significance of embodied energy in certified passive houses. In Proceedings of World Academy of Science, Engineering and Technology (No. 78, p. 453). World Academy of Science, Engineering and Technology (WASET).

De Felice, F., Campagiorni, F., & Petrillo, A. (2013). Economic and environmental evaluation via an integrated method based on LCA and MCDA. *Procedia: Social and Behavioral Sciences, 99*, 1–10. doi:10.1016/j.sbspro.2013.10.465

Dixit, M. K., Culp, C. H., & Fernandez-Solis, J. L. (2015). Embodied energy of construction materials: Integrating human and capital energy into an IO-based hybrid model. *Environmental Science & Technology, 49*(3), 1936–1945. doi:10.1021/es503896v PMID:25561008

Dixit, M. K., Fernández-Solís, J. L., Lavy, S., & Culp, C. H. (2010). Identification of parameters for embodied energy measurement: A literature review. *Energy and Buildings, 42*(8), 1238-1247.

Dixit, M. K., Fernández-Solís, J. L., Lavy, S., & Culp, C. H. (2012). Need for an embodied energy measurement protocol for buildings: A review paper. *Renewable & Sustainable Energy Reviews, 16*(6), 3730–3743. doi:10.1016/j.rser.2012.03.021

Doh, J. H., & Panuwatwanich, K. (2014). Variations in embodied energy and carbon emission intensities of construction materials. *Environmental Impact Assessment Review, 49*, 31–48. doi:10.1016/j.eiar.2014.06.003

Dove, C. A., Bradley, F. F., & Patwardhan, S. V. (2019). A material characterization and embodied energy study of novel clay-alginate composite aerogels. *Energy and Building, 184*, 88–98. doi:10.1016/j.enbuild.2018.10.045

Dowden, M. (2013). Climate change and sustainable development: Law, policy and practice. *Estates Gazette*.

European Committee for Standarization. (n.d.). *CEN/TC 350- Sustainability of construction works.* Accessed June 24, 2019. https://standards.cen.eu/dyn/www/f?p=204:7:0:FSP_ORG_ID:481830&cs=181BD0E0E925FA84EC4B8BCCC284577F8

Brummer, G., & Pienaar, C. (2008). Embodied Carbon in Construction. *Green Design*. Retrieved from http://www.davislangdon.com/upload/Africa%20Research%20Publications/Article%20-%20Embodied%20Carbon%20Jnl%203%2008.pdf

Eaton, K. J., & Amato, A. (1998). A comparative life cycle assessment of steel and concrete framed office buildings. *Journal of Constructional Steel Research, 1*(46), 286–287. doi:10.1016/S0143-974X(98)00074-1

Foraboschi, P., Mercanzin, M., & Trabucco, D. (2014). Sustainable structural design of tall buildings based on embodied energy. *Energy and Building*, *68*, 254–269. doi:10.1016/j.enbuild.2013.09.003

Gertsev, V. I., & Gertseva, V. V. (2004). Classification of mathematical models in ecology. *Ecological Modelling*, *178*(3-4), 329–334. doi:10.1016/j.ecolmodel.2004.03.009

Getter, K. L., Rowe, D. B., Robertson, G. P., Cregg, B. M., & Andresen, J. A. (2009). Carbon sequestration potential of extensive green roofs. *Environmental Science & Technology*, *43*(19), 7564–7570. doi:10.1021/es901539x PMID:19848177

Goggins, J., Keane, T., & Kelly, A. (2010). The assessment of embodied energy in typical reinforced concrete building structures in Ireland. *Energy and Building*, *42*(5), 735–744. doi:10.1016/j.enbuild.2009.11.013

Government, H. M. (2010). *Innovation and growth team, in: Low Carbon Construction-Final Report*. Retrieved from http://www.bis.gov.uk/assets/biscore/business-sectors/docs/l/10-1266-low-carbon-construction-igt-ðnal-report.pdf

Griffin, C. T., Bynum, L., Green, A., Marandyuk, S., Namgung, J., Burkhardt, A., & Hoffman, M. (2013, June). Comparing the embodied energy of structural systems in parking garages. In *2nd International Conference on Structures and Architecture, ICSA 2013* (pp. 945-952). 10.1201/b15267-133

Hu, M. (2017). Balance between energy conservation and environmental impact: Life-cycle energy analysis and life-cycle environmental impact analysis. *Energy and Building*, *140*, 131–139. doi:10.1016/j.enbuild.2017.01.076

Hamilton-MacLaren, F., Loveday, D. L., & Mourshed, M. (2009). *The calculation of embodied energy in new build UK housing*. Academic Press.

International Energy Agency. (n.d.). *The future of cooling*. Accessed March 18, 2019. https://webstore.iea.org/the-future-of-cooling

Dixit, M. K., Fernandez-Solis, J. L., Lavy, S., & Culp, C. H. (2010). Identiðcation of parameters for embodied energy measurement—A literature review. *Journal of Energy and Buildings*, *42*(8), 1238–1247. doi:10.1016/j.enbuild.2010.02.016

Ibn-Mohammed, T., Greenough, R., Taylor, S., Ozawa-Meida, L., & Acquaye, A. (2013). Operational vs. embodied emissions in buildings—A review of current trends. *Energy and Building*, *66*, 232–245. doi:10.1016/j.enbuild.2013.07.026

Iddon, C. R., & Firth, S. K. (2013). Embodied and operational energy for new-build housing: A case study of construction methods in the UK. *Energy and Building*, *67*, 479–488. doi:10.1016/j.enbuild.2013.08.041

Jaffal, I., Ouldboukhitine, S. E., & Belarbi, R. (2012). A comprehensive study of the impact of green roofs on building energy performance. *Renewable Energy*, *43*, 157–164. doi:10.1016/j.renene.2011.12.004

Kabir, M. R., Rooke, B., Dassanayake, G. M., & Fleck, B. A. (2012). Comparative life cycle energy, emission, and economic analysis of 100 kW nameplate wind power generation. *Renewable Energy*, *37*(1), 133–141. doi:10.1016/j.renene.2011.06.003

Krantz, J., Larsson, J., Lu, W., & Olofsson, T. (2015). Assessing embodied energy and greenhouse gas emissions in infrastructure projects. *Buildings*, *5*(4), 1156–1170. doi:10.3390/buildings5041156

Krantz, J., Larsson, J., Lu, W., & Olofsson, T. (2015). Assessing embodied energy and greenhouse gas emissions in infrastructure projects. *Buildings*, *5*(4), 1156–1170. doi:10.3390/buildings5041156

Kua, H. W., & Maghimai, M. (2017). Steel-versus-Concrete Debate Revisited: Global Warming Potential and Embodied Energy Analyses based on Attributional and Consequential Life Cycle Perspectives. *Journal of Industrial Ecology*, *21*(1), 82–100. doi:10.1111/jiec.12409

Lee, B., Trcka, M., & Hensen, J. L. (2011). Embodied energy of building materials and green building rating systems—A case study for industrial halls. *Sustainable Cities and Society*, *1*(2), 67–71. doi:10.1016/j.scs.2011.02.002

Lenzen, M. (2000). Errors in conventional and Input-Output—based Life—Cycle inventories. *Journal of Industrial Ecology*, *4*(4), 127–148. doi:10.1162/10881980052541981

Lenzen, M., & Treloar, G. (2002). Embodied energy in buildings: Wood versus concrete—reply to Börjesson and Gustavsson. *Energy Policy*, *30*(3), 249–255. doi:10.1016/S0301-4215(01)00142-2

Lenzen, M., Wood, R., & Wiedmann, T. (2010). Uncertainty analysis for multi-region input–output models–a case study of the UK's carbon footprint. *Economic Systems Research*, *22*(1), 43–63. doi:10.1080/09535311003661226

Li, J., & Colombier, M. (2009). Managing carbon emissions in China through building energy efficiency. *Journal of Environmental Management*, *90*(8), 2436–2447. doi:10.1016/j.jenvman.2008.12.015 PMID:19344996

Lotteau, M., Loubet, P., & Sonnemann, G. (2017). An analysis to understand how the shape of a concrete residential building influences its embodied energy and embodied carbon. *Energy and Building*, *154*, 1–11. doi:10.1016/j.enbuild.2017.08.048

Menzies, G. F. (2011). *Embodied energy considerations for existing buildings*. Historic Scotland Technical Paper, 13.

Miller, D., Doh, J. H., & Mulvey, M. (2015). Concrete slab comparison and embodied energy optimisation for alternate design and construction techniques. *Construction & Building Materials*, *80*, 329–338. doi:10.1016/j.conbuildmat.2015.01.071

Mithraratne, N., & Vale, B. (2004). Life cycle analysis model for New Zealand houses. *Building and Environment*, *39*(4), 483–492. doi:10.1016/j.buildenv.2003.09.008

Moncaster, A. M., & Symons, K. E. (2013). A method and tool for 'cradle to grave' embodied carbon and energy impacts of UK buildings in compliance with the new TC350 standards. *Energy and Building*, *66*, 514–523. doi:10.1016/j.enbuild.2013.07.046

Myllyviita, T., Holma, A., Antikainen, R., Lähtinen, K., & Leskinen, P. (2012). Assessing environmental impacts of biomass production chains–application of life cycle assessment (LCA) and multi-criteria decision analysis (MCDA). *Journal of Cleaner Production*, *29*, 238–245. doi:10.1016/j.jclepro.2012.01.019

Nässén, J., Holmberg, J., Wadeskog, A., & Nyman, M. (2007). Direct and indirect energy use and carbon emissions in the production phase of buildings: An input–output analysis. *Energy*, *32*(9), 1593–1602. doi:10.1016/j.energy.2007.01.002

Ngo, T., Mirza, A., Gammampila, R., Aye, L., Crawford, R., & Mendis, P. (2009, September). Life cycle energy of steel and concrete framed commercial buildings. In *Solar09, Proceedings of the 47th ANZSES Annual Conference* (*Vol. 29*, pp. 1-10). Academic Press.

Organization, I. (2006a). *ISO 14040:2006-Environmental Management - Life Cycle Assessment - Principles and Framework, 2006*. Author.

Organization, I. (2006b). *ISO 14044:2006-Environmental Management - Life Cycle Assessment e Requirements and Guidelines, 2006*. Academic Press.

Ozoemena, M., Cheung, W. M., Hasan, R., & Hackney, P. (2014). *A hybrid Data Quality Indicator and statistical method for improving uncertainty analysis in LCA of a small off-grid wind turbine*. Academic Press.

Öztaş, S. K., & Ipekci, C. A. (2013). Evaluation of the embodied energy for building materials in Turkey. *Advanced Materials Research, 689*, 273–277. doi:10.4028/www.scientific.net/AMR.689.273

Peuportier, B. L. P. (2001). Life cycle assessment applied to the comparative evaluation of single family houses in the French context. *Energy and Buildings, 33*(5), 443-450.

Pomponi, F., & Moncaster, A. (2016). Embodied carbon mitigation and reduction in the built environment–What does the evidence say? *Journal of Environmental Management, 181*, 687–700. doi:10.1016/j.jenvman.2016.08.036 PMID:27558830

Portland State University. (n.d.). *Embodied energy + Carbon Calculators*. Accessed June 17, 2019. http://web.pdx.edu/~cgriffin/eecc/

Pullen, S. (1996). *Data quality of embodied energy methods* (Doctoral dissertation). Deakin University.

Pullen, S. (2000). Estimating the embodied energy of timber building products. Journal of the institute of wood. *Science, 15*(3), 147–151.

Ramesh, T., Prakash, R., & Shukla, K. K. (2010). Life cycle energy analysis of buildings: An overview. *Energy and Buildings, 42*(10), 1592-1600.

Reddy, B. V., & Jagadish, K. S. (2003). Embodied energy of common and alternative building materials and technologies. *Energy and Buildings, 35*(2), 129-137.

Robati, M., Daly, D., & Kokogiannakis, G. (2019). A method of uncertainty analysis for whole-life embodied carbon emissions (CO2-e) of building materials of a net-zero energy building in Australia. *Journal of Cleaner Production, 225*, 541–553. doi:10.1016/j.jclepro.2019.03.339

Rowley, H. V., Peters, G. M., Lundie, S., & Moore, S. J. (2012). Aggregating sustainability indicators: Beyond the weighted sum. *Journal of Environmental Management, 111*, 24–33. doi:10.1016/j.jenvman.2012.05.004 PMID:22813856

Saadatian, O., Sopian, K., Salleh, E., Lim, C. H., Riffat, S., Saadatian, E., ... Sulaiman, M. Y. (2013). A review of energy aspects of green roofs. *Renewable & Sustainable Energy Reviews*, *23*, 155–168. doi:10.1016/j.rser.2013.02.022

Sazedj, S., Morais, A. J., & Jalali, S. (2014). *Comparison of embodied energy and carbon dioxide emissions of brick and concrete based on functional units*. Academic Press.

Scheuer, C., Keoleian, G. A., & Reppe, P. (2003). Life cycle energy and environmental performance of a new university building: modeling challenges and design implications. *Energy and Buildings, 35*(10), 1049-1064.

Chapter 2
Building Codes Don't Measure Up:
A Case for Urban Material Performance Standards

Jeana Ripple
University of Virginia, USA

ABSTRACT

It is only in the case of fire that materials are considered by the American International Building Codes across an aggregation of scales (i.e., a building, a block, a district) leaving many other essential factors of material performance neglected. Mostly ignored are environmental and social parameters that also present forms of risk. This chapter uses the cities of New York and Chicago and three performance characteristics as case studies to examine additional material impacts at the city-scale. Case studies analyze material maintenance requirements against urban disinvestment, moisture absorption capacity against mold rates within flood-prone communities, and embodied carbon against material lifespan averages across cities. Findings reveal connections between material performance and economic, health, and energy implications across the city, suggesting the need for more broadly defined urban material performance standards.

INTRODUCTION

Early American city builders developed material regulations that define where and when specific building materials can be used based on a singular urban risk,

DOI: 10.4018/978-1-7998-2426-8.ch002

conflagration. Over the next century, building codes translated fire protection goals into rules addressing vulnerabilities at the building scale—including occupancy, building height, and property line proximity—to define the range of allowable building materials in specific locations. The results are twofold. First, the codes produced a product-scale material performance mentality in the construction industry with a myopic set of criteria. Second, the system gave rise to urban neighborhoods defined by a dominant building material with correlating delineations of socioeconomic vulnerability.

The current model of material-defining building codes succeeded in bolstering fire protection in American cities but also established a restricted system that overlooks less obvious performance outcomes. For example, the current regulations assume a static environment, do not account for intersectional urban risks, consider hours-long timeframes only, and ignore the impact of cumulative performance across neighborhoods. The term "intersectional" is used here as defined by legal scholar Kimberle Crenshaw to refer to the compounded impact of disadvantage from multiple sources (Crenshaw, 1989). By limiting the complexity of building codes, we also limit their capacity to adapt to unforeseen social or environmental impacts.

This chapter uses the cities of New York and Chicago, and three performance characteristics as case studies to examine material impacts beyond combustibility at the city-scale. The first case study investigates material maintenance requirements and urban disinvestment. Findings include a correlation between high-maintenance material concentrations and urban disinvestment indicators, particularly within economically vulnerable communities. Secondly, the chapter analyzes the evolution of moisture absorption capacity within combustible construction over the twentieth century and studies mold rates within wood-frame neighborhoods. Results suggest a link between highly absorptive material areas, flood-prone communities, and health risks. Finally, the chapter considers each construction type according to embodied energy and material lifespan averages across cities. This final case study also raises the issue of local resource availability and variations in embodied energy according to regional geology and fire-code requirements.

Overall results suggest the need for further analysis across intersectional material and urban risks and increased attention to material performance at the city scale. The structure of building codes and the visibility of their potential impacts must become more legible to foster experimentation, public conversation, and to improve the adaptive capacity of urban material stock.

Methodology: Scaling Up and Over Time

In recent years, significant efforts have been made to expand the scale of material performance measurement to cities, regions, and systems, particularly in analyzing or predicting energy use (Moe, 2013; Reinhart & Davila, 2016). However, drawing clear connections between building technology choices and urban consequences remains challenging. This chapter presents a blended methodology drawing from building technology, material science, cartography, and design. This study translates between varied scales of material measurement using a specific element of the United States building codes, called construction type assemblies, as both a proxy and a potential pathway to regulatory change.

Life Cycle Analysis (LCA) is one prominent methodology used to expand the timeline and scope of material performance in architecture and related disciplines. LCA is the measurement of total energy or total carbon used (and sequestered) in production, construction, and disposal of a material (Simonen, 2014). Several publicly available LCA databases compare the embodied energy and global warming potential (i.e., carbon dioxide equivalents) of building components (DeWolf & Ochendorf, 2014; Quartz, 2015; Pharos Project, 2019; DeWolf, 2019). Additionally, the International Standards Organization now includes carbon footprint calculation methods and environmental product declarations (International Standards Organization [ISO], 2015; ISO, 2017). However, inconsistencies in embodied energy and carbon measurements across sources still present data reliability challenges (Simonen et al., 2017).

Related methodologies such as Material Flow Analysis [MFA] extend this concept to the scale of the city (Kennedy et al., 2009). However, MFA studies often rely upon city-wide economic data to account for energy and material use, making it difficult to connect to specific building materials, technologies, or design strategies. Research teams like those led by structural engineer, Catherine DeWolf (2017) and building scientist, Christoph F.Reinhart (2016) bridge the scalar gap by creating "building archetypes" generated from building age, shape, height, and local observation to examine both embodied and operational energy across urban neighborhoods. In the field of industrial ecology, Reyna and Chester (2015) extend this methodology further by using municipal material data instead of archetypes to analyze urban building material stock according to residence time, materials, and associated environmental impact. While Reyna and Chester's material data set removes the need for archetypal speculation, it perhaps also reinstates a disconnect between urban performance and design decisions at the scale of material assembly.

This study builds upon the research described above by examining socioeconomic measures of sustainable urban life, such as health and economic stability along with energy outcomes. The primary source material for this research is residential building material data acquired from two case study cities, New York and Chicago. The municipal datasets include exterior wall material, parcel and building footprint, building height, and building age. Using Geographic Information Systems (GIS), this data is compared to additional geographic and social information, primarily from census and open city data sources, including topographic elevation, poverty rates, foreclosure rates, building abandonment reports, mold-related building violations, and asthma rates.

Construction Types I-V (Figure 1) then act as a proxy for the further quantification and translation of exterior wall information into specific assemblies. Construction types are defined by the International Building Code [IBC] according to fire resistance ratings, with the highest (Type I-A, one-A) consisting of protected non-combustible materials and the lowest (Type V-B, five-B) consisting of unprotected combustible materials (International Code Council [ICC], 2017). Types II (two) through IV (four) slowly decrease in fire resistance. For example, Type III (three) construction, known as "ordinary construction," most often consists of masonry exterior walls, wood floors, and a wood roof. Type IV construction is heavy timber, originally used to describe typical manufacturing and industrial buildings of the early twentieth century. Today, Type IV construction often consists of engineered lumber systems. Type IV is sufficiently rare in residential development as to be virtually excluded from public exterior wall data. However, due to growing interest in multi-story mass-timber construction, Type IV systems are expected to become increasingly common. International Building Code, table 721.1 prescribes a variety of material assemblies and specifies material thickness to satisfy each of the construction types. The most common selections from this table are the basis of this study's material

Figure 1. Construction types regulate fire resistance focusing on load-bearing elements, floor, and roof construction (Table 721.1, ICC, 2018). The most common selections from each type form the basis of analysis.
Source: Jeana Ripple, 2019

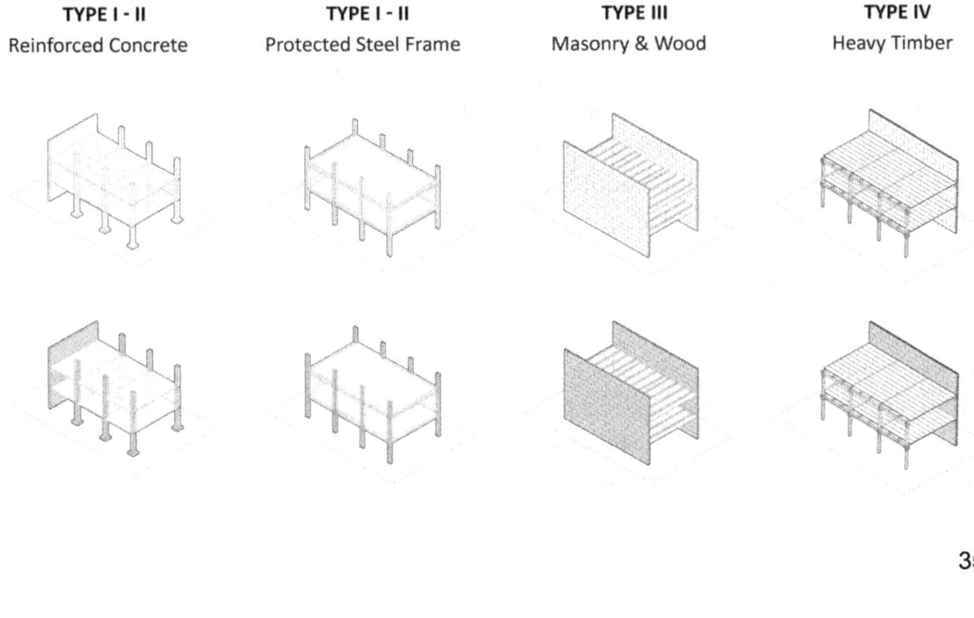

quantities. Because Type V construction is not rated for fire protection, the codes do not specify material components, thicknesses, and quantities. In order to compare Type V more precisely, several wood-frame wall assemblies, including wood, vinyl, and aluminum cladding were chosen from the industry-leading source used to quantify construction costs for typical assemblies (Mewis, 2019).

Cities require specification of construction type information during permitting. However, the author has yet to find a city that digitizes construction type data. In its absence, exterior wall data is sufficient to make the distinction broadly between non-combustible (Types I – III) vs. combustible (Type V) construction. Ongoing research by the author (not yet included in the findings below) uses building footprint and height information to translate exterior wall data into more specific construction type subsets across the city.

In each case study, construction type assemblies are quantified according to mass per square foot of envelope or occupiable floor area and compared according to multiple critical performance factors such as embodied carbon versus lifespan, service life versus cost, and water absorption versus decay rate. A graphic method common to material science, known as an Ashby chart (Ashby, 2017), uses a scatter plot grouped by material families to compare two simultaneous material properties. In this case, construction type assembly groups are used in place of material families to compare the relative performance of each according to concurrent measures.

Finally, this study relies on normative design and cartographic techniques of mapping (Figure 2, Figure 3) and site analysis to situate material conditions in a spatial, environmental context. Mapping and site analysis are considered standard within the design disciplines. They are also well-recognized as inherently subjective practices. Cartography has been historically used to enforce power and influence (Crampton & Krygier, 2018) and site analysis is often intentionally expressive of personal impressions, as described in Kevin Lynch's canonical, "Image of the City" (Lynch, 1960). In this case, perception is used as fodder for speculation and visual data comparison to guide more in-depth quantitative and material-scale analysis. By combining the observations enabled through mapping with quantitative material accounting, this multi-scalar and transdisciplinary research draws from design and stands apart from those methodologies derived purely from geography, building technology, or material sciences.

Construction Types: Anomalies in RANGE and Impact

Building codes generally regulate within the boundary of the single building on an individual site. Within the systematic, product-scale of American building codes, construction types are unique because they bridge scales, protecting not only the individual building occupants but also the city itself.

Figure 2. Types I-III versus Type V neighborhoods, New York City
Source: Jeana Ripple, 2019

Figure 3. Types I-III versus Type V neighborhoods, Chicago
Source: Jeana Ripple, 2019

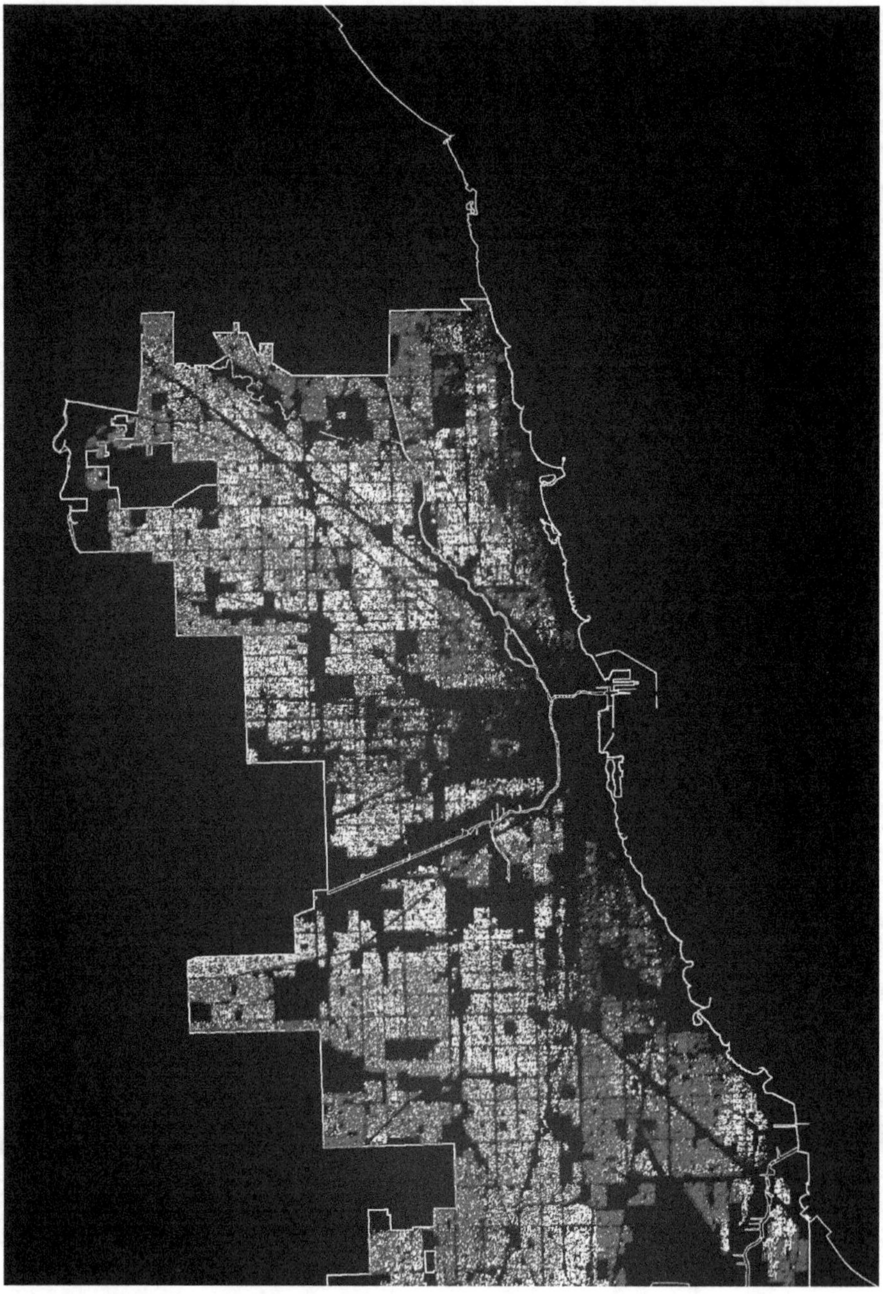

In their historical and theoretical analysis of American building codes, *Questioning Architectural Judgement: The Problem of Codes in the United States*, Moore and Wilson (2013) describe the dominant "prescriptive economic model" of building codes guiding North American and Western European code paradigms. "Prescriptive codes" differ from "performance codes" by dictating component-based performance standards rather than whole-building performance standards. Moore and Wilson characterize prescriptive economic codes as seeking civic economy, universal standards, and "see[ing] the built environment as an assembly of technological objects and spaces in which the relationship between humans and their environment is ordered at a distance, by experts." They go on to state that the "principal problem with prescriptive codes is, however, that code-makers—perhaps owing to their training—tend to be unaware of the complexity of the systems they hope to regulate. Simply put, prescriptive codes tend to have unintended consequences" (Moore & Wilson, 2013).

In support of this argument, Listokin and Hattis describe the historical development of American building codes resting on four primary influences: the insurance industry, the tenement and housing movements, the engineering profession, and the construction industry (Listokin & Hattis, 2005). They describe the combination of these social, economic, and technical perspectives leading to a consensus method of standard-making representing the interests of the building production economy equally to technical or social outcomes (Listokin & Hattis, 2005).

Municipal zoning ordinances focus on the city-scale through land-use, and therefore significantly impact the cumulative performance of buildings in a city. However, zoning ordinances are not explicitly measuring the performance of buildings, rather defining zones of allowable building uses (retail, industrial, residential, etc). As a result, the cumulative impacts of building technologies across cities can be considered the "unintended consequences" referenced by Moore and Wilson. There is one notable exception in the building codes, and it is particularly relevant in light of increasing environmental, health, and social concerns within cities.

Construction types within building codes are tied as closely to urban performance as to an individual building's safety. City builders developed construction types within building codes at the turn of the century as American cities were struggling to survive in the face of frequent fires. In the years that followed, building codes translated fire protection goals into a matrix of rules dictating the material that could be built in specific locations within the city, on a site, and within a building. The matrix considers occupancy, building height, and property line proximity to define and apply allowable material assemblies based on fire-resistance rating, i.e., the number of hours a material will survive in a fire. In one hundred years, construction types remained mostly unchanged, even in the specifics of their language. The focus of regulated material choice remains fixed on combustibility. However, urban conflagrations have drastically decreased over time, and fire-fighting infrastructure has improved (Wermiel, 2000). There is a noticeable decline in attention to construction material in the real estate market and a related relaxing of some construction type specifics such as height limits for combustible construction.

Meanwhile, the model and implementation of building codes have evolved more generally. Increasingly, performance-based codes allow for a more holistic analysis of building performance and opportunities for locally-specified criteria (Moore & Wilson, 2013). Performance-based codes allow testing of a proposed building system through simulation, therefore moving away from prescribing pieces and parts and toward an analysis of cumulative results. Although fire resistance can now be defined through performance testing, there is no cumulative urban scale performance measurement used to analyze the impact of material patterns across cities. Rather, construction types still operate solely on the pieces and parts, failing to study overall effects and potential unintended consequences.

This blind spot toward analysis and lack of evolution of construction types is surprising for several reasons. First among them, the data exists. Insurance companies, most notably Sanborn, began recording and mapping building material patterns in 1867 and continued until 1967 (Mueller, 2015). Sanborn map collections now reside in the Library of Congress for public use. More recently, cities began to digitize geographic data, including exterior wall materials. Still, material data remains mostly invisible and absent from city open data sources as well as most analysis of urban health, housing, and energy use. Secondly, construction types dictate the most environmentally and perhaps socially impactful components of a building and a city. The significant effect of exterior wall and roof materials on a building's lifespan and operational energy is well known (Lovell, 2010; Moghtadernejad, 2018). These same components, along with primary load-bearing elements, are also drivers of embodied energy and embodied carbon for buildings (Thormark, 2006). These same elements are the focus of construction type regulation. Considering the need to understand and account for the holistic impact of building design, construction types' role in shaping American cities offers a unique opportunity to develop communities according to energy investment, health, and economic goals.

THE RISE AND FALL OF MATERIAL AWARENESS

Sara Wermiel documents the transformational influence of fire concerns on the development of American buildings and cities in her historical book, "The Fireproof Building" (Wermiel, 2000). Wermiel also points out the distinctly American nature of the issue of urban material choice at the turn of the century. Almost all major conflagrations globally between 1815 and 1915 occur in North America (Wermiel,

2000). Wermiel attributes this imbalance to the substantial fireproof European building stock at that time, compared to the wood-framed construction that became popular in turn-of-the-century American cities.

A series of issues led to the American tendency toward wood-framed construction within cities. First, as American cities were built, new material technologies became available. For example, dimensioned lumber, mechanization, and "a rapidly growing distribution system of railways and water routes thoroughly entered the domain of building [through] wooden one-family homes" (Korvenmaa, 1990). Wood construction in new and growing cities offered an affordable material alternative to masonry or heavy timber construction because it involved simple techniques and relatively unskilled labor (Cronon, 1992). Second, American cities grew to include and protect single-family detached housing. Sonia Hirt (2013) describes the zoning of single-family detached areas within cities as a "uniquely American phenomenon." With less density, and smaller scale structures, light wood-framed construction remains a feasible option in many American cities today.

However, the question of wood construction safety and urban sustainability featured prominently in historical public debate. In Chicago following the Great Fire of 1871, disputes over allowable building material resulted in the creation of a new political party for the mayoral election called the "fireproofer's union" while workers marched in protest of wood construction bans (Sawislak, 1995). Like many American cities, Chicago's fire limits eventually outlawed wood construction within the city center and became a line of compromise between public safety and the social mobility enabled by affordable construction (Rosen, 1986).

The demand for fireproof material in the early twentieth century created economic opportunity through new building products. For example, entrepreneurs like Daniel Badger and James Bogardus produced catalogs of load-bearing cast iron, helping to mark the birth of the "fireproof" skeleton frame in the United States (Badger, 1982; Gayle, 1998). Cast iron systems eventually gave way to steel and terracotta in major cities like NY and Chicago (Leslie, 2013). The focus on combustibility outweighed other performance considerations, occasionally even usurping essential human comfort. For example, Sarah Wermiel (2000) references the "no tree rule," once established by mill insurance companies to outlaw the presence of trees near buildings.

Over the next one hundred years, awareness and concern over the choice of building material and its connection to urban geography appear to fade. For example, the 1940 Census of Housing was the last census to account for material composition—masonry versus frame—of urban dwellings across United States cities despite continued prevalence of urban wood-framed construction (Census, 1940; Census, 1950).

By the 1950s, advances in building technology and post-war housing demands shifted the public focus toward construction speed and affordability rather than urban material performance (Chow, 2002). This trend toward the primacy of affordability over performance characteristics continues today, most notably in the rapid expansion of "five over one" construction (Fox, 2019). Five over one is an apt term used to describe "podium construction" in building codes that often results in five stories of Type V (wood frame) construction over one story of Type I (noncombustible) construction (Podesto, 2015). The introduction of five over one first emerged on the west coast (conflicting accounts credit California and Seattle, (Hinshaw, 2015; Fox, 2019)), first using treated lumber to qualify as Type III multistory construction. In 2017, the IBC in increased the allowable height of Type V construction, now permitting up to five stories of combustible construction over a podium (ICC, 2017). Durability challenges in five over one construction are common, in part due to compounding multi-story expansion and contraction of wood (Podesto, 2015). The pervasive adoption of five-over-one is defining the next wave of material choice and real-estate investment trends in the American city, demonstrating a continued shift away from sensitivity to material in urban development.

In summary, the current conceptual framework of building code construction types effectively address threats to urban sustainability arising from fire. However, construction types also developed as an arcane and esoteric material regulation system that is largely invisible to the general public. The codes have become more relaxed over time while census or similar analysis of building material relative to other housing statistics has become scarce. Diminished public awareness of material vulnerabilities is currently combined with a narrow scope of performance criteria based solely on fire resistance. The following case studies compare neighborhoods defined by a dominant material, focusing on the socioeconomic and spatial impacts they produce.

CASE STUDIES

Case 1: Durability and Disinvestment

The first case assesses the impact of material durability on neighborhood disinvestment. Urban disinvestment, which includes economic decline, depopulation, and building abandonment, is an ongoing concern in many U.S. Cities (Raliegh & Galster, 2015). As neighborhoods experience symptoms of disinvestment, property values sink, and "property owners face increasing pressure to defer maintenance, mortgage payments, and property tax payments" (Raliegh & Galster, 2015). Myriad complex factors contribute to neighborhood disinvestment. However, research generally agrees that disinvestment perpetuates the deterioration of physical buildings (Ferriera et al., 2018). This case study investigates the degree to which material durability is a relevant performance indicator of neighborhood vulnerability to disinvestment by comparing building material with foreclosure and abandonment patterns throughout New York and Chicago.

Durability is used to refer to a building component's ability to withstand change, decay, or wear. Many factors can influence an assembly's longevity, including climatic conditions and susceptibility to moisture and mold, as will be discussed further in the next case study. However, this case study will focus on the average service life of the assembly's outer-most layers to anticipate the durability of the entire structure and a corresponding pace of maintenance cost requirements.

Figure 4 compares upfront construction costs and the service life of materials typically used to build and enclose construction Types I-III versus Type V construction assemblies. R.S. Means Company's annual publication of construction industry costs (R.S. Means, 2019) is the source of material and labor cost estimates for each element. Building component service life is best measured by analyzing the warranties of particular building products. This study uses a list of service life averages across products provided by the International Association of Certified Home Inspectors [InterNACHI] (InterNACHI, 2019).

Type V materials offer the lowest upfront investment, but also the shortest service life of weather barrier sealants and cladding. Type V systems, therefore, require the highest ongoing maintenance or replacement costs, requiring double or triple the replacement rate of Types I-III systems on average. Also, Type V load-bearing elements below these barriers are more susceptible to moisture and decay as will be discussed further in the next case study.

Figure 4 also compares maintenance timelines with residential ownership timelines, including tax and financial structures that may influence ownership patterns. For example, the National Association of Realtors reports that the 2018 average ownership tenure of American homes is nine years, a generally high number compared to a six-to-seven-year historical average (Mendenhall & Yun, 2018). The short average ownership may explain a lack of prioritization of material durability during construction. However, one would still expect to see a more significant emphasis on long-term durability within the resale market.

To test the correlation between material maintenance cycles and urban disinvestment patterns, New York and Chicago census tracts are compared according to dominant material, abandonment rates, and foreclosure rates (Figures 6 and 7). In Chicago, the median density of Type V housing is 33%. Chicago neighborhoods with over 50% Type V housing were considered high-density Type V neighborhoods. Chicago's high-density Type V neighborhoods demonstrate higher rates of abandonment (Chicago Open Data, 2019) and foreclosure (U.S. Census, 2015) than the city's

Figure 4. Upfront cost vs. service life of construction type building components
Source: Jeana Ripple, 2019

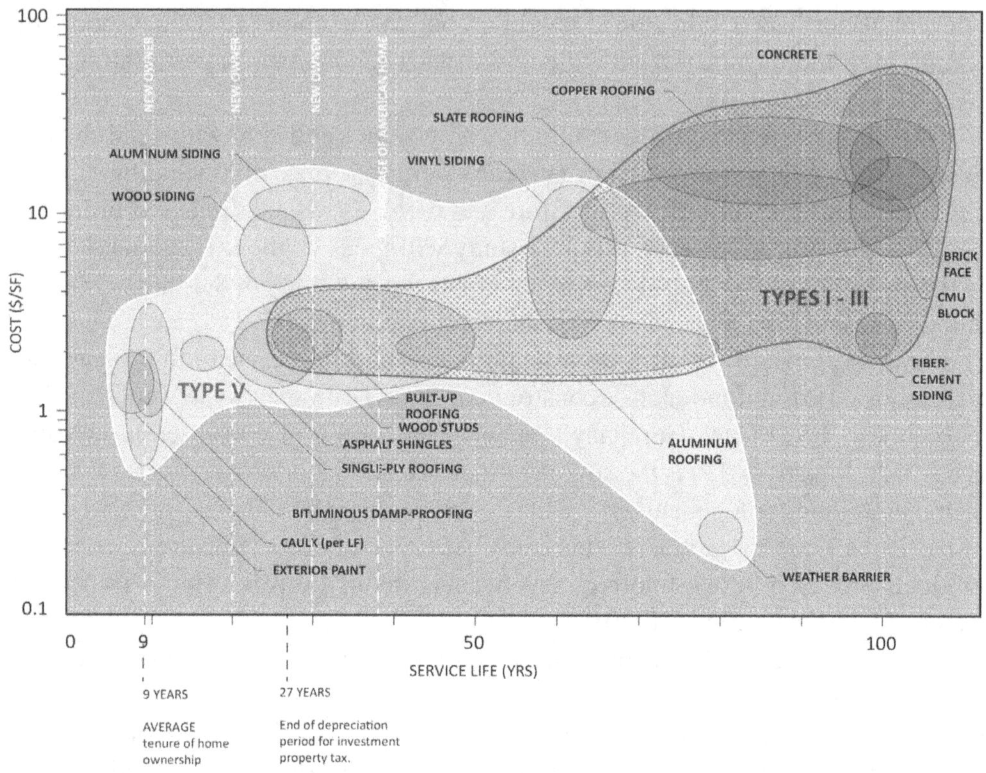

average, at 173% and 132% percent respectively. To contextualize these percentage increases, Chicago neighborhoods with an above-average family poverty rate (U.S. Census, 2015) were analyzed according to disinvestment correlation. The results (see Figure 5) indicate that building material is a reliable indicator of neglect and disinvestment, although not as strong an indicator as neighborhood poverty rate. Those neighborhoods with both high Type V density and above-average poverty rates have the highest percentage of abandonment and foreclosure over the city's averages, at 360% and 190% respectively. New York analysis demonstrated similar findings, with an overall median density of Type V housing per neighborhood at 31%. Communities with over 50% Type V housing showed higher rates of abandonment and foreclosure over the city's averages, at 152% and 175% respectively.

Urban disinvestment, including residential property abandonment, generate unsafe conditions, deter development, reduce property values, and create tax burdens for other surrounding owners (Han, 2014). An individual owner may be ill-equipped to counteract a phenomenon of neighborhood-level devaluation particularly when

Figure 5. foreclosure and abandonment rate trends according to poverty or Type V construction density
Source: Jeana Ripple, 2019

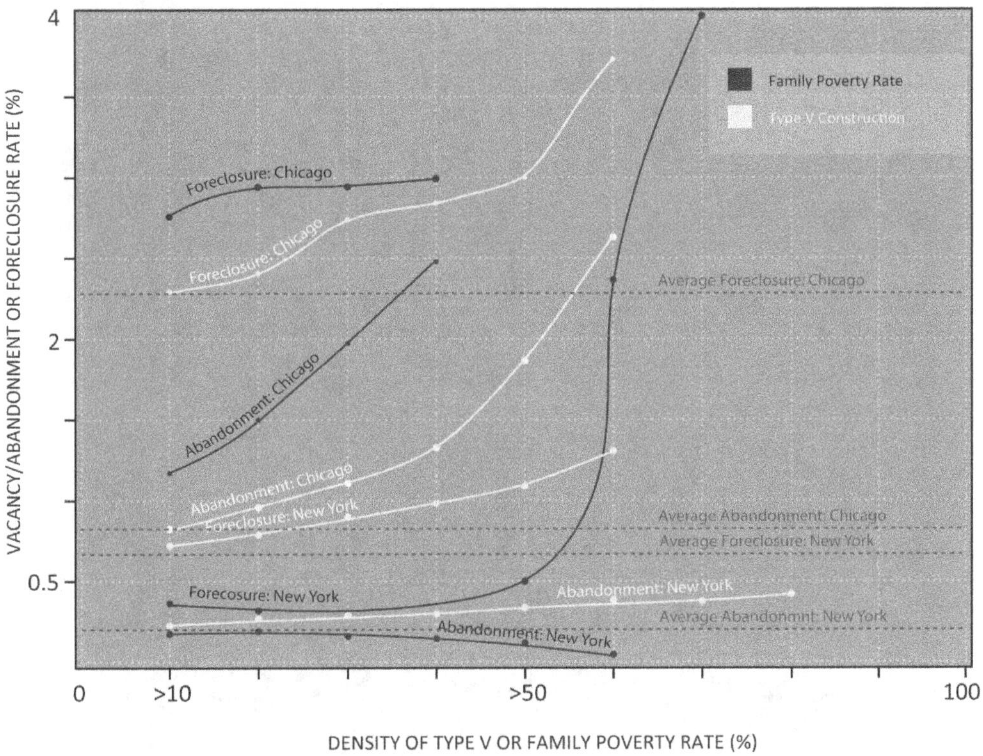

this is combined with ongoing high-maintenance costs. In addition, neighborhood conditions and urban decay are significant indicators for well-being, quality of life, and mental health (Geronimus, 2000). Environmental justice studies of neighborhood quality and health also establish the disproportionate impact of poor neighborhood conditions on communities of color (Wilson et al., 2008). A public health study of New York City neighborhoods in the 1990s relates urban disinvestment to famine, stating that it causes the "standard consequence of any famine, a raised death rate among those most susceptible" (Wallace & Wallace, 1990).

Findings of this case study do not point to wood frame construction as a sole cause of urban disinvestment. Rather, these findings raise the possibility that material properties may compound existing social and economic risks. Building code construction types are intended to protect collective vulnerability beyond the domain of an individual owner or the level demanded by the market. It is, therefore worth considering the risk that multiplying high-maintenance materials in urban neighborhoods may produce or exacerbate economic risk.

Figure 6. The density of Type V housing within high foreclosure zones in New York City
Source: Jeana Ripple, 2019

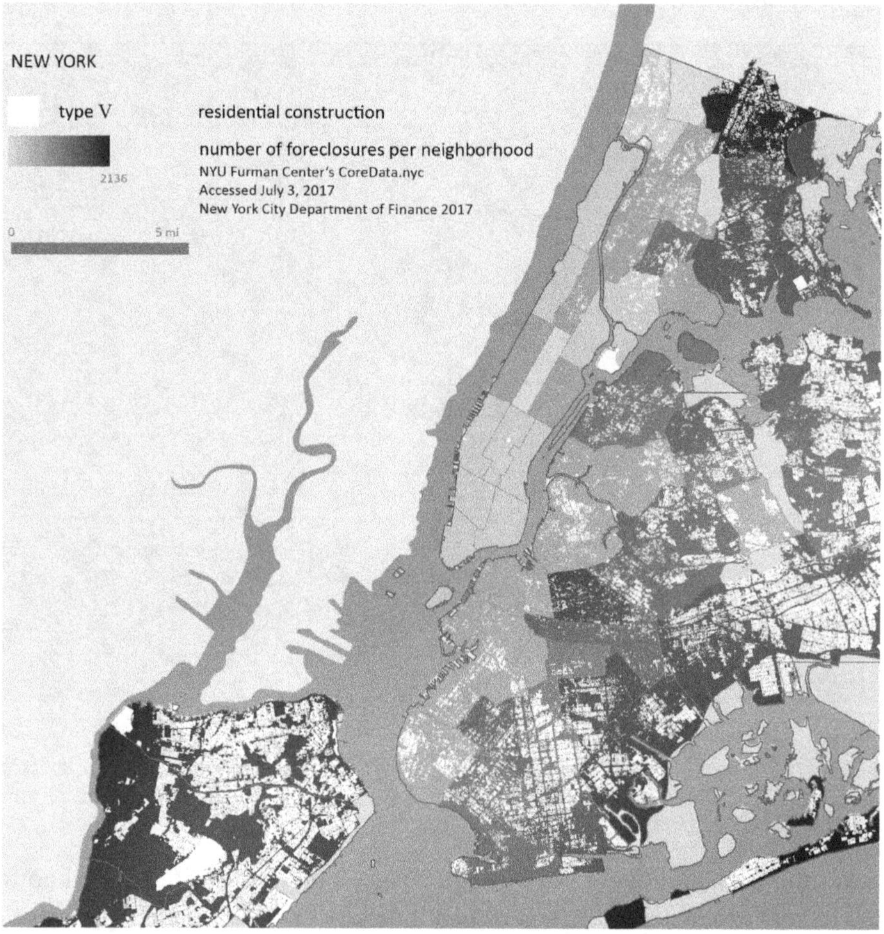

Case 2: Moisture Absorption and Health

The second case study investigates construction type assemblies according to water absorption capacity and health impacts according to material properties and urban data in New York City. This study considers two types of potential urban health implications related to material absorption. First, to what degree has the evolution of Type V construction over the last century increased its associated health risks? Second, can Type V construction be considered a health indicator at an urban scale? Specifically, this case study will focus on the relationship between combustible materials, water absorption, mold, and decay. Susceptibility to mold is a concern

Figure 7. The density of Type V housing within high foreclosure zones in Chicago
Source: Jeana Ripple, 2019

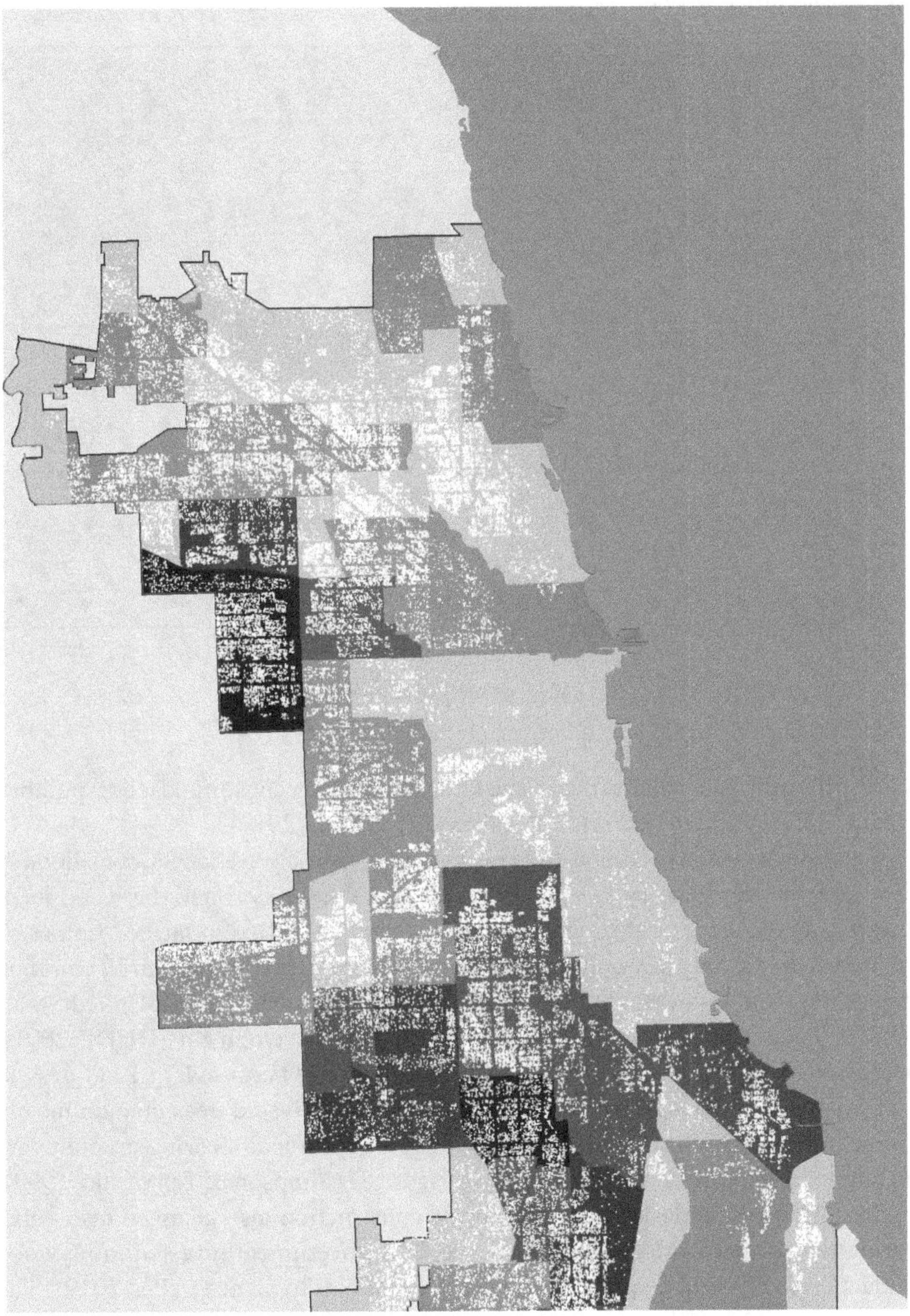

Figure 8. Moisture content vs. decay resistance comparison between species used early and late century Type V construction
Source: Jeana Ripple, 2019

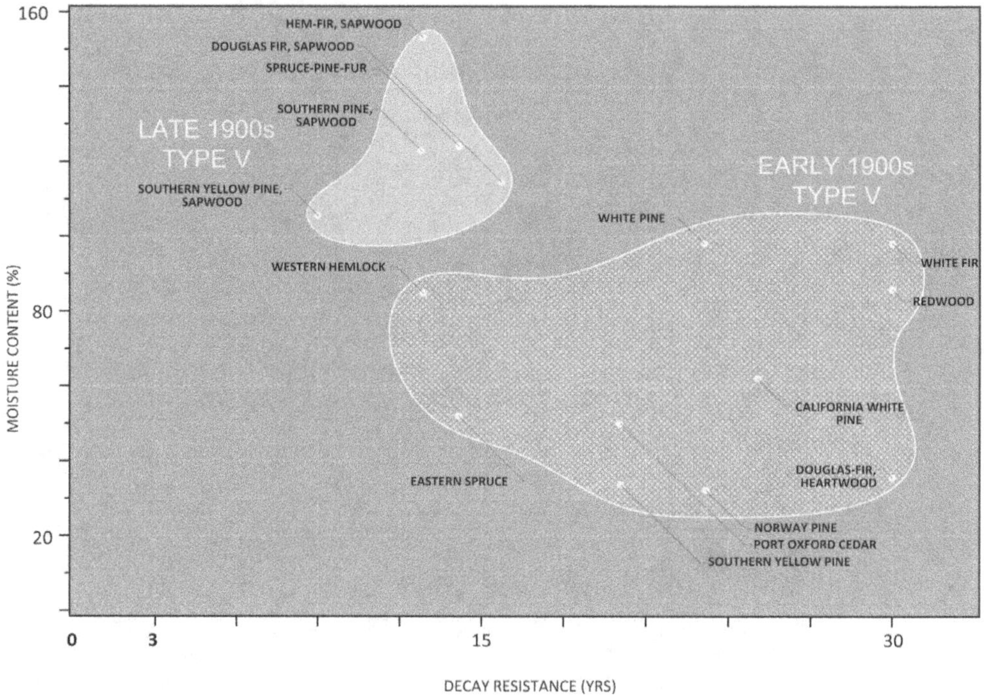

because it can lead to material decay and failure and also because mold is a respiratory irritant, linked to increased rates of asthma (Fisk et al., 2017).

The materials used in combustible residential construction changed considerably over the twentieth century in ways that decrease durability. First, the wood itself changed, as the use of old-growth timber necessarily shifted to farmed timber. A decrease in the variety of wood species common to construction—from 17 common species, down to 4—was a byproduct of this production shift in the United States (United States Department of Housing and Urban Development [HUD], 2001). Decreased lumber density is associated with this change because larger old-growth trees contain more dense heartwood while smaller cultivated trees contain mostly porous sapwood. Sapwood and, therefore, today's Type V load-bearing structures are thus more susceptible to mold and decay (Figure 8) (Simpson & TenWolde, 1999).

Sheathing and underlayment in Type V construction also changed over time. During the second world war (WWII), the U.S. government invested in plywood production to construct equipment ranging from boats to planes (APA, 2019). The engineered lumber industry was then poised to grow further to meet the post-war boom in housing demand. Plywood production nearly tripled in the United States from 1944 to 1954 (APA, 2019). In the 1970s, the industry developed a new form of sheathing known as Oriented Strand Board [OSB] (Engineered Wood Association [APA], 2019). OSB is created from small wood strands rather than sheets of veneer. Both plywood and to a higher degree, OSB, increase the ability for mold to reach the wood's cellulose fibers on which it feeds (Lstiburek, 2007).

The industry shift toward more insulation in frame construction also created moisture and mold challenges. In the 1930s, home builders observed that peeling exterior paint was a byproduct of insulated exterior walls, prompting the misinterpretation that, "insulation draws water" (Rose et al., 2011). In fact, wet cavity wall interiors were not uncommon before insulation. However, the introduction of both insulation less porous cladding material decreased the ability of wall cavities to sufficiently dry (Rose et al., 2011). This issue led to the use of plastic air and vapor barriers in wood frame construction. However, building scientists disagree about the extent to which these barriers create or ameliorate moisture problems within cavity walls (Lstiburek, 2007; Rose, 2005).

Figure 8 shows water absorption (Glass & Zelinka, 2010) and susceptibility to decay (Highley, 1995) for both the early-1900s and late-1900s Type V construction. The overall shift from heartwood to sapwood and from boards to engineered sheathing increases absorption rates, mold-susceptibility, and decay rate if exposed to moisture (Highley, 1995; Rose et al. 2011).

Using exterior wall material data from New York City, the neighborhood rate of Type V construction was tested against mold-violation complaints to determine whether a high density of Type V construction is an indicator of neighborhood mold or asthma risk. This analysis used mold-violation reports made to the city via 311 reporting (NYC Open Data, 2019), meaning owner-occupied units will be unlikely to register. This potential inconsistency requires further study to control for the rate of rental units per neighborhood. Chicago's publicly available 311 data did not include mold violations and was therefore not tested.

In particular, this inquiry focuses on the "inundation zones" impacted by flooding during Hurricane Sandy in 2012 (NYC Open Data, 2019). Municipal Housing and Urban Development analysis after the hurricane demonstrated that combustible construction was more susceptible to structural damage than other construction types. "Single story combustible construction represented 18 percent of the buildings in the inundated areas of the city, but 73 percent of all structurally damaged or destroyed buildings in the city" (Yanowitz, 2013).

Findings from this analysis indicate that the density of Type V construction is an indicator of higher mold risk in the inundation zones. Census tracts with more than 50% Type V construction within the inundation zone show a rate of 121% of mold violation reports compared to the city's median rate. Communities with less

Figure 9. The density of Type V housing within zones of predicted sea-level rise at an increase of 4 degrees Celsius.
Source: Ripple, 2019

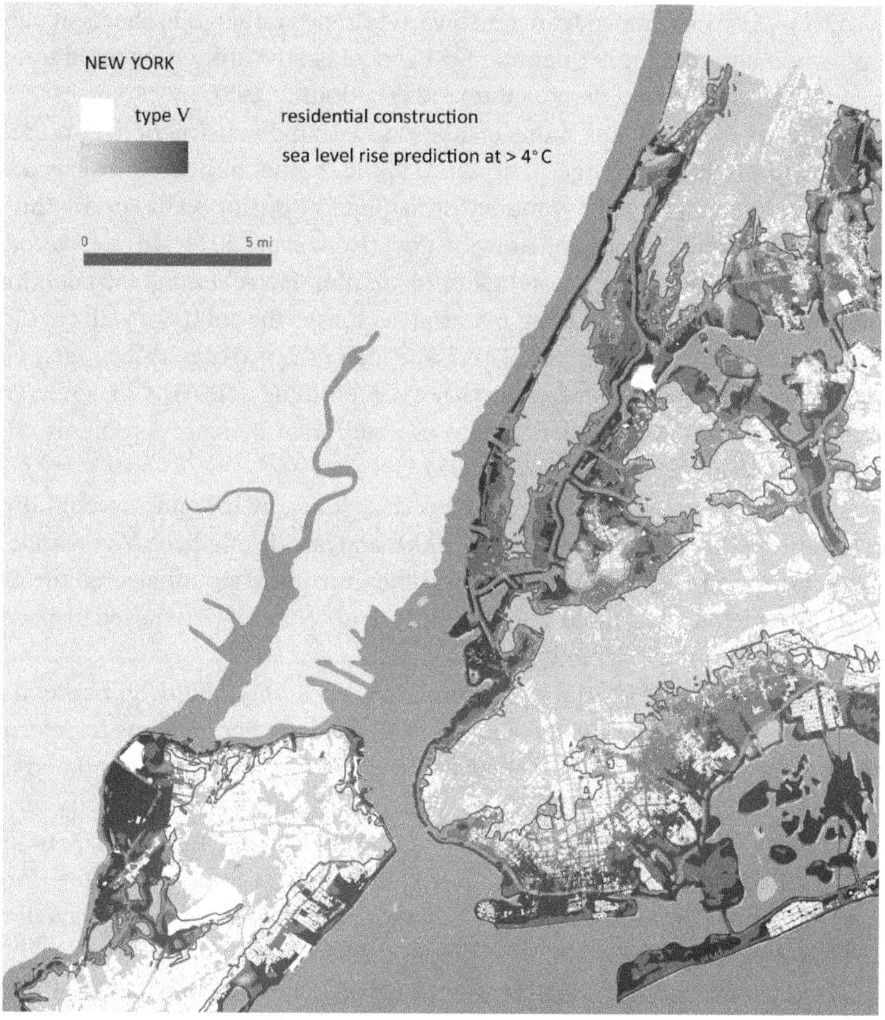

than 50% Type V construction tracts show a rate of 72% compared to the city's median. The poverty rate is a more significant indicator of the incidence of mold violation reports at a 156% mold report rate in high-poverty census tracts. Type V construction was not, however, an indicator of increased asthma rates or mold violation rates for the city overall. These latter findings may reflect the complexity of the housing market in New York and the variety of environmental factors that contribute to asthma risk. Research in other coastal cities is currently underway by the author to provide comparison data.

The Hurricane Sandy inundation area is relatively small compared to the coastal zone that is threatened by sea-level rise predictions. Figure 9 shows the density of Type V construction within the area that will be submerged if global temperatures increase by four degrees Celsius (NOAA, 2019). This map does not include the expanded area that will be threatened by storm surge, indicating that structural and moisture vulnerability of Type V construction methods is an increasingly urgent risk.

American building codes are not unique in their product-scale or combustibility-focus. Many other international building codes have similar fire-protection measures built into their structural requirements for buildings. However, the US *is* more limited in its discussion of material durability than some international codes, such as the Eurocodes. For example, American building codes only regulate wood durability in reference to materials proximate to the ground (ICC, 2017). By contrast, Eurocodes build in a broadly applicable classification system for various European wood species according to decay resistance and moisture content (Porteous, 2007).

Case 3: Embodied Carbon and Lifespan

The first two case studies demonstrate heightened economic and health vulnerabilities in Type V urban neighborhoods. However, the construction market will continue to integrate the use of wood products for affordability reasons. Wood is also a renewable resource and environmental sustainability is often cited as another benefit of wood construction. The final case study compares the relative embodied carbon across construction type assemblies in each case study city. Embodied carbon is used as a unit of energy measurement rather than embodied energy because the former takes into account the carbon sequestration benefit of wood construction.

Several recent studies in the fields of architecture and engineering have quantified both operational and embodied energy based on archetypal buildings and materials at the city scale (DeWolf et al. 2017; Reinhart & Davila, 2016). This case study examines New York and Chicago and adds to the growing body of energy research in two ways. First, by using building code assembly requirements, it is possible to quantify a minimum material-thickness for each construction type and to compare the relative embodied values per square foot of the wall, floor, and roof area. This calculation also takes into account the impact of local resources, specifically aggregate, on material thickness requirements according to fire codes. Second, using municipal material data and associated building ages across each city, this study can factor the average rate of replacement into embodied carbon calculations. This study is limited by the high degree of LCA data inconsistency, a well-documented issue under development in the industry (Simonen et al., 2017). However, LCA still represents one of the most promising measures in seeking to understand comprehensive ecological impacts of building material.

Figure 10 and Table 1 compare the embodied carbon and average service life of each construction type component. These analyses used an average building size to normalize results across assembly types. All construction types are allowable up to five stories above-grade, and 60'. Five stories and 60' was therefore used as the case study building height with a 60' x 25' footprint (standard city lot size). Standard selections and required material thickness for each construction type are based on IBC table 721.1 for Types I – IV. Because Type V is not fire-rated, specific material quantities are not prescribed by the code. Typical wood-frame assemblies and associated material quantities were sourced from the industry-leading reference for cost estimating (Mewis, 2019). For these embodied energy calculations, a wood-cladding and wood-sheathing assembly was assumed for Type V, including wood studs, insulation, and a vapor barrier for exterior walls. Although wood-cladding is less common in contemporary construction, it was chosen to look at the conservative, low-end of embodied energy and carbon calculations for Type V construction.

Building age and component lifespan also play an essential role in understanding a building's embodied energy. Both New York and Chicago have an older building stock and significantly older residential building ages than the country's average. In New York, the average Type I-III home is 82 years old, and the average Type V home is 75 years old. In Chicago, likely due to Chicago's increasingly strict fire codes, the average Type I-III residence is 83 years old, and the average Type V residence is 93 years old.

Additional case study cities around the country (Denver, Seattle, Tampa) were analyzed according to material and building age averages to further generalize an estimate of building age according to construction type. The compiled data across all five cities yields an average building age for Types I-III of 77.6 years and an average building age for Type V of 57.6 years. A 20-year difference in age is conservative relative to potential service lifespan estimates of each construction type. However, building age is dictated by more than the service lifespan of its pieces and parts. Proper maintenance can extend a less-durable material system, and functional obsolescence can cut a sound building's age short. Therefore, this empirical evidence across cities offers a mechanism to balance material durability with other harder-to-predict factors.

Table 1 shows the average embodied carbon and embodied energy calculated for each construction type using a percentage measurement of roof, wall, and floor components to provide values per square foot of occupiable floor area. Although Types IV and V have the highest embodied energy (primarily owing to insulation and vapor barrier values), they are the lowest in embodied carbon due to wood's sequestration of carbon during a tree's growth. High embodied rates for Type IV construction result from kiln drying and the lamination processes used to make contemporary heavy timber members. However, the potential for use of mass timber in

Table 1. Embodied energy and embodied carbon per construction type and lifespan averages

	values / SF of occupiable floor area*			
	Type I/II	**Type III**	**Type IV**	**Type V**
Embodied Energy [MJ]	111.2	107.9	141.7	126.9
Lifespan Adjusted Embodied Energy [MJ]	**111.2**	**107.9**	**141.7**	**171.3**
Embodied Carbon [kg CO2]	10.0	7.9	8.1	6.4
Lifespan Adjusted Embodied Carbon [kg CO2]	**10.0**	**7.9**	**8.1**	**8.6**
	These calculations use a 6-story building for each type, sized for an urban lot (25' x 60' x 60') and standard selections for each construction type (based on IBC table 721.1) for material measurement and proportions of roof, wall, and floor per sf of occupiable floor area;			

(LCA Data Sources: Quartz Open Data (2019))

taller buildings may significantly change its relative embodied energy and embodied carbon per square foot. Type III construction demonstrates the lowest overall value in embodied energy and the second lowest in embodied carbon.

When the embodied carbon and embodied energy estimates for are adjusted to account for a more frequent replacement rate of Type V construction by 35%, the relative carbon advantage of wood construction reverses to become the highest in embodied energy per square foot of all construction types (Table 1). After adjusting for building age and replacement rate, Types III represents the lowest embodied energy and embodied carbon values per square foot.

Current trends in the construction industry are utilizing Type V in multistory construction and Type IV in mass timber applications. The maintenance requirements of multistory wood frame construction due to compounded expansion and contraction rates may impact overall building service lifespan and affect or negate the energy advantages of wood. Types III and IV construction show potential for an advantageous balance of carbon sequestration and longevity, assuming the future evolution of engineered heavy timber can achieve the same lifespan average to that of Types I-III.

Figure 10 plots typical components for each assembly based on embodied energy and service lifespan. It also shows an increase in embodied carbon based on the type of aggregate used in concrete assemblies. This increase is not due to any significant energy distinction between aggregates, rather the different thickness and overall material quantity required by fire codes due to aggregate type. The most common aggregates defined by the fire code are siliceous and carbonate, with approximately

Figure 10. Embodied carbon vs. service life of construction type building components
Source: (Anonymized), 2019

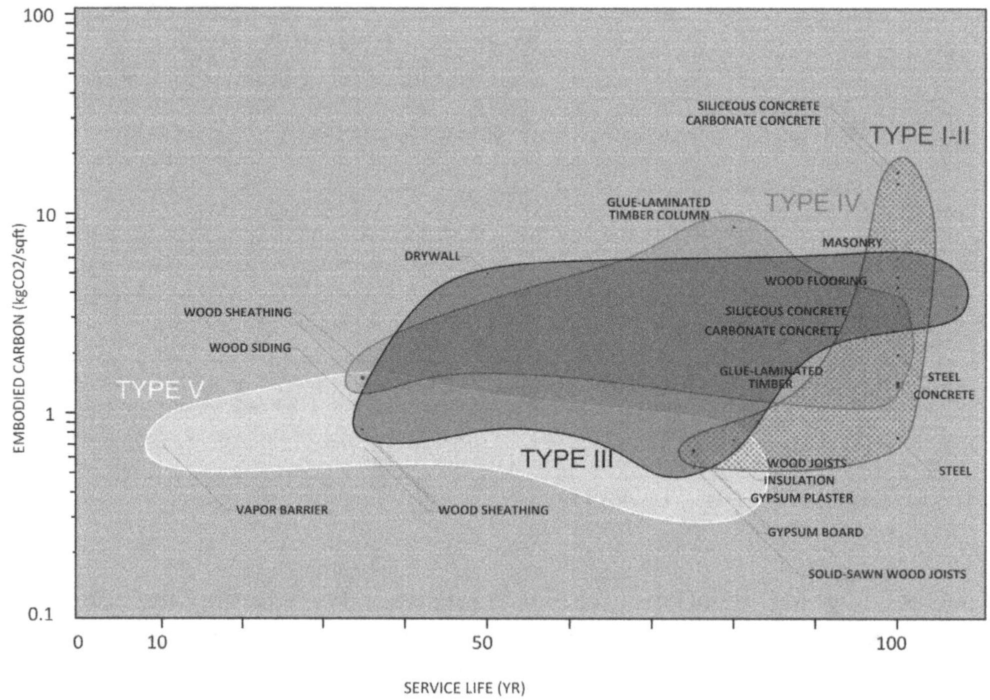

an inch more of siliceous concrete thickness required to meet carbonate concrete's fire resistance in every regulated component. Siliceous aggregate is primarily sandstone and granite, while carbonate consists of dolomite and limestone (Bilow & Kamara, 2008). According to the United States Geological Survey, it is financially impractical to transport aggregate beyond a roughly thirty-minute travel distance (Langer, 2011). Therefore, the local geology surrounding any city dictates its concrete makeup to some degree and impacts the overall calculation of energy implications.

Figure 11 shows the locations of siliceous and carbonate aggregate across the United States. New York's proximity to siliceous resource creates an embodied carbon multiplier of 8.77% per occupiable square foot of a building (when calculated using the generic building size outlined for this study) over the same concrete building in carbonate-proximate Chicago. The increase of 8-9% of embodied carbon per square foot is significant when considering the entirety of urban building stock. This example illustrates the local specificity with which construction types and their use should be considered, and is in direct conflict with the current one-rule-fits-all model of American building codes.

Figure 11. Siliceous and carbonate resource deposits
Source: (Anonymized), 2019

FINDINGS AND RECOMMENDATIONS

This study builds upon existing research seeking to quantify the ecological impacts of building material by integrating social, economic, and energy factors that impact the wellbeing of urban communities. The study relies upon the assumption that material-scale performance properties, such as the water absorption of wood might scale up to explain urban phenomenon such as prevalent mold in wood frame, flood-proneneighborhoods. While these assumptions drastically simplify the complex dynamics of urban communities, they are not fundamentally different from the current logic of building code material regulations. Current codes make a similar assumption that the behavior of a single building component is a determining factor in urban fire risk, regardless of complicating external factors.

Findings in this study can be summarized as follows. Construction type is an indicator of health and economic vulnerabilities at the neighborhood scale, including increased risk of disinvestment and increased susceptibility to mold within flood zones. The average building age of Type V construction within cities is lower than other construction types (with the exception of Chicago), which eliminates the relative embodied carbon advantages of Type V construction. Types III and IV (masonry envelope and heavy timber) show promise when considering a balance of durability, lifespan, and embodied carbon. The average age of urban residential buildings is significantly older than the national average of only 37 years (Zhao, 2017). While frame construction is unlikely to disappear completely, longer building age within cities should prompt consideration of regulatory demands on material investment according to service life, health, and energy.

CONCLUSION

The implications of these findings present complex challenges. For example, revising codes to address the risks discussed in this chapter might lead to further limitations of wood frame construction in urban areas. This would most certainly drive housing costs up, exacerbating affordable housing crisis and threatening the use of renewable resources for construction in many cities. Instead, the answers may lie in developing broader incentives and resources with these findings in mind. For example, the definition of "affordable" might be revised to take into account ongoing maintenance costs. Financial development incentives might encourage the construction or preservation of durable material structures. Information resources might clarify the advantages or disadvantages of local material resources. Information resources might also be used to raise public awareness of more diverse material performance considerations according to local urban environments and risks beyond combustibility. Whether or not the current regulations change, more diverse analysis of their consequences is needed.

Social scientist Ulrich Beck (2006) warned that "the experience of the past encourages anticipation of the wrong kind of risk, the one we believe we can calculate and control, whereas the disaster arises from what we do not know and cannot calculate." Urban scholar Eran Ben-Joseph (2005) argues that, over time, we internalize rules and codes to the point of forgetting the reason for their making in the first place. Analysis of the products and patterns generated by American material codes seem to fulfill both warnings, demonstrating both narrow performance objectives and diminishing public awareness that the choice of building material in cities is at all important. Encoded in these material choices and the patterns they establish, one can find a direct link between building codes, construction materials, financial policy, and overall quality of life, marking an essential arena for social and economic debate in the built environment.

ACKNOWLEDGMENT

The author thanks researchers Joshua Aronson, Jessica Smith, and Zazu Swistel for their essential contributions to this body of work. This work was supported by grants from the University of Virginia, School of Architecture.

REFERENCES

Ashby, M. F. (2017). *Materials selection in mechanical design*. Academic Press.

Badger, D. D., & Gayle, M. (1982). Badger's Illustrated Catalogue of Cast-Iron Architecture. New York: Dover Publications.

Bath Inventory of Carbon and Energy (ICE) | Greenhouse Gas Protocol. (n.d.). Retrieved August 6, 2019, from https://ghgprotocol.org/Third-Party-Databases/Bath-ICE

Beck, U. (2006). Living in the world risk society. *Economy and Society*, *35*(3), 329–345. doi:10.1080/03085140600844902

Ben-Joseph, E. (2005). *The Code of the City: Standards and the Hidden Language of Place Making*. Cambridge, MA: The MIT Press.

Bilow, D. N., & Kamara, M. E. (2008). Fire and Concrete Structures. *Structures Congress 2008*, 1–10.

Braubach, M., & Fairburn, J. (2010). Social inequities in environmental risks associated with housing and residential locations: A review of evidence. *European Journal of Public Health*, *20*(1), 36–42. doi:10.1093/eurpub/ckp221 PMID:20047933

Census Bureau. (1940). *1940 Census of Housing: Volume 2. General Characteristics*. Retrieved from United States Department of Commerce website: https://www.census.gov/library/publications/1943/dec/housing-vol02.html

Census Bureau. (1950). *1950 Census of Housing: Volume 1. General Characteristics*. Retrieved from United States Department of Commerce website: https://www.census.gov/library/publications/1953/dec/housing-vol-01.html

Chow, R. (2002). *Suburban Space: The Fabric of Dwelling* (1st ed.). Berkeley, CA: University of California Press.

Crampton, J. W., & Krygier, J. (2018). *An Introduction to Critical Cartography*. Retrieved from http://beu.extension.unicen.edu.ar/xmlui/handle/123456789/359

Crenshaw, K. (1989). Demarginalizing the Intersection of Race and Sex: A Black Feminist Critique of Antidiscrimination Doctrine, Feminist Theory and Antiracist Politics. *University of Chicago Legal Forum, 1989*, 139–168.

Cronon, W. (1992). *Nature's Metropolis: Chicago and the Great West (Reprint edition)*. New York: W. W. Norton & Company.

De Wolf, C. (n.d.). *deQo: Database of Embodied Quantity Outputs*. Retrieved August 5, 2019, from deQo: Database of Embodied Quantity Outputs website: https://www.carbondeqo.com/

De Wolf, C., Cerezo, C., Murtadhawi, Z., Hajiah, A., Al Mumin, A., Ochsendorf, J., & Reinhart, C. (2017). Life cycle building impact of a Middle Eastern residential neighborhood. *Energy, 134*, 336–348. doi:10.1016/j.energy.2017.06.026

Duthinh, D. (2014). Structural Design for Fire: A Survey of Building Codes and Standards. *Technical Note (NIST TN) - 1842*. Retrieved from https://www.nist.gov/publications/structural-design-fire-survey-building-codes-and-standards

Ferreira, F. A. F., Spahr, R. W., Sunderman, M. A., & Jalali, M. S. (2018). A prioritisation index for blight intervention strategies in residential real estate. *The Journal of the Operational Research Society, 69*(8), 1269–1285. doi:10.1080/01605682.2017.1390535

Fisk, W. J., Lei-Gomez, Q., & Mendell, M. (2007). Meta-analyses of the associations of respiratory health effects with dampness and mold in homes. *Indoor Air, 17*(4), 284–296. doi:10.1111/j.1600-0668.2007.00475.x PMID:17661925

Fox, J. (2019, February 13). Why America's New Apartment Buildings All Look the Same. *Bloomberg.Com*. Retrieved from https://www.bloomberg.com/news/features/2019-02-13/why-america-s-new-apartment-buildings-all-look-the-same

Gayle, C., & Gayle, M. (1998). *Cast-Iron Architecture in America: The Significance of James Bogardus* (1st ed.). New York: W. W. Norton & Company.

Geronimus, A. T. (2000). To mitigate, resist, or undo: Addressing structural influences on the health of urban populations. *American Journal of Public Health, 90*(6), 867–872. doi:10.2105/AJPH.90.6.867 PMID:10846503

Gibson, L. J., & Ashby, M. F. (1999). *Cellular Solids: Structure and Properties*. Cambridge University Press.

Glass, S., & Zelinka, S. (2010). *Moisture Relations and Physical Properties of Wood*. Academic Press.

Han, H.-S. (2014). The Impact of Abandoned Properties on Nearby Property Values. *Housing Policy Debate*, *24*(2), 311–334. doi:10.1080/10511482.2013.832350

Highley, T. L. (1995). Comparative durability of untreated wood in use above ground. *International Biodeterioration & Biodegradation*, *35*(4), 409–419. doi:10.1016/0964-8305(95)00063-1

Hinshaw, M. (2015, April 19). *The new Seattle, where everything looks the same*. Retrieved August 5, 2019, from https://crosscut.com/2015/04/the-new-seattle-where-everything-looks-the-same

Hirt, S. (2013). Home, Sweet Home: American Residential Zoning in Comparative Perspective. *Journal of Planning Education and Research*, *33*(3), 292–309. doi:10.1177/0739456X13494242

History of APA, Plywood, and Engineered Wood. (n.d.). Retrieved July 30, 2019, from APA – The Engineered Wood Association website: https://www.apawood.org/apas-history

International Association of Certified Home Inspectors (InterNACHI). (2019). *InterNACHI's Standard Estimated Life Expectancy Chart for Homes, InterNACHI*. Retrieved from https://www.nachi.org/life-expectancy.htm

International Code Council (ICC). (2017). *2018 International Building Code*. ICC.

International Organization for Standardization (ISO). (2015). *ISO 14001:2015 "Environmental management systems: Requirements with guidance for use."* Retrieved from https://www.iso.org/cms/render/live/en/sites/isoorg/contents/data/standard/06/08/60857.html

International Organization for Standardization (ISO). (2017). *ISO 21930:2017, "Sustainability in buildings and civil engineering works — Core rules for environmental product declarations of construction products and services."* Retrieved from https://www.iso.org/obp/ui/#iso:std:iso:21930:ed-2:v1:en

Kennedy, C., Steinberger, J., Gasson, B., Hansen, Y., Hillman, T., Havránek, M., ... Mendez, G. V. (2009). Greenhouse Gas Emissions from Global Cities. *Environmental Science & Technology*, *43*(19), 7297–7302. doi:10.1021/es900213p PMID:19848137

Korvenmaa, P. (1990). The Finnish Wooden House Transformed: American prefabrication, war-time housing and Alvar Aalto. *Construction History, 6*, 47–61.

Langer, W. (2011). *Aggregate Resource Availability in the Conterminous United States, Including Suggestions for Addressing Shortages, Quality, and Environmental Concerns* (United States Geological Survey No. Open-File Report 2011-1119). United States Department of the Interior.

Legget, R. F. (1958). *The Durability of Buildings*. doi:10.1520/STP39480S

Leslie, T. (2013). *Chicago Skyscrapers, 1871-1934*. University of Illinois Press. doi:10.5406/illinois/9780252037542.001.0001

Listokin, D., & Hattis, D. B. (2005). Building Codes and Housing. *Cityscape (Washington, D.C.), 8*(1), 21–67.

Lovell, J. (2010). *Building Envelopes: An Integrated Approach* (1st ed.). New York: Princeton Architectural Press.

Lstiburek, J. W. (2007a). The material view of mold. *ASHRAE Journal*.

Lstiburek, J. W. (2007b). The Perfect Wall. ASHRAE Journal, 49(5), 74,76,78.

Lynch, K. (1960). *The Image of the City*. Cambridge, MA: MIT Press.

Mendenhall, E., & Yun, L. (2018). *2018 Profile of Home Buyers and Sellers*. Retrieved from National Association of Realtors website: https://www.nar.realtor/sites/default/files/documents/2018%20HBS%20Highlights.pdf

Mewis, R. (Ed.). (2018). Building Construction Costs With RSMeans Data 2019 (Annual edition). Gordian.

Moe, K. (2013). *Convergence: An Architectural Agenda for Energy* (1st ed.). London: Routledge. doi:10.4324/9780203489444

Moghtadernejad, S., Chouinard, L. E., & Mirza, M. S. (2018). Multi-criteria decision-making methods for preliminary design of sustainable facades. *Journal of Building Engineering, 19*, 181–190. doi:10.1016/j.jobe.2018.05.006

Moore, S. A., & Wilson, B. B. (2013). *Questioning Architectural Judgment* (1st ed.). London: Routledge. doi:10.4324/9780203796658

Mueller, L. (2015). Sanborn Fire Insurance Maps: History, Use, Availability. *The Primary Source, 26*(2).

National Oceanic and Atmospheric Administration Office for Coastal Management (NOAA). (2019). *Sea Level Rise Viewer*. Retrieved August 6, 2019, from https://coast.noaa.gov/digitalcoast/tools/slr

NYC Open Data. (2019). *311 Service Requests from 2010 to Present*. Retrieved from New York City Department of Information Technology & Telecommunications website: /Social-Services/311-Service-Requests-from-2010-to-Present/erm2-nwe9, /Social-Services/311-Service-Requests-from-2010-to-Present/erm2-nwe9

Pharos Project. (n.d.). Retrieved August 5, 2019, from https://www.pharosproject.net/dashboard/

Podesto, L. (2015). Maximizing Value with Mid-Rise Construction. Wood Products Council.

Porteous, J. (2007). *Structural timber design to Eurocode 5*. Malden, MA: Blackwell Pub. doi:10.1002/9780470697818

Quartz: open data for a healthier, more sustainable future. (2015). Retrieved August 5, 2019, from Quartz website: http://quartzproject.org/

Raleigh, E., & Galster, G. (2015). Neighborhood Disinvestment, Abandonment, and Crime Dynamics. *Journal of Urban Affairs*, *37*(4), 367–396. doi:10.1111/juaf.12102

Reinhart, C. F., & Cerezo Davila, C. (2016). Urban building energy modeling – A review of a nascent field. *Building and Environment*, *97*, 196–202. doi:10.1016/j.buildenv.2015.12.001

Reyna, J. L., & Chester, M. V. (2015). The Growth of Urban Building Stock: Unintended Lock-in and Embedded Environmental Effects. *Journal of Industrial Ecology*, *19*(4), 524–537. doi:10.1111/jiec.12211

Ristow, W. W. (1968). United States Fire Insurance and Underwriters Maps · 1852—1968. *Quarterly Journal. Library of Congress*, *25*(3), 194–218.

Rose, W. B. (2005). *Water in Buildings: An Architect's Guide to Moisture and Mold* (1st ed.). Hoboken, NJ: Wiley.

Rose, W. B., Kaskel, B. S., Kudder, R. J., Mitchell, M. R., & Link, R. E. (2011). Insulation Draws Water. *Journal of Testing and Evaluation*, *39*(1), 102972. doi:10.1520/JTE102972

Sandy Inundation Zone | NYC Open Data. (n.d.). Retrieved August 4, 2019, from https://data.cityofnewyork.us/Environment/Sandy-Inundation-Zone/uyj8-7rv5

Sawislak, K. (1996). *Smoldering City: Chicagoans and the Great Fire, 1871-1874*. University of Chicago Press.

Simonen, K. (2014). *Life Cycle Assessment* (1st ed.). London: Routledge. doi:10.4324/9781315778730

Simonen, K., Rodriguez, B. X., & Wolf, C. D. (2017). Benchmarking the Embodied Carbon of Buildings. *Technology|Architecture + Design, 1*(2), 208–218.

Simpson, W., & TenWolde, A. (1999). Physical Properties and Moisture Relations of Wood. In *Wood handbook—Wood as an engineering material* (p. 463). Forest Products Laboratory.

Thormark, C. (2006). The effect of material choice on the total energy need and recycling potential of a building. *Building and Environment, 41*(8), 1019–1026. doi:10.1016/j.buildenv.2005.04.026

United States Department of Commerce. (2015). *2015 United States Census via Social Explorer*. Retrieved from https://www.socialexplorer.com/explore-maps

United States Department of Housing and Urban Development. (2001). *Review of Structural Materials and Methods for Home Building in the United States: 1900 to 2000*. Retrieved from https://www.huduser.gov/portal/publications/destech/revstruc.html

Wallace, R., & Wallace, D. (1990). Origins of public health collapse in New York City: The dynamics of planned shrinkage, contagious urban decay and social disintegration. *Bulletin of the New York Academy of Medicine, 66*(5), 391–434. PMID:2257377

Wermiel, S. E. (2000). *The Fireproof Building: Technology and Public Safety in the Nineteenth-Century American City*. Baltimore, MD: The Johns Hopkins University Press.

Wilson, S., Hutson, M., & Mujahid, M. (2008). How Planning and Zoning Contribute to Inequitable Development, Neighborhood Health, and Environmental Injustice. *Environmental Justice, 1*(4), 211–216. doi:10.1089/env.2008.0506

Yanowitz, B. (2013). *A Stronger, More Resilient New York*. Retrieved from New York City Economic Development Corporation website: https://www.nycedc.com/resource/stronger-more-resilient-new-york

Zhao, N. (2017, January 5). *The Aging Housing Stock*. Retrieved August 5, 2019, from Eye On Housing website: http://eyeonhousing.org/2017/01/the-aging-housing-stock-3/

KEY TERMS AND DEFINITIONS

Building Abandonment: The evacuation, relinquishment, and often neglect of a property by its owner.

Building Codes: Municipal regulation defining construction and design guidelines. The International Code Council describes its mission as the creation of model codes to protect building occupant health, safety, and welfare.

Building Lifespan: A building's expected useful service life. This is often up to double the average building age.

Building Stock: The collection of buildings in a city or region.

Construction Types: Material assemblies defined by building codes according to their fire-resistance rating.

Flashing: Part of a drainage system, typically protecting a joint between materials.

Intersectionality: The compounded impact of disadvantage from multiple sources.

Life-Cycle Analysis: Measurement of total energy or total carbon used (and sequestered) in production, construction, and disposal of a material.

Performance Codes: Giving directions that the building as a whole perform to a certain standard.

Prescriptive Codes: Giving directions that each component is built to a certain standard.

Service Life: Measurement of material durability; the period during which a material component of a building is functioning correctly, withstanding wear and decay.

Sheathing: Protective covering, typically between primary wall structure and exterior cladding.

Underlayment: Similar to sheathing, but operating between floor structure and floor covering.

Urban Disinvestment: The economic decline of a neighborhood.

Chapter 3
Teardown Index:
Emissions of Single-Family Homes in Vancouver

Joseph Dahmen
University of British Columbia, Canada

Jens von Bergmann
MountainMath Software and Analytics, Canada

Misha Das
University of British Columbia, Canada

ABSTRACT

Replacing older homes with new ones constructed to higher efficiency standards is one way to raise the operating efficiency of building stocks. However, new buildings require large amounts of embodied energy to construct, and it can take years before more efficient operations offset carbon emissions associated with new construction. This chapter looks at the carbon dioxide emission payback period of newly constructed, efficient single-family homes in Vancouver, British Columbia, where the authors find that it takes over 150 years for the operation to equal the embodied carbon associated with the of a typical high-efficiency new home. The findings suggest that current policies aimed at reducing emissions by replacing older homes with new high-efficiency buildings should be reconsidered.

DOI: 10.4018/978-1-7998-2426-8.ch003

INTRODUCTION

Buildings are significant drivers of climate change, generating one-third of global greenhouse gas emissions. Despite their widely acknowledged contribution to climate change, the design of green buildings does not always rest on evidence-based practices, and the policies governing green design frequently lack sound scientific judgment that delivers measurable environmental outcomes. For example, increased efficiency is often used as a justification for new green buildings. Replacing older, inefficient buildings with architecture designed to meet more stringent performance standards can indeed raise the overall operating efficiency of building stocks. However, it can take years before the embodied greenhouse gas emissions incurred by the new construction are offset by their more efficient operation. This chapter investigates the carbon dioxide emission payback period of newly constructed, efficient single-family homes in Vancouver, Canada, using it as a case study. The methods of the study can be applied to different contexts. The chapter finds that the carbon dioxide emission payback period for new homes meeting current efficiency standards in Vancouver averages 168 years, despite forty percent increases in operational efficiency over existing single-family homes. The length of this carbon payback period suggests that replacing older single-family homes with high-efficiency homes in Vancouver adds to—rather than reduces—overall emissions.

Rising carbon emissions in the face of aggressive performance standards can be explained by the fact that environmental performance is only one of many competing factors influencing decision-makers. Rising property values often compel property owners to tear down and replace older buildings as they seek to capitalize on growing market demand. In this scenario, more efficient building operation is, at best, a positive externality. A statistical model called the Teardown Index is presented, which correlates rising land value to increases in carbon emissions. This model indicates that replacing older poorly performing homes with new high-efficiency homes in Vancouver will result in 1.3–2.8 million tonnes of additional carbon dioxide equivalent emissions between 2017 and 2050. For each percent increase to the compound annual growth rate of property values, an additional 150 thousand tonnes of CO_2 equivalents will be released between 2017-2050. The findings suggest that current policies in Vancouver aimed at reducing emissions by replacing functional buildings with new high-efficiency buildings, while well-intentioned, should be reconsidered.

The research presented suggests that land economics have a significant impact on overall emissions from the built fabric due to their influence on construction cycles. The methods used in the study can be applied to other contexts to determine the carbon dioxide emission payback period when replacing existing structures. However, the carbon payback period of buildings is highly context-specific, and the results will vary greatly from region to region. Many factors influence initial embodied carbon emissions, operating carbon emissions, and the longevity of structures. These factors include the type of construction materials and methods (e.g., wood-frame or masonry), envelope design, morphology (single-family vs. multi-unit residential), climate (severe or mild), and the available energy sources (e.g., coal-fired versus hydroelectricity), each of which can affect the carbon payback period. The chapter concludes by looking at the relative effect of these factors on the conclusions reached for Vancouver and their influence on the applicability of the findings to different contexts.

BACKGROUND AND CONTEXT

Carbon dioxide (CO_2) emissions are significant drivers of climate change, which constitutes an intensifying threat to human and natural systems (IPCC 2014c). Buildings generate one-third of global greenhouse gas (GHG) emissions (Ürge-Vorsatz et al. 2007; Pérez-Lombard et al. 2008; IEA and OECD 2016; De Wolf et al. 2017) and consume approximately 30% of secondary energy. GHG emissions from buildings have more than doubled since 1970, and are projected to double or triple by 2050 (IPCC 2014a). Reducing CO_2 emissions of buildings will be a key component to meeting the Intergovernmental Panel on Climate Change (IPCC) recommendations, which call for the reduction of CO_2 emissions by a factor of 4 to prevent irreversible changes to climate (IPCC 2014b).

Operating and Embodied Impacts

Buildings cause environmental impacts at every stage of their life cycle, from construction to operation and maintenance, to decommissioning at end-of-life (Cabeza et al. 2014). Operating impacts refer to environmental impacts produced by basic building operations, such as heating and cooling, lighting, appliances and equipment, and plug loads. Embodied impacts refer to the environmental impacts associated with constructing the buildings themselves. Embodied impacts include upstream effects associated with producing construction materials, such as extracting raw materials, processing, and manufacturing, as well as the impacts of the construction itself, which include transporting materials to a site and assembling them. Finally, embodied impacts include maintenance or replacement of components during the useful life of the building, as well as decommissioning, disassembly, and recycling or disposal at the end of life (Reddy and Jagadish 2003).

Quantifying operating and embodied impacts requires a consistent methodology to ensure accurate assessments and comparisons. Methods for measuring the embodied impacts of buildings have been reviewed by many authors (Treloar et al. 2003; Menzies et al. 2007; Dixit et al. 2010; Chang et al. 2010; Abanda et al. 2013) and can be broadly categorized into (1) input-output, (2) process and (3) hybrid-based analysis. Life Cycle Assessment (LCA), which is a leading internationally accepted method that can be used to compare the environmental impacts of similar products or services (ISO 2006), is becoming among the most commonplace metric used for buildings. LCA has been employed to compare the cumulative operating and embodied impacts of buildings on the environment during all lifecycle stages, from products and assemblies to construction, use, end-of-life and beyond (Buyle et al. 2013; Pomponi and Moncaster 2016).

Although LCA is governed by ISO standards, comparisons across studies can be complicated due to a lack of agreement about system boundaries. These unresolved debates include whether to measure primary versus secondary energy or account for indirect as well as direct effects of resource use (Huijbregts et al. 2004; Lloyd and Ries 2007; Bribián et al. 2009; Dixit et al. 2012; Ibn-Mohammed et al. 2013). Also, Pomponi and Moncaster (2016) demonstrate that the duration investigated in LCAs varies. 90% of existing LCAs cover less than 60% of a building's life cycle stages, and often little data is available on the use and end-of-life stages. The same study also demonstrates that the lack of consistency in system boundaries (Matthews et al. 2008) and varied assumptions yield strong variations in results that prevent effective benchmarking across studies. Haapio and Viitaniemi's (2008) analysis of 17 environmental assessment tools concludes that comparing different LCAs is "difficult, if not impossible" due to variations in building types or lifecycle stages between studies, and a general lack of methodological transparency. Despite these challenges, LCA studies of buildings can provide a useful approximation of the impacts of buildings on the environment, particularly as regards the relationship of operating to embodied impacts. A survey of 73 LCA's of conventional (as opposed to high-performance) residential and commercial building indicates that operating impacts typically account for between 75-90% of environmental effects over a period of 50 years (Ramesh et al. 2010; Cabeza et al. 2014).

The Interrelation of Operating, Embodied, and Total Impacts

Current policies aimed at reducing the environmental impacts of buildings in North America and Europe have focused exclusively on operating impacts (Dixit 2017), in keeping with their dominant contribution to overall impacts. However, focusing exclusively on operating impacts remains, at best, an incomplete strategy, which fails to acknowledge the interrelationships between operating and embodied impacts. For example, a common method of reducing operating energy is to increase insulation, which adds to the embodied impacts of the building (Sartori and Hestnes 2007).

Research has shown that embodied impacts can reach up to 60% of total impacts in high-efficiency houses (Thormark 2006; Sartori and Hestnes 2007; De Wolf 2014). As building operations become more efficient, reducing the total impacts of buildings will require a thorough understanding of the relationship of operating to embodied impacts.

Emission Payback Period

The relationship between operating and embodied impacts can be understood in terms of its payback period, a term borrowed from finance that refers to the length of time required to recover the cost of an initial investment. We use the term "emission payback period" to refer to the length of time required for the reductions to operating CO_2e emissions to equal the initial "investment" of embodied CO_2e (carbon dioxide equivalent) emissions released during upgrades required to improve efficiency. The emission payback period is a function of the amount of embodied CO_2e emissions released, the time it takes for the consequent reductions to operating CO_2e emissions to equal the embodied impacts, and the service life of the new building or renovation. For example, Verbeeck and Hens (2010b) show that in the Belgian context, the embodied impacts of upgrading building envelopes with additional insulation are paid back in less than three years by reduced energy consumption. Once the initial investment of embodied CO_2e emissions are recovered, reductions to overall CO_2e emissions result. The CO_2e emission payback period can also illustrate the ineffectiveness of certain measures. For instance, Huang et al. demonstrate that shading devices on a particular Hong Kong building facade have a minimal effect on energy consumption and never achieve payback, adding to CO_2e emissions over the life of the building, rather than reducing them (Huang et al. 2012). A growing body of literature attests to the importance of assessing the emission payback period of environmental impact mitigation strategies (IPCC 2014a), which makes it possible to assess the viability and effectiveness of potential environmental impact mitigation strategies.

Case: Vancouver, BC

We use the emission payback period to assess the environmental effects of the recent development of single-family homes (SFHs) in Vancouver, Canada. Between 2005-2017, residential property prices in Vancouver increased by 261%, with a 28% price increase in the last three years alone (Real Estate Board of Vancouver 2017). This dramatic rise in property values has produced Canada's most expensive real estate (Holzhey and Skoczek 2016). One result of rapidly escalating property values is that increasing numbers of older SFH are torn down and replaced as owners seek to maximize property values in areas where nearly all available lots are already built out. Between 1985 to 2017, 26,800 SFH, or 40% of all SFHs in residentially (RS) zoned areas, were demolished and replaced with new SFH (BC Assessment 2015b, von Bergmann and Dahmen 2016). BC Assessment standards dictate that when 65% or more of a home is replaced, it is considered new construction; less than 65% is considered a renovation. During the period of 2005-2011, 97% of all permitted construction activity in BC was above the 65% standard; only 3% of construction activity consisted of renovations or 206 of 6521 permits issued (BC Assessment 2015b). The data shows that complete teardown and replacement of existing SFH is currently the dominant approach in the Vancouver context.

Replacing older building stock with newer buildings constructed to higher efficiency standards in Vancouver is often defended by local authorities on the grounds that it reduces operating emissions, despite numerous studies that have shown renovation to be an effective strategy for reducing operating emissions (Ma et al., 2012; and Rysanek and Choudhary 2013). It is generally (though not always) true that new buildings built to higher energy standards reduce operating emissions. In the case of RS-zoned areas of Vancouver, new SFH must meet aggressive efficiency performance standards that restrict the CO_2e emissions of new homes to 12 kg $CO_2e/m^2/yr$, from the 2007 baseline of 23 kg $CO_2e/m^2/yr$ (Vancouver 2016). This represents a nearly 50% gain in efficiency, which is substantial. At least one study has shown that similar gains could also be realized through renovation instead of total replacement (Mohammadpourkarbasi and Sharples 2017). However, approximately 97% of all SFHs in Vancouver consist of wood-frame construction, with older homes insulted minimally or constructed without insulation altogether. The relatively poor quality of typical wood-frame construction here results in renovations that preserve little more than the shell of the building, and frequently include pouring a new concrete foundation. While establishing the impacts of these deep renovations is beyond the scope of the current research, they likely approach that of new construction. When embodied emissions are considered, the environmental justification for both total replacement or deep renovation is questionable. What is the overall effect of teardown and replacement on total emissions for SFH in Vancouver? Answering this question requires calculating the emission payback period for the structures in question, at the scale of the individual building lot, and the city as a whole.

Research Contribution

This paper presents a method for assessing total CO_2e emissions resulting from SFHs in Vancouver, at the scale of a single building lot and the entire city. The method includes operating and embodied impacts, and the payback period required to offset embodied impacts through reductions to operating impacts is calculated. At the scale of the lot, the research compares the typical average environmental impact of a new SFH against the existing SFH it replaces, by calculating the emission payback period and comparing it to the impacts of the original structure. At the scale of the city, the research establishes a relationship between rising real estate values and changes to the replacement rate of SFHs in Vancouver. Finally, by applying the environmental impacts established in 1) to the correlation of real estate values with replacement rates in 2) enables us to 3) assess the impact of changing land values on the total CO_2e emissions of the city's SFH stock.

The research focuses exclusively on RS (Residential Single) areas of Vancouver, which at the time of writing are zoned for detached single-family and duplex homes with a secondary basement suite. New construction has only slightly higher rates of secondary suites than the existing stock, so new SFH construction adds few additional dwelling units to the overall supply in the city. Laneway homes can also be added to existing lots without impacting the main house, but we ignore them in this discussion, as their construction does not typically require demolition of an existing home. With few exceptions, available building lots in RS-zoned areas of Vancouver are completely built out. As a consequence, almost all new SFH construction in these areas follows the teardown of an existing SFH. As such, the RS-zoned areas of the city offer a unique opportunity to compare the environmental consequences of replacing existing buildings with more efficient buildings of the same type.

The research is motivated by a desire to create a method capable of establishing the total impacts of building replacement that accounts for embodied as well as operating impacts. The methods presented enable a more accurate assessment of impacts and outcomes associated with replacing older buildings with newer, more efficient structures, as well as the time required to realize any projected savings. Sound policymaking must rest on accurate information. The research will be of value to municipalities seeking evidence-based environmental policies to meet building emissions targets currently in place and, ultimately, to strengthen climate change mitigation efforts.

Materials and Methods

Assessing the environmental impacts corresponding to different rates of replacement of existing single-family homes requires three elements: 1) a method for establishing operating and embodied CO_2e emissions of new and existing SFH building stock; 2) a method for establishing the period of time, referred to as the "emission payback period" required to recover the CO_2e emissions released to build the new, more efficient building and 3) a model capable of predicting how many SFH will be replaced, to enable projected outcomes into the future at the scale of the city.

Although operating and embodied impacts include impacts on human health, ecosystem quality, natural resource use, and climate change and others, for the sake of simplicity, the study employs global warming potential (GWP), which indicates the contribution to climate change expressed in CO_2 equivalents (CO_2e) (Pomponi and Moncaster 2016). As defined by RS zoning, SFHs refer to detached homes, which includes duplex homes and auxiliary dwelling units where they occur on the same property, but excludes low- and high-rise multi-unit residential buildings.

Overview of Operating Emissions

Operating CO_2e emissions from SFHs in Vancouver come from two primary sources: electricity and natural gas. The latter is used primarily for space and water heating in homes where electricity is not used for those purposes. Estimating operating CO_2e emissions requires establishing usage rates for electricity and natural gas, and then converting energy usage in each category to associated CO_2e emissions based on geographically specific conversion factors determined by the methods used to produce the energy from primary sources. Approximately 66% of SFH energy demand in Vancouver is met by electricity, of which 98% is produced from low-carbon renewable sources such as hydroelectricity (BC Hydro 2017). The uncommonly high percentage of renewable energy in Vancouver means that emissions per unit of energy consumed in the city are lower than regions that rely on non-renewable, high-carbon sources of energy. Determining energy consumption for existing buildings or new construction requires different methods, as indicated below.

Operating Emissions of Existing Buildings

Establishing the operating CO_2e emissions associated with existing SFHs of different vintages in Vancouver is difficult due to a lack of specificity about the age and location of buildings in available data sets documenting actual energy use. However, a reasonably accurate account of operating CO_2e emissions of SFH by vintage can be achieved through comparison and extrapolation of available consumption data sets. Energy consumption by end-use (e.g., space heating, lighting, etc.) and typical energy source (gas, electricity) for SFHs in Lower Mainland, (the region surrounding and including Vancouver), was obtained from BC Hydro and Power Authority (BC Hydro), which supplies electricity to Vancouver (Appendix: Table 4, Young 2017). BC Assessment data was used to calculate the areas of SFHs in the Lower Mainland, (BC Assessment 2015b), so that energy consumption data by end-use and energy source could be converted to consumption per area (Appendix: Table 5). This yielded a general picture of consumption rates across all vintages.

Space heating and ventilation energy requirements per square meter account for 48% of SFH energy demand in Vancouver. To generate vintage-specific operating CO_2e emissions, emissions from these sources were adjusted using gross thermal output by vintage cohort and by building type data. The raw data was drawn from the Natural Resources Canada Comprehensive Energy Use Database (CEUD) (Natural Resources Canada 2014) for SFHs across BC (Appendix Table 6). The CEUD database indicates that gross thermal output across Canada varies widely by building vintage, with older buildings generally consuming more thermal energy. The remaining end-use categories (hot water, space cooling, appliances, lighting, electronics, and other) were held constant across vintages, as it was assumed that consumption in these categories does not vary significantly by the age of the building. Finally, the average consumption of natural gas and electricity per square meter was calculated by vintage for SFH in the Lower Mainland. (Appendix: Table 7). This secondary energy data was converted to primary energy, scaling up consumption by factors of 2.05 for electricity and 1.02 for natural gas (Energy Star 2013) to account for the energy required to extract, process, and deliver the fuel to the building (Deru and Torcellini 2007). Carbon emissions associated with electricity and gas use were calculated using regionally-specific environmental impact conversion factors provided by the British Columbia Ministry of Environment (British Columbia 2016). As a final step, the data was checked for accuracy by comparing it with the Vancouver Community Energy and Emissions Inventory (British Columbia, 2012), which measures aggregate CO_2e emissions at the municipal scale, but does not differentiate by building type or age. The aggregate operating CO_2e emissions calculated were found to be within 10% of the Vancouver Community Energy and Emissions Inventory, which was deemed acceptable.

The operating energy data sets considered demonstrate that operating energy consumption, and the corresponding CO_2e emissions, generally vary with building age in the Canadian context. This finding is supported by previous studies (by Swan and Ugursal 2009; Kavgic et al. 2010; Tooke et al. 2014a; Tooke et al. 2014b), which have demonstrated the viability of using building age as a predictor of energy consumption and GHG emission rates for bottom-up and archetypical analyses of multiple buildings. The operating emission rates of Vancouver SFHs by vintage (using NRCan's CEUD vintage cohorts) are summarized in Table 1.

Operating Emissions of New Buildings

Operating CO_2e emissions for newly built SFHs are assumed to follow guidelines established by the Zero Emissions Building Plan (ZEBP, Vancouver 2016), a municipal policy first proposed in Vancouver in 2016. While the policy is still under review, it establishes an aspirational municipal emission target that will exceed

Table 1. Vancouver SFH primary operating CO_2e rates and area metrics by vintage

Vintage	Operating Carbon Emissions ($kgCO_2e/m^2$/ year)	Number of SFH	SFH Internal Floor Area (m^2)	Average Lot Area (m^2)	Average Unused Buildable Area (m^2)
Prior to 1946	**15.42**	20,147	266	499	108
1946–1960	**14.17**	10,606	220	499	154
1961–1977	**12.43**	9,014	259	467	91
1978–1983	**11.00**	3,180	304	432	20
1984–1995	**9.78**	11,909	411	502	-34
1996–2000	**8.74**	3,227	357	487	8
2001–2005	**8.56**	3,447	316	475	40
2006–2010	**7.99**	2,938	368	493	1
2011–2017	**7.74**	3,468	392	520	-2

Source: Dahmen, von Bergmann and Das (2017), derived from data from the following sources: BC Assessment (2015b), Natural Resources Canada (2014), Energy Star (2013) British Columbia (2012), Eddie Young (BC Hydro).

provincial requirements in British Columbia. The ZEBP mandates that SFH built from 2020 to 2025 emit at most 7 kg CO_2e/m^2/year, or approximately 50% less than SFH constructed prior to 1946. Allowable operating carbon emissions of SFH will be reduced to 0 kg CO_2e/m^2/year after 2025. The research assumes that SFHs built from 2017 to 2020 will emit at the same rate as SFHs built from 2011 to 2017.

Although the ZEBP reduces operating CO_2e emissions for newly built SFHs on a per square meter basis, the average single-family home is growing in size. The average single-family homes built since 2000 has an internal space of 359m^2, a gain of 16% when compared to the average internal size of SFH's constructed prior to 2000. Homes constructed between 2011-216 contained 32% more internal space than homes constructed prior to 1946. Figure 1 below shows the rise in floor space ratio (FSR), which is the ratio of the total floor area of the home to the lot size of single-family homes constructed in Vancouver from 1905 - 2015. Almost all new single-family homes are now built to the maximum FSR, which is the main metric the city uses to regulate the maximum size of a building. The rise in FSR beginning in 2009 corresponds to the laneway house legislation, which enabled property owners to build secondary structures on RS lots. Single-family homes are getting larger as a result of increased FSR. Gross floor area (GFA) is an absolute measure of building size. The increase in overall size mitigates somewhat the reductions to overall operating impacts of new construction brought about by the ZEBP and also increases the embodied impacts of new construction.

Figure 1. Floor space ratio and gross floor area in Vancouver 1905-2016
Source: Authors, derived from data provided by BC Assessment (2015b).

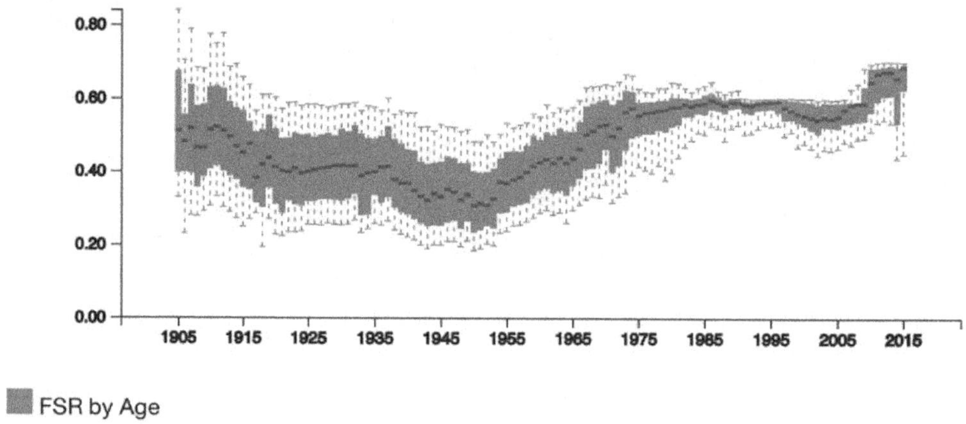

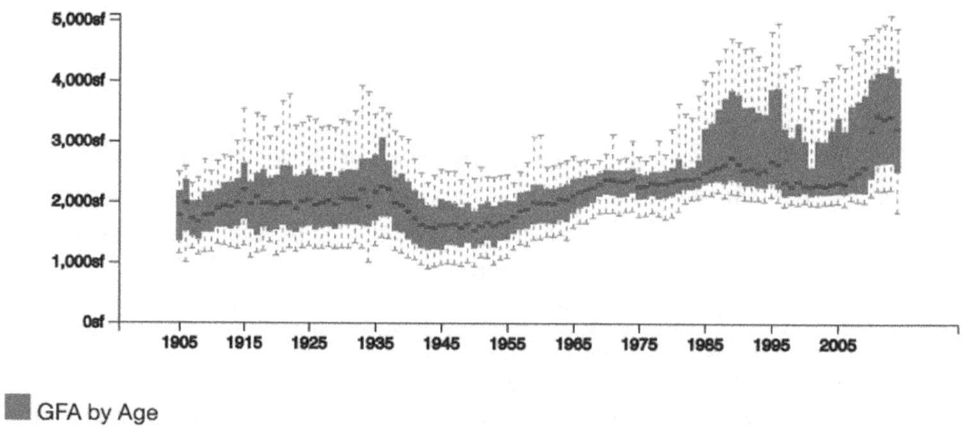

Embodied Emissions

Embodied CO_2e emissions can be divided into two categories: initial and recurring (Ibn-Mohammed et al. 2013). Initial embodied CO_2e emissions do not repeat and are predominantly concentrated in lifecycle stages preceding the use stage. Recurring embodied CO_2e emissions reflect the emissions resulting from maintaining, repairing, and replacing building components that have shorter lifespans than the overall building itself, such as the roof of exterior cladding.

Embodied Emissions of Existing SFHs in Vancouver

The materials required to replace existing SFHs with new buildings produce embodied CO_2e emissions that would not otherwise be incurred. Embodied CO_2e emissions are thus calculated for new construction, whereas the embodied CO_2e emissions of existing SFHs are considered to be a "sunk costs" and are not calculated, as they do not affect the CO_2e emission payback period. There is one exception, however. Recurring embodied CO_2e emissions have been applied to existing SFH. Verbeeck and Hens (2010a) establish that embodied CO_2e emissions increase with building age due to recurring embodied CO_2e emissions. An LCA by Treloar et al. (2000) investigates a two-story semi-detached residence building in Australia constructed with brick veneer with a timber upper floor, finding that recurring embodied energy added 32% to the initial embodied energy outlay after 30 years, while the LCA by Zhang et al. (2014) of a wood-framed SFH in Vancouver revealed that 33% of the total embodied CO_2e emissions over the 60-year lifespan of the building were attributed to recurring embodied carbon, representing an annualized rate of 1 kg $CO_2e/m^2/$year. Thus, a rate of 1 kg CO_2e/m^2/year of recurring embodied CO_2e emissions has attributed to existing SFHs in Vancouver.

Embodied Emissions of New SFHs in Vancouver

Embodied emissions of buildings vary widely by building type (e.g., residential vs. commercial) and construction method (e.g., wood-frame versus masonry). An analysis of sales data since 2000 by the authors demonstrates that 96% of new SFH built since 2000 in Vancouver consist of two-and-one-half story light wood platform framing on concrete foundations. Accordingly, 48 LCAs of North American residential buildings under 465 m^2 between 1 and 6 stories were reviewed from an embodied CO_2e benchmark study by Simonen et al. (2017), in addition to 3 LCAs of Canadian wood-frame residential structures (Norman et al. 2006; Zhang et al. 2014; Salazar and Meil 2009) and 2 Passivhaus LCAs (Brunklaus et al. 2010; Dahlstrom et al. 2012). LCAs which did not directly account for Stage A (cradle through construction), as well as those with outlying embodied CO_2e emissions above 1000 kg CO_2e/m^2 or below 100 kg CO_2e/m^2, were removed from the data set provided by Simonen et al. The average embodied CO_2e emissions during Stage A for the 34 remaining LCAs is 232 kg CO_2e/m^2 (see Appendix: Table 8). An LCA by Dahlstrom et al. (2012) comparing an SFH built to conventional versus PassiveHouse standards finds that the building to PassiveHouse standards results in an increase in embodied CO_2e emissions from 216 kg CO_2e/m^2 to 251 kg CO_2e/m^2, or 16%. Because the ZEBP energy and CO_2e emission guidelines accord with or surpass PassiveHouse standards, the 232 kg CO_2e/m^2 average for North American residential construction was increased by

16% to account for the additional embodied CO_2e impacts of building to a higher standard. This yields a benchmark initial embodied CO_2e emission rate for new SFH in Vancouver at 269 kg CO_2e/m^2. This figure lies within the range of 201 kg CO_2e/m^2, the median value of the 53 LCAs North American residential buildings under six stories and less than 465m^2 filtered from the review by Simonen et al. (2016), and 400 kg CO_2e/m^2, determined by Salazar and Meil's (2009) LCA of a high-efficiency wood-framed SFH in Canada. The size of new SFH was assumed to be the maximum allowable FSR, in keeping with the trends indicated in Figure 1 above. To the initial embodied CO_2e of new SFHs was added 1 kg CO_2e/m^2/year in recurring embodied CO_2e emissions. This rate is also applied to existing SFHs.

It could be argued that the embodied CO_2e emissions of demolition should be counted toward the overall embodied CO_2e emissions of new SFHs because these emissions are released as part of the site preparation required for new construction. However, embodied CO_2e emissions of the demolition of existing SFH have been omitted from the study due to a lack of reliable data. This omission was deemed acceptable, as numerous studies find that demolition accounts for less than 2% of overall embodied impacts of buildings (Junnila and Horvath 2003; Norman et al. 2006; Ramesh et al. 2010; Cabeza et al. 2014; Zhang et al. 2014; Vilches et al. 2017).

Total Emissions, Building Longevity and Payback periods

Once operating and embodied CO_2e emissions are determined, it is possible to calculate the total CO_2e emissions. Total CO_2e emissions are the sum of operating emissions and embodied emissions, expressed as CO_2e. The total CO_2e emissions of a newly constructed SFH are evaluated against the operating emissions of the older SFH it replaces, using a payback period calculation. The payback period represents the number of years required for the total CO_2e emissions of the new construction to equal the annual operating CO_2e emissions of the existing SFH, had it not been replaced. The payback period calculation is expressed in Equation 1 below:

Equation 1: SFH payback period formula

$$Y = ECi / [(ECr' + OC') - (ECr + OC)]$$

Legend: (Y) Payback period; (EC) embodied CO_2e emissions; (OC) operating CO_2e emissions; (i) initial; (r) recurring; (') existing SFH.

Replacing an existing building with a new building with lower operating CO_2e emissions results in a positive CO_2e payback period, measured in years. If the replacement of an existing building has the same overall operating carbon emissions, the denominator is zero and the payback time is infinite. Embodied emissions can never be paid back because no additional efficiencies were introduced. The payback period is negative if the new building has higher operating emissions than the building it replaces.

The payback period is the relative yardstick by which emission savings or debt of new SFHs can be measured, compared to a control state that consists of the operating emissions and recurring embodied emissions of the original building. If a replacement SFH is younger than its payback period, the total emissions of that lot exceed the emissions that would have been incurred had the SFH been left in place (CO_2e debt). If the new SFH is older than its payback period, the total emissions of the lot are lower than the emissions of the original SFH (CO_2e savings). Since operating CO_2e rates vary by SFH vintage cohort, and replacement SFHs maximize their allowable buildable square footage per lot, payback periods will vary by vintage, as well as by any potential increase in the size of new construction.

Teardown Index

Assessing the environmental benefits of replacing different vintages of SFH with more efficient buildings requires estimating the life expectancy of the new construction. If the payback period exceeds life expectancy, the structure efficiency gains are unlikely to be realized. Rosenthal et al. (1994) developed a land and building value assessment model that concludes that the relative building value, which refers to the quotient of the assessed building value over the total value of the property, can be used to predict reliably whether a building will be demolished when it is sold. The lower the relative building value, the more likely a building is to be demolished, as new owners seek to balance the value of the building with the overall value of the property.

Building on the work of Rosenthal et al., we developed a statistical model called the Teardown Index to assess the duration of life of individual SFH's in Vancouver. This general linear logistic model uses the relative building value at the time of sale to predict the probability that an SFH will be replaced when it is transacted (bought or sold). Historic land transfer data and land value assessments from BC Assessment were used to identify the determinants of SFH teardown/rebuilds within RS zoning. Confirming the earlier research by Rosenthal et al., all but 1.3% of historic teardowns of SFH could be associated with the property being transacted (BC Assessment 2015a, 2015b; BC Assessment et al. 2016). This represents a binary outcome (i.e., the building will be torn down, or will survive). We have expanded on this model by also considering in floor area, total building value, lot size, building age, and year of last major improvement as other potential predictors. To fit and evaluate the model, we used single-family homes in single-family zoned areas transacted between 2006 and 2015. To account for homes being torn down some time before or after the transaction, we used a moving buffer of up to four years around the transaction time, which narrowed the time window considered. We found little variation in model parameters or accuracy across time windows.

We experimented with several improvements to the model because of high dependence among some of the predictors. However, the accuracy of the model was not significantly improved by adding variables like lot area or the total building value, nor by using more sophisticated models. We added an L1 and L2 regularization penalty to diminish the effect of co-linearity. We also build a random forest model, in which a large number of individual decision trees operate as an ensemble, which yielded an improvement in predictive power by 3% when compared to the naive single variable logistic regression. We decided that the added complexity and lack of transparency did not justify the inclusion of the random forest method in the final model, based on the relatively small increase in accuracy it produced. Prediction accuracy of the logistic model was constant across test and training data. The final analysis was fit on 23,101 homes transacted between 2006 and 2012 in order to allow for enough time for a transacted property to get torn down and rebuilt. The model is 82% accurate.

The fitted logistic model yields the following equation, which uses the relative building value to predicts the teardown probability in the event of a property transaction. The modeled probability of a transacted SFH with relative building value x to be torn down is given by Equation 2:

Equation 2. Teardown index formula

$$P(x) = \frac{1}{1+e^{-(0.53-20.23*x)}}$$

This work relied on BC Assessment's valuations of land and buildings, which we found to be an unbiased estimator of building values, although at times with high variance on individual buildings. The teardown probability tops out at 63% for essentially worthless buildings, which cannot fully be explained by inconsistencies in property assessments, and indicates that speculation on future land value gains is an important factor in some of these transactions.

Projecting Total Emissions of Vancouver's SFH stock to 2050

Correlating the predictions with emission payback periods for different vintages of SFHs makes it possible to estimate the total CO_2e emissions of different scenarios at the scale of the city. Six scenarios are modeled to study their effect on total CO_2e

Teardown Index

Figure 2. Teardown Index graph

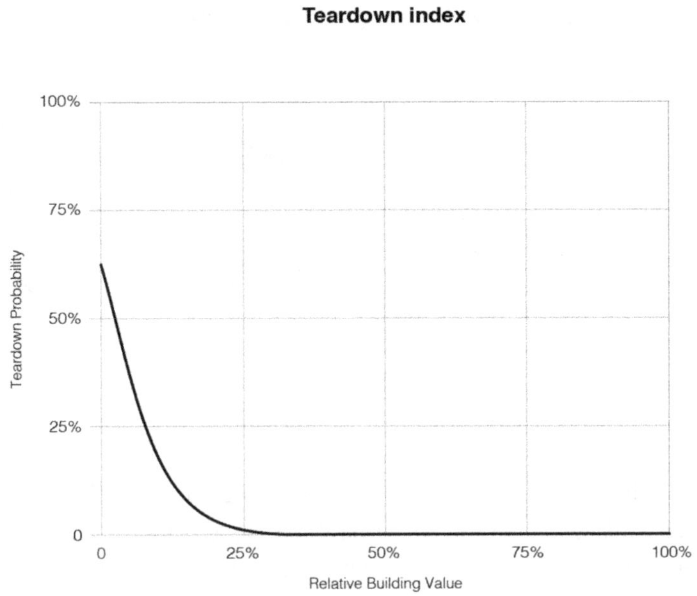

emissions of SFH in Vancouver from 2017 to 2050. The first scenario is a control, which is defined by no change in property prices and an artificial mechanism whereby no SFHs are replaced, resulting in emissions produced entirely building operation, with no new embodied emissions. The remaining scenarios project the evolution of the SFH stock to 2050 using average price compound annual growth rates (CAGR) from -2% to 8.24%, which reflect different CAGR in Vancouver over the past 8 years. The effect these different scenarios produce on total CO_2e emissions is presented below.

RESULTS

Emission Payback Periods by SFH Vintage

The average CO_2e payback periods by SFH vintage cohort calculations are summarized in Figure 3 and Table 2 below. Because operating emission rates of existing SFHs increase with age, new SFH homes that replace older SFH demonstrate shorter payback periods. For example, a new SFH replacing an SFH built prior to 1946 has a payback period of 73 years, if constructed between 2020 and 2025, and 25 years if built after 2025, when emissions standards are tightened for new construction. In contrast, replacing an average SFH built after 2011 has a CO_2e payback period

Figure 3. CO_2e Payback Period for a new SFH replacing pre-1946 SFH

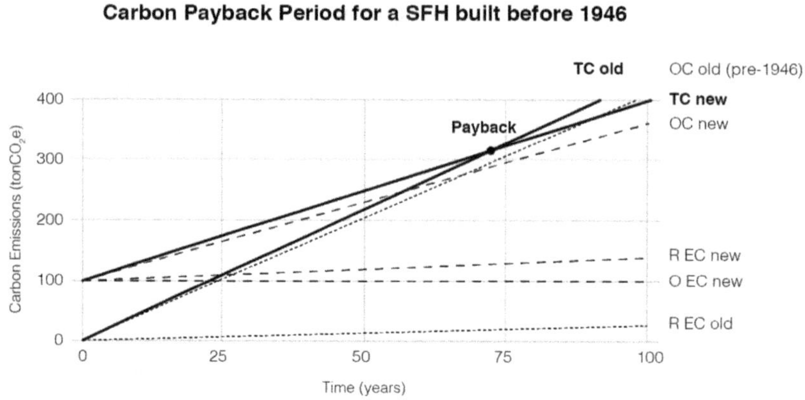

of 344 years, which is reduced to 34.6 years if built after 2025. For all SFHs in Vancouver, the weighted average CO_2e payback period is 168 years, which drops to 29 years for SFH built between 2020 and 2025 and those built after 2025, when operating emissions effectively drop to zero.

Scenarios of Real Estate Appreciation

Table 3 summarizes the CO_2e emissions from Vancouver SFHs for different real estate price compound annual growth rates (CAGR). The growth rates of 2.23%, 5.08%, and 8.24% correspond to the CAGR of SFHs prices in Vancouver over the

Table 2. Replacement SFH CO_2e payback periods by vintage

Vintage	Payback period of SFH built from 2020 to 2025 (years)	Payback period of SFH built after 2025 (years)
Before 1946	73.1	25.2
1946–1960	291.3	34.0
1961–1977	138.6	30.1
1978–1983	82.6	26.2
1984–1995	71.8	25.0
1996–2000	176.3	31.6
2001–2005	548.4	35.9
2006–2010	277.5	33.8
2011–2017	343.9	34.6

last one, five, and three years, respectively. The 2% reduction to the CAGR scenario is added to provide insight into the likely effect of a loss of value on replacement rates. The six scenarios test the sensitivity of CO_2e emissions toward different land value change scenarios, holding building values constant. This changes the relative building value over time, which in turn impacts the probability of a building getting torn down when transacted. We then run simulations to randomly transact properties at historical rates, apply our model to predict teardowns, and compute the resulting CO_2e emissions. Given the probabilistic nature of this estimate, we repeat this process for each scenario and average over the predicted CO_2e missions until they converge. The results are listed in Table 3.

In the control scenario, in which no SFHs are replaced and the current stock of SFHs is conserved, the total CO_2e emissions released by operating and maintaining these homes between 2017 and 2050 amount to 8.9M mt CO_2e.

Running the Teardown Index for an SFH price CAGR of 2.23% over the same time period projects that an estimated 32,366 SFHs will be demolished and replaced. Exploiting maximum FSR buildable area during rebuilding in keeping with the trends documented in Figure 1 above will add 3.2M m^2 of interior space to those structures. From 2017-2050, the total emissions from this scenario will amount to 10.9M mt CO_2e, which represents an increase of 1.9M mt CO_2e emissions when compared to the control scenario mentioned above.

At a CAGR of 5.08%, the Teardown Index projects that 39,929 SFHs will be replaced, and 3.3M m^2 of interior space will be added to the housing stock, causing total emissions from 2017-2050 to reach 11.3M mt CO_2e. This represents a 26.5% increase from the control scenario.

In the extreme growth scenario, based on the 8.24% CAGR rise recorded in SFH prices from 2015-2017, 47,276 SFH would be replaced between 2017-2050, resulting in a net increase of 3,390,215 m^2 of interior space, and 11.7M mt CO_2e of total emissions, a 31.8% increase over the control scenario.

Notably, the model shows that for a CAGR of -2%, 21,921 SFH will be replaced, producing 10.2M mt CO_2e emissions. A 2% reduction in property value does not affect the low relative building values significantly enough to cause a major change in replacement rates.

The Real Estate Price Elasticity of SFH CO_2e Emissions

Price elasticity refers to the responsiveness of one variable to changes in another. Generated by the median value of 100 teardown index simulations, the real estate price elasticity of SFH CO_2e emissions is presented in Figure 4. This curve represents the change in total, operating, and embodied CO_2e emissions from the total stock of SFHs over the 2017-2050 period given a change in the CAGR of SFH land values

Table 3. Total CO_2e emissions by scenario

Scenario	Real estate CAGR (%)	SFH replaced	Net increase of SFH stock internal area (m^2)	Total carbon emissions (metric tonnes of CO_2e)
1	Moratorium	0	0	8,913,000
2	-2%	21,921	2,627,278	10,214,801
3	0%	26,183	2,836,140	10,441,025
4	2.23%	32,366	3,178,000	10,856,000
5	5.08%	39,929	3,262,000	11,275,000
6	8.24%	47,276	3,390,000	11,744,000

Source: Authors, 2019

over that period. For each percentage point increase in CAGR land value price gains, an additional estimated 130M tonnes of CO_2e are added between 2017 to 2050.

DISCUSSION

The results show that the environmental benefits of tearing down and replacing even very poorly performing buildings are dubious at best in Vancouver. The average CO_2e payback period of 168 years for a typical home built to current standards renders it unlikely that emission savings will be realized before the structure is replaced. Even the shortest CO_2e emission payback time of 71 years for new SFH constructed to current emission standards may easily exceed the life of the home

Figure 4. Real estate price elasticity of SFH CO_2e emissions

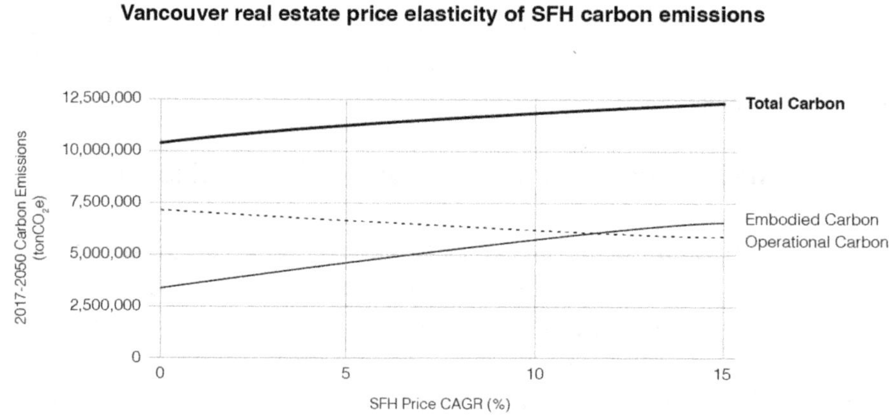

in the current economic climate, in which buildings are torn down and replaced at a frenetic pace. Increasingly stringent net-zero energy standards after 2025 bring down the payback period significantly, although the CO_2e emission payback period ranges between 25 and 36 years even under the zero operating consumption scenario. It is as yet uncertain whether these aspirational guidelines will be achieved on the proposed schedule. It is certain that the current practice of tearing down functional homes and replacing them with new homes built to present performance standards is a losing proposition environmentally, at the scale of both the individual building lot and the city.

The long payback times documented give rise to large CO_2e emission debts for all of the real estate growth scenarios considered at the scale of the city. For all of the scenarios considered, replacing existing SFH with more efficient buildings causes a significant net increase in CO_2e emissions for the period between 2017-2050. It is especially notable that the increases to overall emissions occur despite 50% reductions to operating emissions on a per square meter basis, which can be considered very significant, although reductions are mitigated by the overall increase in GFA as homeowners seek to maximize FSR. This finding can be considered conservative in terms of the environmental effects of replacement. Carbon emissions associated with the demolition of existing SFHs were not considered, which would add additional CO_2e emissions to new construction. More significantly, among the many impacts of building construction, only CO_2e emissions were considered. The overall impacts of new construction would increase considerably were impacts on human health, ecosystems, water quality, and overall resource use to be taken into account.

It should be noted that the findings reported are highly specific to Vancouver. The predominance of wood-frame SFH construction in Vancouver is typical of most North American cities, but Vancouver benefits from a climate considerably milder than the rest of Canada. The mild climate reduces operating energy requirements, which in turn reduces operating emissions, lengthening the payback period for building replacement. A more severe climate would increase operating energy consumption, which would also reduce the CO_2e payback periods for renovation or replacement. Vancouver also benefits from an unusually high percentage of renewable energy, reducing the emissions per unit of energy that is consumed. Hydroelectricity and natural gas, the predominant sources of operating energy in Vancouver, have low emission factors per unit of energy consumed. CO_2e payback periods would diminish considerably for localities where operating energy demand is met by non-renewable energy sources with higher emission factors, such as coal-fired electricity or oil heat. Put another way, an SFH constructed in Vancouver, with its mild climate and renewable sources of energy, will have a significantly longer CO_2e payback period than an identical SFH constructed in a city such as Calgary, with its severe climate and energy grid based on non-renewable sources of energy.

The payback times for SFH are also influenced by the type of materials and assemblies used in new construction, which have a significant impact on initial impacts, ongoing maintenance, and associated energy costs. The predominant structural method used to construct new SFH in Vancouver is wood-frame with a concrete foundation, which has a relatively low initial embodied impacts. Other construction types, such as masonry, would likely lengthen payback times, further due to higher initial embodied impacts. Unlike the low carbon sources of energy used to operate buildings in Vancouver, many of the construction materials used to build homes in the city are the product of global supply chains that do not benefit from renewable energy or sustainable practices. The carbon-intensive sources of energy used to manufacture many construction materials increase the initial carbon investment of new construction, lengthening the payback time.

The site-by-site analysis shows that in a mild climate with ample sources of low-carbon energy, the current practice of demolishing functional SFH and replacing them with more efficient SFHs of the same type does not make environmental sense. However, measures can be taken to reduce the payback periods of new construction, which currently exceed the life expectancy of the new buildings. The most straightforward method for reducing payback periods is simply to adopt more stringent energy performance standards. If Vancouver adopts PassivHaus standards in 2025, payback periods will be reduced to between 25-35 years, depending on the vintage of existing SFH. Realizing the environmental benefits of this approach would require adopting a "one and done" replacement policy that would incentivize the construction of high-efficiency buildings while ensuring that the new structures themselves are not torn down and replaced prematurely due to low RBV.

Preventing reconstruction of efficient SFH could perhaps be done by downzoning, such that reconstruction after initial replacement would have an associated GFA/FAR penalty that would create a powerful disincentive to rebuild once efficiency was achieved. Another alternative could be to adopt policies that would require owners or developers to provide carbon accounting at the building permit stage. In this scenario, new construction that could be shown to diminish overall emissions within a reasonable payback period could proceed without additional fees; where this was not the case, owners would pay the penalty in the form of a carbon tax. This would require reliable methods of establishing impacts of buildings as well as setting a price on carbon in a manner similar to that of cap and trade systems. The inconsistency of current carbon accounting, as well as the difficulty of adopting carbon pricing in Canada to date, suggests that this method, although perhaps the most efficient and just, would be challenging to implement.

Although the policy measures outlined above give some sense of how carbon emissions could be dealt with on a site-by-site basis in Vancouver, these alternatives do little to address the underlying reason that so many SFH are torn down. The decision to replace SFHs is typically made by individual owners on economic grounds, as the relative building value used to calculate the probability of replacement suggests. In Vancouver, the average relative building value of existing SFH currently hovers around 12%. At this rate, the Teardown Index model predicts that one in four homes will be torn down when the property changes hands. All but the most expensive new homes constructed in Vancouver are below 30% relative building value, well under the 70% typical of a healthy housing market. The problem is getting worse. Property values are rising faster than the cost of construction, driving the relative building value of both new and existing SFH ever lower. This produces a teardown cycle, as new owners demolish existing homes and rebuild commensurate with the overall value of their property. When property values rise so high that even high-end SFH are unable to maintain sufficient value to avoid becoming tomorrow's teardown, it points to the need to rethink land use more broadly through zoning revisions.

The teardown cycle is unlikely to be broken until RS zoning is loosened to allow alternatives to SFH. Revising zoning so that SFH can be selectively replaced with row-housing up to four or five stories built to PassivHaus standards would offer improved energy efficiency, decreasing emissions and shortening carbon payback to acceptable time frames. The higher initial cost of these high-density residential structures will increase their relative building value, decreasing the likelihood that they will be torn down before their carbon debt is repaid. Although affordability will still present a challenge, the cost of construction will be shared among more residents than an SFH. Finally, the row-house typology will increase the overall supply of housing to inhabitants in close proximity to the downtown core, reducing transportation and related impacts associated with neighborhoods comprised of SFH.

Assumptions and Limitations

The research focuses on CO_2e emissions as the most closely linked to global warming. Relying on a review of prior LCAs to determine embodied CO_2e emissions means that the research inherits their shortcomings in the form of scenario, boundary, and other model uncertainties (Lloyd and Ries 2007). Moreover, many other types of environmental impacts are associated with the construction, maintenance, operation, and replacement of SFHs, such as ozone depletion, respiratory inorganics, human toxicity, ionizing radiation, ecotoxicity, ozone formation, acidification, terrestrial eutrophication, land use, waste, and resource consumption. When these are considered, it is likely that the impacts could increase unevenly between operating and embodied impacts. In addition, the scope of the study is limited to CO_2e emissions directly associated with SFHs. Indirect emissions associated with infrastructural requirements of SFHs, such as roads and parking are not considered, nor are the indirect emissions due to transportation, or from consumption-related emissions (Kellett et al. 2013).

Emissions are considered on a per-area basis, which does not account for the number of people served by spaces, and as such may be a less accurate method of accounting for emissions than a per capita basis. It is likely that replacing SFHs with denser forms of inhabitation, such as mid-rise, ground-oriented development, would lead to a profound reduction of CO_2e emissions per capita.

Projecting SFH total CO_2e emissions to 2050 requires making assumptions about current policy while disregarding the possibility of potentially significant policy changes. The Zero Emissions Building Plan is used to predict future energy consumption, despite uncertainty about its prospects for approval and anticipated enforcement and compliance. The projections also assume that the current distribution of energy sources will be maintained in the future. Natural gas produces over sixteen times more CO_2e per unit of energy than hydroelectricity. If homeowners move toward replacing natural gas with electric heat pumps for space and water heating that account for more than half of the energy demands of SFH, the projections will overstate operating CO_2e emission rates. This would lengthen CO_2e emission payback periods further than currently predicted.

Finally, the research presented accounts only for the environmental impacts of SFH construction and operation at the scale of individual buildings. Where the environmental effects are indicated at the scale of the city, they are calculated by aggregating the effects of individual buildings. This method does not account for the additional environmental impacts of carbon-intensive patterns of living commonly associated with suburban configurations. Reserving large areas of land in proximity to the central business district for SFH land use also carries an opportunity cost: it requires a large portion of the metropolitan population to live farther away from jobs, forcing them into longer commutes. Accounting for these additional impacts would have an effect of the carbon payback time of SFH.

FUTURE DIRECTIONS

Operating energy efficiency gains are likely to be realized across the stock of SFH in Vancouver in the future, as the city progresses toward net-zero operational goals. Operating emissions will diminish as a result, making embodied impacts a larger proportion of total building impacts. The growing importance of embodied impacts will make it essential to account for and regulate embodied CO_2e emissions in future policy. Moving forward, establishing clear accounting practices (Basbagill et al. 2013; De Wolf et al. 2017), strengthening access to LCA data for building components and processes, such as a localized Environmental Product Declarations database, and mandating institutional oversight are likely prerequisites to that end. Such measures could facilitate the introduction of embodied CO_2e emission accounting in the design phase of construction (Häkkinen et al. 2015) at the scale of the SFH, rather than on an ad hoc post-occupancy basis, as is currently the case. Fortunately, the City of Vancouver is requiring developers to track embodied carbon in new commercial construction, the first step toward a stated goal of reducing embodied carbon in new buildings in the city by forty percent by 2030 (Pander, 2019).

The results also point to the relevance of the renovation of poorly performing existing structures, as opposed to outright replacement. Mohammadpourkarbasi and Sharples (2017) note that bringing two Victorian houses in the UK to near Passive-House standards results in a 75% reduction in operating emissions, with a CO_2e payback period of 6 years. Ma et al. (2012) and Rysanek and Choudhary (2013) note that renovation can be an effective method of increasing the sustainability of buildings, but that complex factors influence adoption. In the case of Vancouver, where older single-family homes typically consist of light wood framing on foundations of marginal quality, the impacts of extensive renovations are likely to be comparable to the impacts of outright replacement. Future research should apply similar methods to quantify the carbon payback of renovation to assess its viability as an alternative to replacing buildings entirely. It may be that simply upgrading building systems would offer the shortest payback times in the Vancouver context. For example, replacing gas-fired space heating equipment with electric heat pumps powered by renewable hydroelectricity would address the consumption of natural gas, which is the largest source of CO_2e emissions for many SFH. Quantifying the impacts of different renovation strategies and their associated payback times is beyond the scope of the present study but would provide important information to policymakers.

The research strongly suggests a need to rethink land-use strategies in Vancouver through changes to zoning in the future to permit buildings that are more commensurate with property values than SFH. A natural extension of this research would be to quantify the most carbon-efficient way to grow the metropolitan region, accounting for embodied and operating carbon, as well as associated emissions due to transportation and land-use changes.

Finally, the research presented could be extended by accounting for regional effects. The abundance of low-carbon electricity sources makes it possible to adhere to stringent carbon emission requirements in Vancouver without reducing operating energy consumption significantly. However, the grid infrastructure that provides low-carbon electricity to Vancouver is also connected to other regions throughout the Pacific Northwest. Reducing electricity consumed in Vancouver would potentially provide more low-carbon electricity that could be used to offset more carbon-intensive electricity in Washington, Oregon, and California. A systems-based investigation of energy savings in Vancouver could identify potentially significant regional reductions to overall emissions.

CONCLUSION

Buildings emit up to a third of CO_2e emissions globally. Most emission mitigation strategies to date have focused on reducing operating impacts due to their large share of overall impacts. However, as the proportion of embodied impacts increases due to operational efficiencies, accounting for embodied impacts is imperative. Embodied impacts can be understood in the context of the CO_2e emission payback period, which refers to the period of time required for operational savings to recover the amount of embodied CO_2e emitted by constructing a new building. This study uses CO_2e payback periods to assess total CO_2e emissions for the stock of single-family houses in Vancouver Canada in the context of the city's plan to reduce the CO_2e emissions of building by 80% by 2050. The amount of time for the total emissions of replacement SFHs to equal those of replaced SFHs ranges from 73 to 548 years for SFHs built to 2020 efficiency standards, and 25 to 36 years for SFHs built to 2025 efficiency standards. A statistical tool called the Teardown Index was used to project future demolition and replacement of single-family homes in Vancouver. Four scenarios of real estate capital appreciation growth rates were investigated, ranging between -2% to 8.24% CAGR. Generally, higher rates of real estate growth translate to higher rates of demolition and replacement of SFH. Each percentage point increase in annual land value price gains corresponds to an additional 130 million tonnes of CO_2e released between 2017 to 2050 as a result of operating and embodied impacts. Despite significant operating efficiency increases, replacing existing buildings with new efficient buildings will emit between 2-3.2 million tonnes of additional CO_2e emissions. These findings suggest that replacing functional older SFH with SFH constructed in accordance with current efficiency standards in Vancouver is not an effective strategy to achieve municipal CO_2e emission mitigation goals when all impacts are considered. Rather, increased efficiency standards should be adopted to decrease carbon payback times, and changes to land use should be considered to increase building longevity.

ACKNOWLEDGMENT

The chapter was adapted from an article originally published as "Teardown Index: Impact of property values on carbon dioxide emissions of single-family housing in Vancouver" in the journal Energy & Buildings 170 (2018) 95–106.

An interactive data visualization of the Teardown Index is available online here: https://mountainmath.ca/teardowns

The authors are grateful to the Peter Wall Institute for Advanced Studies for supporting the research.

REFERENCES

Abanda, F. H., Tah, J. H. M., & Cheung, F. K. T. (2013). Mathematical Modelling of Embodied Energy, Greenhouse Gases, Waste, Time? Cost Parameters of Building Projects: A Review. *Building and Environment*, *59*(January), 23–37. doi:10.1016/j.buildenv.2012.07.014

BC Assessment. (2015b). *BIR - A09 - 2014-2017. BC* Assessment and University of British Columbia.

Basbagill, J., Flager, F., Lepech, M., & Fischer, M. (2013). Application of Life-Cycle Assessment to Early Stage Building Design for Reduced Embodied Environmental Impacts. *Building and Environment*, *60*(February), 81–92. doi:10.1016/j.buildenv.2012.11.009

BC Assessment, . (2015a). *Bulk Extract for KPI Sales Detail, A09*, 2005–2015. BC Assessment and University of British Columbia.

BC Assessment & the Revenue Services Vancouver, and Community Services Vancouver. (2016). Property Tax Report Data. *City of Vancouver Open Data Catalogue*. Retrieved from https://data.vancouver.ca/datacatalogue/propertyTax.htm

BC Hydro. (2017). *About BC Hydro*. Accessed July 18. https://www.bchydro.com/about.html

Bribián, I., Usón, A. A., & Scarpellini, S. (2009). Life Cycle Assessment in Buildings: State-of-the-Art and Simplified LCA Methodology as a Complement for Building Certification. *Building and Environment*, *44*(12), 2510–2520. doi:10.1016/j.buildenv.2009.05.001

British Columbia. Government of. (2012). *Vancouver Community Energy & Emissions Inventory*. Retrieved from https://www2.gov.bc.ca/gov/content/environment/climate-change/data/ceei

British Columbia, Ministry of Environment. (2016). *2016/17 B.C. Best Practices Methodology for Quantifying Greenhouse Gas Emissions: Including Guidance for Public Sector Organizations, Local Governments and Community Emissions*. Author.

Brunklaus, B., Thormark, C., & Baumann, H. (2010). Illustrating Limitations of Energy Studies of Buildings with LCA and Actor Analysis. *Building Research and Information*, *38*(3), 265–279. doi:10.1080/09613211003654871

Buyle, M., Braet, J., & Audenaert, A. (2013). Life Cycle Assessment in the Construction Sector: A Review. *Renewable & Sustainable Energy Reviews*, *26*(October), 379–388. doi:10.1016/j.rser.2013.05.001

Cabeza, L. F., Rincón, L., Vilariño, V., Pérez, G., & Castell, A. (2014). Life Cycle Assessment (LCA) and Life Cycle Energy Analysis (LCEA) of Buildings and the Building Sector: A Review. *Renewable & Sustainable Energy Reviews*, *29*(January), 394–416. doi:10.1016/j.rser.2013.08.037

California Building Standards Commission. (2017). *California Building Standards Code (California Code of Regulations, Title 24)*. Retrieved from http://www.bsc.ca.gov/codes.aspx

Canadian Commission on Building and Fire Codes. (2015). *National Building Code of Canada, 2015*. Ottawa: National Research Council of Canada. Retrieved from http://ubc.summon.serialssolutions.com/ 2.0.0/link/0/ eLvHCXMwfV09C8Iw ED0UF8FBq2K1QiYnK2nahmQWiz_AvbT5mKQu-v-91 KqlVCFDSOBIQpLHS-7 uAcTsQMPOnRBppsrUSqtYaiy ThaFUKa5TLbDUGn7_XX W6oXa9vzMIPVJw JPBDJB0fNQMEVbc1 I6cY2wqtd_i RTcH7atoSvNcfrjKD ganmsGtSU19J2UhU ExdnTm6W1KkDij1 B-EwXsM5Ol-M5RNt5 8-iSN0NhS5gUzl W9utchbXoFpHAp AHWMqMDiRAoth VGWcpNoHllhmQ9 ejyUfgnbre_Y5 MliKHOZXN3Jfjmd 23WtzA2M3gde LQgAjizvebOtFegJy03-a

Canadian Mortgage and Housing Corporation. (2016). *Rental Market Report: Vancouver CMA 2016.* Author.

Chang, Y., Ries, R. J., & Wang, Y. (2010). The Embodied Energy and Environmental Emissions of Construction Projects in China: An Economic Input–output LCA Model. *Energy Policy*, *38*(11), 6597–6603. doi:10.1016/j.enpol.2010.06.030

Dahlstrom, O., Sornes, K., Eriksen, S. T., & Hertwich, E. G. (2012). Life Cycle Assessment of a Single-Family Residence Built to Either Conventional- or Passive House Standard. *Energy and Building*, *54*(November), 470–479. doi:10.1016/j.enbuild.2012.07.029

Dahmen, J., von Bergmann, J., & Das, M. (2018). Teardown Index: Impact of property values on carbon dioxide emissions of single-family housing in Vancouver. *Energy and Building*, *170*, 95–106. doi:10.1016/j.enbuild.2018.03.012

De Wolf, C., Pomponi, F., & Moncaster, A. (2017). Measuring Embodied Carbon Dioxide Equivalent of Buildings: A Review and Critique of Current Industry Practice. *Energy and Building*, *140*(April), 68–80. doi:10.1016/j.enbuild.2017.01.075

De Wolf, C. C. E. L. (2014). *Material Quantities in Building Structures and Their Environmental Impact*. Massachusetts Institute of Technology. Retrieved from https://dspace.mit.edu/handle/1721.1/91298

Deru, M. P., & Torcellini, P. A. (2007). *Source Energy and Emission Factors for Energy Use in Buildings*. National Renewable Energy Laboratory. Retrieved from https://www.nrel.gov/docs/fy07osti/38617.pdf

Dixit, M. K. (2017). Life Cycle Embodied Energy Analysis of Residential Buildings: A Review of Literature to Investigate Embodied Energy Parameters. *Renewable & Sustainable Energy Reviews*, *79*(November), 390–413. doi:10.1016/j.rser.2017.05.051

Dixit, M. K., Fernández-Solís, J. L., Lavy, S., & Culp, C. H. (2010). Identification of Parameters for Embodied Energy Measurement: A Literature Review. *Energy and Building*, *42*(8), 1238–1247. doi:10.1016/j.enbuild.2010.02.016

Dixit, M. K., Fernández-Solís, J. L., Lavy, S., & Culp, C. H. (2012). Need for an Embodied Energy Measurement Protocol for Buildings: A Review Paper. *Renewable & Sustainable Energy Reviews*, *16*(6), 3730–3743. doi:10.1016/j.rser.2012.03.021

European Parliament, Council of the European Union. (2010). *Directive 2010/31/Eu on the Energy Performance of Buildings*. Official Journal of the European Union.

European Parliament, Council of the European Union. (2012). *Directive 2012/27/Eu on Energy Efficiency*. Official Journal of the European Union.

Gold, K. (2017). Amid a Housing Crisis, Vancouver Buys Houses and Leaves Them Empty. *The Globe and Mail*. Retrieved from https://www.theglobeandmail.com/real-estate/vancouver/amid-a-housing-crisis-vancouver-buys-houses-and-leaves-them-empty/article35513761/

Haapio, A., & Viitaniemi, P. (2008). A Critical Review of Building Environmental Assessment Tools. *Environmental Impact Assessment Review*, *28*(7), 469–482. doi:10.1016/j.eiar.2008.01.002

Häkkinen, T., Kuittinen, M., Ruuska, A., & Jung, N. (2015). Reducing Embodied Carbon during the Design Process of Buildings. *Journal of Building Engineering*, *4*(December), 1–13. doi:10.1016/j.jobe.2015.06.005

Holzhey, M., & Skoczek, M. (2016). *UBS Global Real Estate Bubble Index: For Housing Markets of Select Cities*. UBS.

Huang, Y., Niu, J., & Tse-ming, C. (2012). Energy and Carbon Emission Payback Analysis for Energy-Efficient Retrofitting in Buildings: Overhang Shading Option. *Energy and Building*, *44*(January), 94–103. doi:10.1016/j.enbuild.2011.10.027

Huijbregts, M., Heijungs, R., & Hellweg, S. (2004). Complexity and Integrated Resource Management: Uncertainty in LCA. *The International Journal of Life Cycle Assessment*, *9*(5), 341–342. doi:10.1007/BF02979428

Ibn-Mohammed, T., Greenough, R., Taylor, S., Ozawa-Meida, L., & Acquaye, A. (2013). Operational vs. Embodied Emissions in buildings—A Review of Current Trends. *Energy and Building*, *66*(November), 232–245. doi:10.1016/j.enbuild.2013.07.026

IEA & OECD. (2016). *Energy Technology Perspectives 2016*. Paris: International Energy Agency.

IPCC. (2014a). Buildings. *In Climate Change 2014: Mitigation of Climate Change. Contribution of Working Group III to the Fifth Assessment Report of the Intergovernmental Panel on Climate Change*, (pp. 671–738). Cambridge, UK: Cambridge University Press. Retrieved from http://pure.iiasa.ac.at/11117

IPCC. (2014b). Climate Change 2014: Synthesis Report. Contribution of Working Groups I, II and III to the Fifth Assessment Report of the Intergovernmental Panel on Climate Change. *IPCC*. Retrieved from https://epic.awi.de/37530/

IPCC. (2014c). Summary for Policy Makers. In *Climate Change 2014: Impacts, Adaptations, and Vulnerability. Part A: Global and Sectoral Aspects. Contribution of Working Group II to the Fifth Assessment Report of the Intergovernmental Panel on Climate Change* (pp. 1–32). Cambridge, UK: Cambridge University Press.

ISO. (2006). *ISO 14040:2006. Environmental Management—Life Cycle Assessment — Principles and Framework*. International Organization for Standardization. Retrieved from https://www.iso.org/obp/ui/#iso:std:iso:14040:ed-2:v1:en

JRC European Commission & Institute for Environmental Sustainability. (2010). *Analysis of Existing Environmental Impact Assessment Methodologies for Use in Life Cycle Assessment*. Author.

Junnila, S., & Horvath, A. (2003). Life-Cycle Environmental Effects of an Office Building. *Journal of Infrastructure Systems*, *9*(4), 157–166. doi:10.1061/(ASCE)1076-0342(2003)9:4(157)

Kavgic, M., Mavrogianni, A., Mumovic, D., Summerfield, A., Stevanovic, Z., & Djurovic-Petrovic, M. (2010). A Review of Bottom-up Building Stock Models for Energy Consumption in the Residential Sector. *Building and Environment*, *45*(7), 1683–1697. doi:10.1016/j.buildenv.2010.01.021

Kellett, R., Christen, A., Coops, N. C., van der Laan, M., Ben Crawford, T. R. T., & Olchovski, I. (2013). A Systems Approach to Carbon Cycling and Emissions Modeling at an Urban Neighborhood Scale. *Landscape and Urban Planning*, *110*(February), 48–58. doi:10.1016/j.landurbplan.2012.10.002

Lloyd, S. M., & Ries, R. (2007). Characterizing, Propagating, and Analyzing Uncertainty in Life-Cycle Assessment: A Survey of Quantitative Approaches. *Journal of Industrial Ecology*, *11*(1), 161–179. http://onlinelibrary.wiley.com/doi/10.1162/jiec.2007.1136/full. doi:10.1162/jiec.2007.1136

Ma, Z., Cooper, P., Daly, D., & Ledo, L. (2012). Existing building retrofits: Methodology and state-of-the-art. *Energy and Buildings, 55*, 889–902.

Marr, G. (2017). Vancouver Real Estate Market Heating up Again as Sales and Prices Recover from Buyers Tax. *Financial Post*. Retrieved from https://business.financialpost.com/personal-finance/mortgages-real-estate/vancouver-real-estate-market-heating-up-again-as-sales-and-prices-rise

Matthews, H. S., Hendrickson, C. T., & Weber, C. L. (2008). The Importance of Carbon Footprint Estimation Boundaries. *Environmental Science & Technology*, *42*(16), 5839–5842. doi:10.1021/es703112w PMID:18767634

Menzies, G. F., Turan, S., & Banfill, P. F. G. (2007). Life-Cycle Assessment and Embodied Energy: A Review. *Proceedings of the Institution of Civil Engineers - Construction Materials, 160*(4), 135–43. 10.1680/coma.2007.160.4.135

Mohammadpourkarbasi, H., & Sharples, S. (2017). *Eco-Retrofitting Very Old Dwellings*. Accessed July 18. https://mediatum.ub.tum.de/doc/1169276/1169276.pdf

Natural Resources Canada. (2014). *Comprehensive Energy Use Database - Residential Sector – British Columbia*. Retrieved from https://oee.nrcan.gc.ca/corporate/statistics/neud/dpa/menus/trends/comprehensive/trends_res_bc.cfm

Norman, J., MacLean, H. L., & Kennedy, C. A. (2006). Comparing High and Low Residential Density: Life-Cycle Analysis of Energy Use and Greenhouse Gas Emissions. *Journal of Urban Planning and Development*, *132*(1), 10–21. doi:10.1061/(ASCE)0733-9488(2006)132:1(10)

Pander, S. (2019). How Vancouver is pushing to build a city of zero emissions buildings. *Fast Company*. Accessed Dec. 3, 2019. https://www.fastcompany.com/90421205/how-vancouver-is-pushing-for-all-its-new-buildings-to-be-zero-emissions

Pérez-Lombard, L., Ortiz, J., & Pout, C. (2008). A Review on Buildings Energy Consumption Information. *Energy and Building*, *40*(3), 394–398. doi:10.1016/j.enbuild.2007.03.007

Pomponi, F., & Moncaster, A. (2016). Embodied Carbon Mitigation and Reduction in the Built Environment? What Does the Evidence Say? *Journal of Environmental Management, 181*(October), 687–700. doi:10.1016/j.jenvman.2016.08.036 PMID:27558830

Ramesh, T., Prakash, R., & Shukla, K. K. (2010). Life Cycle Energy Analysis of Buildings: An Overview. *Energy and Building, 42*(10), 1592–1600. doi:10.1016/j.enbuild.2010.05.007

Real Estate Board of Vancouver. (2017). *June 2017 MLS® Home Price Index.* Accessed July 9. https://www.rebgv.org/home-price-index?region=Greater+Vancouver&type=all&date=2017-06-01

Reddy, Venkatarama, & Jagadish. (2003). Embodied Energy of Common and Alternative Building Materials and Technologies. *Energy and Buildings, 35*(2), 129–137. Retrieved from https://www.sciencedirect.com/science/article/pii/S0378778801001414

Rosenthal, S. S., & Helsley, R. W. (1994). Redevelopment and the Urban Land Price Gradient. *Journal of Urban Economics, 35*(2), 182–200. doi:10.1006/juec.1994.1012

Rysanek, A., & Choudhary, R. (2013). Optimum building energy retrofits under technical and economic uncertainty. *Energy and Building, 57*, 324–337.

Salazar, J., & Meil, J. (2009). Prospects for Carbon-Neutral Housing: The Influence of Greater Wood Use on the Carbon Footprint of a Single-Family Residence. *Journal of Cleaner Production, 17*(17), 1563–1571. doi:10.1016/j.jclepro.2009.06.006

Sartori, I., & Hestnes, A. G. (2007). Energy Use in the Life Cycle of Conventional and Low-Energy Buildings: A Review Article. *Energy and Building, 39*(3), 249–257. doi:10.1016/j.enbuild.2006.07.001

Sathre, R., & Gustavsson, L. (2009). Using Wood Products to Mitigate Climate Change: External Costs and Structural Change. *Applied Energy, 86*(2), 251–257. doi:10.1016/j.apenergy.2008.04.007

Simonen, K., Droguett, B. R., Strain, L., & McDade, E. (2017). *Embodied Carbon Benchmark Study: LCA for Low Carbon Construction.* University of Washington. Retrieved from https://digital.lib.washington.edu/researchworks/handle/1773/38017

Sousa, G., Jones, B. M., Mirzaei, P. A., & Robinson, D. (2017). A Review and Critique of UK Housing Stock Energy Models, Modelling Approaches and Data Sources. *Energy and Building, 151*(September), 66–80. doi:10.1016/j.enbuild.2017.06.043

Star, E. (2013). *Source Energy.* Portfolio Manager - Technical Reference.

Swan, L. G., & Ismet Ugursal, V. (2009). Modeling of End-Use Energy Consumption in the Residential Sector: A Review of Modeling Techniques. *Renewable & Sustainable Energy Reviews*, *13*(8), 1819–1835. doi:10.1016/j.rser.2008.09.033

Thormark, C. (2006). The Effect of Material Choice on the Total Energy Need and Recycling Potential of a Building. *Building and Environment*, *41*(8), 1019–1026. doi:10.1016/j.buildenv.2005.04.026

Tooke, T. R., Coops, N. C., & Webster, J. (2014a). Predicting Building Ages from LiDAR Data with Random Forests for Building Energy Modeling. *Energy and Building*, *68*(January), 603–610. doi:10.1016/j.enbuild.2013.10.004

Tooke, T. R., van der Laan, M., & Coops, N. C. (2014b). Mapping Demand for Residential Building Thermal Energy Services Using Airborne LiDAR. *Applied Energy*, *127*(August), 125–134. doi:10.1016/j.apenergy.2014.03.035

Treloar, G., Fay, R., Love, P. E. D., & Iyer-Raniga, U. (2000). Analysing the Life-Cycle Energy of an Australian Residential Building and Its Householders. *Building Research and Information*, *28*(3), 184–195. doi:10.1080/096132100368957

Treloar, G. J., Gupta, H., Love, P. E. D., & Nguyen, B. (2003). An Analysis of Factors Influencing Waste Minimisation and Use of Recycled Materials for the Construction of Residential Buildings. *Management of Environmental Quality*, *14*(1), 134–145. doi:10.1108/14777830310460432

Upton, B., Miner, R., Spinney, M., & Heath, L. S. (2008). The Greenhouse Gas and Energy Impacts of Using Wood instead of Alternatives in Residential Construction in the United States. *Biomass and Bioenergy*, *32*(1), 1–10. doi:10.1016/j.biombioe.2007.07.001

Ürge-Vorsatz, D., Koeppel, S., & Mirasgedis, S. (2007). Appraisal of Policy Instruments for Reducing Buildings' CO_2 Emissions. *Building Research and Information*, *35*(4), 458–477. doi:10.1080/09613210701327384

Vancouver, City of. (2016). *Zero Emissions Building Plan*. Author.

Verbeeck, G., & Hens, H. (2010a). Life Cycle Inventory of Buildings: A Calculation Method. *Building and Environment*, *45*(4), 1037–1041. doi:10.1016/j.buildenv.2009.10.012

Verbeeck, G., & Hens, H. (2010b). Life Cycle Inventory of Buildings: A Contribution Analysis. *Building and Environment*, *45*(4), 964–967. doi:10.1016/j.buildenv.2009.10.003

Vilches, A., Garcia-Martinez, A., & Sanchez-Montañes, B. (2017). Life Cycle Assessment (LCA) of Building Refurbishment: A Literature Review. *Energy and Building*, *135*(January), 286–301. doi:10.1016/j.enbuild.2016.11.042

von Bergmann, J., & Dahmen, J. (2016). *Teardown Index: Vancouver SFH Teardowns*. Retrieved from https://mountainmath.ca/teardowns

Wang, H., & Zhai, Z. J. (2016). Advances in Building Simulation and Computational Techniques: A Review between 1987 and 2014. *Energy and Building*, *128*(September), 319–335. doi:10.1016/j.enbuild.2016.06.080

Wang, Y., Kuckelkorn, J., & Liu, Y. (2017). A State of Art Review on Methodologies for Control Strategies in Low Energy Buildings in the Period from 2006 to 2016. *Energy and Building*, *147*(July), 27–40. doi:10.1016/j.enbuild.2017.04.066

Xing, S., Xu, Z., & Jun, G. (2008). Inventory Analysis of LCA on Steel- and Concrete-Construction Office Buildings. *Energy and Building*, *40*(7), 1188–1193. doi:10.1016/j.enbuild.2007.10.016

Zhang, W., Tan, S., Lei, Y., & Wang, S. (2014). Life Cycle Assessment of a Single-Family Residential Building in Canada: A Case Study. *Building Simulation*, *7*(4), 429–438. doi:10.100712273-013-0159-y

APPENDIX

Operating Impacts

Table 4. Average energy requirements by end-use and fuel source per SDH per year for 2014 in lower Mainland, BC. (Source: Young 2017)

Fuel	Gas (kWh)	Gas (GJ)	Gas %	Electric (kWh)	Electric (GJ)	Electric %	Total (kWh)	Total (GJ)	Total %
Space Heating	8,132.6	29.3	81.44%	1,853	6.7	18.56%	9,985.6	35.95	44.66%
Hot Water	2,380.0	8.6	77.78%	680	2.4	22.22%	3,060.0	11.02	13.68%
Space Cooling	0.0	0.0	0.00%	78	0.3	100.00%	78.0	0.28	0.35%
Appliances	0.0	0.0	0.00%	3,244	11.7	100.00%	3,244.0	11.68	14.51%
Lighting	0.0	0.0	0.00%	1,871	6.7	100.00%	1,871.0	6.74	8.37%
Electronics	0.0	0.0	0.00%	1,784	6.4	100.00%	1,784.0	6.42	7.98%
Other	0.0	0.0	0.00%	1,460	5.3	100.00%	1,460.0	5.26	6.53%
Ventilation	0.0	0.0	0.00%	879	3.2	100.00%	879.0	3.16	3.93%
Total	**10,512.6**	**37.8**	-	**11,849**	**42.7**	-	**22,361.6**	**80.50**	**100.00%**
% of total energy	47.01%	47.01%	47.01%	52.99%	52.99%	52.99%	100.00%	100.00%	100.00%

Embodied Impacts

Table 5. Average energy requirements by end-use and fuel source per SDH m^2 per year for 2014 in Lower Mainland, BC. (Source: Young 2017)

Fuel	Gas (kWh/ m^2)	Gas (GJ/m^2)	Gas % of Use	Electric (kWh/m^2)	Electric (GJ/ m^2)	Electric % of Use	Total (kWh/ m^2)	Total (GJ/m^2)	Use % of Total
Space Heating	36.825	0.133	81.44%	8.390	0.030	18.56%	45.215	0.163	44.66%
Hot Water	10.777	0.039	77.78%	3.079	0.011	22.22%	13.856	0.050	13.68%
Space Cooling	0.000	0.000	0.00%	0.353	0.001	100.00%	0.353	0.001	0.35%
Appliances	0.000	0.000	0.00%	14.689	0.053	100.00%	14.689	0.053	14.51%
Lighting	0.000	0.000	0.00%	8.472	0.030	100.00%	8.472	0.030	8.37%
Electronics	0.000	0.000	0.00%	8.078	0.029	100.00%	8.078	0.029	7.98%
Other	0.000	0.000	0.00%	6.611	0.024	100.00%	6.611	0.024	6.53%
Ventilation	0.000	0.000	0.00%	3.980	0.014	100.00%	3.980	0.014	3.93%
Total	**47.60**	**0.17**	-	**53.65**	**0.19**	-	**101.25**	**0.36**	**100.00%**
% of total	47.01%	47.01%	47.01%	52.99%	52.99%	52.99%	100.00%	100.00%	100.00%

Table 6. Vintage adjusted space heating and ventilation energy requirement (Source: Derived from Natural Resources Canada Comprehensive Energy Use Database, 2014)

Vintage	Total Single Detached Floor Space (in BC, million m²)	Gross Output Thermal Requirements for Single Detached in BC (GJ/household)	Gross Output Thermal Requirements for Single Detached in BC (GJ/m²)	Gross Output Therm. Req. weighted average by m² per vintage (GJ/m²)	Gross Thermal Output Delta from weighted average	Vintage adjusted space heating and ventilation energy requirement (GJ/m²)
Before 1946	8.17	80.20	0.57	-	67.54%	**0.297**
1946–1960	9.97	76.38	0.51	-	50.79%	**0.267**
1961–1977	34.16	70.59	0.43	-	27.67%	**0.226**
1978–1983	23.59	66.89	0.37	-	8.52%	**0.192**
1984–1995	44.77	59.09	0.31	-	-7.75%	**0.163**
1996–2000	16.73	49.33	0.27	-	-21.59%	**0.139**
2001–2005	15.56	51.20	0.26	-	-23.94%	**0.135**
2006–2010	20.48	49.31	0.23	-	-31.55%	**0.121**
2011–2014	13.65	50.12	0.22	-	-34.97%	**0.115**
Average	-	-	-	-	-	-
Total	**187.08**	-	-	**0.339**	-	-
source	*NRCan CEUD 2014: Table 19*	*NRCan CEUD 2014: Table 32*	*NRCan CEUD 2014: Table 33*	*derived*	*derived*	*derived*

Table 7. Primary operating CO_2e emissions per m^2 of lower mainland SFH adjusted by vintage (Source: Derived from data from BC Assessment (2015b), Natural Resources Canada (2014), Energy Star (2013), British Columbia (2012), British Columbia (2016), Young (2017)1.)

Vintage	Space Heating+Ventilation: Total (GJ/m2)	Space Heating+Ventilation: Gas (GJ/m2)	Space Heating+Ventilation: Electricity (GJ/m2)	Water Heating: Total (GJ/m2)	Water Heating: Gas (GJ/m2)	Water Heating: Electricity (GJ/m2)	Appliances: Total - All Elec (GJ/M2)	Lighting: Total - All Elec. (GJ/M2)	Space Cooling: Total - All Elec (GJ/M2)	Electricity total (GJ/m2)	Gas total (GJ/m2)	Electricity Converted to primary in GJ	Gas converted to primary in GJ	Prim. elec in kgCO2e	Prim. Gas in kgCO2e	Total Prim. Operating Energy (kgCO2e/m2/year)
Before 1946	0.297	0.242	0.055	0.050	0.039	0.011	0.106	0.030	0.001	0.204	0.280	0.418	0.286	**1.238**	**14.183**	**15.42**
1946–1960	0.267	0.217	0.050	0.050	0.039	0.011	0.106	0.030	0.001	0.198	0.256	0.406	0.261	**1.204**	**12.961**	**14.17**
1961–1977	0.226	0.184	0.042	0.050	0.039	0.011	0.106	0.030	0.001	0.191	0.223	0.391	0.227	**1.158**	**11.275**	**12.43**
1978–1983	0.192	0.157	0.036	0.050	0.039	0.011	0.106	0.030	0.001	0.184	0.195	0.378	0.199	**1.120**	**9.878**	**11.00**
1984–1995	0.163	0.133	0.030	0.050	0.039	0.011	0.106	0.030	0.001	0.179	0.172	0.367	0.175	**1.087**	**8.691**	**9.78**
1996–2000	0.139	0.113	0.026	0.050	0.039	0.011	0.106	0.030	0.001	0.174	0.152	0.357	0.155	1.060	7.681	**8.74**
2001–2005	0.135	0.110	0.025	0.050	0.039	0.011	0.106	0.030	0.001	0.174	0.148	0.356	0.151	1.055	7.510	**8.56**
2006–2010	0.121	0.099	0.022	0.050	0.039	0.011	0.106	0.030	0.001	0.171	0.138	0.351	0.140	1.040	6.955	**7.99**
2011–2014	0.115	0.094	0.021	0.050	0.039	0.011	0.106	0.030	0.001	0.170	0.133	0.348	0.135	1.033	6.705	**7.74**
Average	0.177	0.144	0.033	0.050	0.039	0.011	0.106	0.030	0.001	0.181	0.183	0.372	0.187	1.103	9.256	10.36
Total	-	-	-	-	-	-	--	-	-	-	-	-	-	-	-	-

Table 8. Source: Data extracted from Simonen et al.

BLDG_PUB	BLDG_TYP + BLDG_US	BLDG_LOC	BLDG_NEW	$BLDG_AREA_M2	$BLDG_AREA_FT2	BLDG_$	BLDG_ST	LCA_YEAR	LCA_RE	LCA_SOUR	LCA_STAG	LCA_BLDG	LCA_MAT_Q	EC_WB_EX	EC_LCAA_PERM2
A05	Residentia Single-fam	North Ame	New	94 to 465	1 001 to 5 000	0	1 to 6	2004			A	SFEI	Y	46826000	**244.31**
A05	Residentia Single-fam	North Ame	New	94 to 465	1 001 to 5 000	0	1 to 6	2004			A	SFEI	Y	37047000	**193.29**
A05	Residentia Single-fam	North Ame	New	94 to 465	1 001 to 5 000	0	1 to 6	2004			A	SFEI	Y	28004000	**139.93**
A05	Residentia Single-fam	North Ame	New	94 to 465	1 001 to 5 000	0	1 to 6	2004			A	SFEI	Y	21367000	**106.77**
A05	Residentia Other	North Ame	New	94 to 465	1 001 to 5 000	0	1 to 6	2011	60		ABC	SEI	N	40	**266.67**
A05	Residentia Other	North Ame	New	94 to 465	1 001 to 5 000	0	1 to 6	2011	60		ABC	SEI	N	28.4	**258.18**
A05	Residentia Other	North Ame	New	94 to 465	1 001 to 5 000	0	1 to 6	2011	60		ABC	SEI	N	25.8	**258**
A05	Residentia Other	North Ame	New	94 to 465	1 001 to 5 000	0	1 to 6	2011	60		ABC	SEI	N	38.4	**256**
A05	Residentia Other	North Ame	New	94 to 465	1 001 to 5 000	0	1 to 6	2011	60		ABC	SEI	N	43.1	**253.53**
A05	Residentia Other	North Ame	New	94 to 465	1 001 to 5 000	0	1 to 6	2011	60		ABC	SEI	N	27.6	**250.91**
A05	Residentia Other	North Ame	New	94 to 465	1 001 to 5 000	0	1 to 6	2011	60		ABC	SEI	N	39.7	**248.13**
A05	Residentia Other	North Ame	New	94 to 465	1 001 to 5 000	0	1 to 6	2011	60		ABC	SEI	N	39.3	**245.63**
A05	Residentia Other	North Ame	New	94 to 465	1 001 to 5 000	0	1 to 6	2011	60		ABC	SEI	N	45.6	**240**
A05	Residentia Other	North Ame	New	94 to 465	1 001 to 5 000	0	1 to 6	2011	60		ABC	SEI	N	30.7	**236.15**
A05	Residentia Other	North Ame	New	94 to 465	1 001 to 5 000	0	1 to 6	2011	60		ABC	SEI	N	23.3	**233**
A05	Residentia Other	North Ame	New	94 to 465	1 001 to 5 000	0	1 to 6	2011	60		ABC	SEI	N	25.6	**232.73**
A05	Residentia Other	North Ame	New	94 to 465	1 001 to 5 000	0	1 to 6	2011	60		ABC	SEI	N	34.4	**229.33**
A05	Residentia Other	North Ame	New	94 to 465	1 001 to 5 000	0	1 to 6	2011	60		ABC	SEI	N	24.8	**225.45**
A05	Residentia Other	North Ame	New	94 to 465	1 001 to 5 000	0	1 to 6	2011	60		ABC	SEI	N	29.2	**224.62**
A05	Residentia Other	North Ame	New	94 to 465	1 001 to 5 000	0	1 to 6	2011	60		ABC	SEI	N	35.6	**222.5**
A05	Residentia Other	North Ame	New	94 to 465	1 001 to 5 000	0	1 to 6	2011	60		ABC	SEI	N	34.9	**218.13**
A05	Residentia Other	North Ame	New	94 to 465	1 001 to 5 000	0	1 to 6	2011	60		ABC	SEI	N	27.6	**212.31**

continued on following page

Table 8 continued

BLDG_PUB	BLDG_TYP + BLDG_US	BLDG_LOC	BLDG_NEW	$BLDG_ AREA_M2	$BLDG_ AREA_FT2	BLDG_$	BLDG_ST	LCA_YEAR	LCA_RE	LCA_SOUR	LCA_STAG	LCA_BLDG	LCA_MAT_Q	EC_WB_EX	EC_LCAA_ PERM2
A05	Residentia Other	North Ame	New	94 to 465	1 001 to 5 000	0	1 to 6	2011	60		ABC	SEI	N	42	**210**
A05	Residentia Other	North Ame	New	94 to 465	1 001 to 5 000	0	1 to 6	2011	60		ABC	SEI	N	22.5	**204.55**
A05	Residentia Other	North Ame	New	94 to 465	1 001 to 5 000	0	1 to 6	2011	60		ABC	SEI	N	42.9	**204.29**
A05	Residentia Other	North Ame	New	94 to 465	1 001 to 5 000	0	1 to 6	2011	60		ABC	SEI	N	26.3	**202.31**
A05	Residentia Other	North Ame	New	94 to 465	1 001 to 5 000	0	1 to 6	2011	60		ABC	SEI	N	30.2	**201.33**
A05	Residentia Other	North Ame	New	94 to 465	1 001 to 5 000	0	1 to 6	2011	60		ABC	SEI	N	29.8	**198.67**
A05	Residentia Other	North Ame	New	94 to 465	1 001 to 5 000	0	1 to 6	2011	60		ABC	SEI	N	27.4	**195.71**
A05	Residentia Other	North Ame	New	94 to 465	1 001 to 5 000	0	1 to 6	2011	60		ABC	SEI	N	31.2	**195**
A05	Residentia Other	North Ame	New	94 to 465	1 001 to 5 000	0	1 to 6	2011	60		ABC	SEI	N	27.3	**182**
A05	Residentia Other	North Ame	New	94 to 465	1 001 to 5 000	0	1 to 6	2011	60		ABC	SEI	N	26.7	**178**
A05	Residentia Other	North Ame	New	94 to 465	1 001 to 5 000	0	1 to 6	2011	60		ABC	SEI	N	24.8	**177.14**
A05	Residentia Single-fam	North Ame	New	94 to 465	1 001 to 5 000	0		2014	60		ABCD	SFEI	Y	184.79	**746.64**

Section 2
Material Topics

This section consists of in-depth studies of various material topics. The chapters offer insights regarding the environmental performance of plastics, cementitious materials, and water in building construction. Approaches include linking disparate fields of inquiry, proposing a new assessment methodology, and developing and testing a new composite. These works demonstrate that in-depth explorations of specific materials can contribute to the broader knowledge base of building construction as a whole.

Chapter 4
Toxicity in Architectural Plastics:
Life-Cycle Index of Human Health in Building (LCI-HHB)

Franca Trubiano
https://orcid.org/0000-0001-6919-7217
University of Pennsylvania, USA

ABSTRACT

The building industry lacks a holistic and integrated method for assessing the possible human health risks attendant to using materials that have been verified as toxic. In particular, it lacks an open-source, interactive interface for measuring the health risks associated with sourcing, manufacturing, selecting, installing, using, maintaining, and disposing of building-based polymers. Because of their high degree of chemical synthesis, polymers are typically more toxic than wood, glass, or concrete; yet architects, engineers, builders, clients, and the general public remain poorly informed about the deadly accumulation of synthetic polymers that originate in the building industry and that pervade our air, water, and bodies. This question should be central to the very definition and practice of life-cycle assessment, and this chapter outlines a process for developing an industry-based life-cycle index of human health in building (LCI-HHB). After all, traditional LCAs are of little help to anyone not healthy enough to enjoy them.

DOI: 10.4018/978-1-7998-2426-8.ch004

DIAGNOSING THE PROBLEM

The AEC Industry: Uniformed, Negligent, or Missing in Action?

The Architecture, Engineering, and Construction (AEC) industry is substantially in support of Life-Cycle Assessments (LCA) and increasingly aware of the very real costs of embodied energy, operational energy, and carbon emissions that contribute to the production of greenhouse gases and the devastating effects of climate change. It has largely embraced LCA methods and metrics, having attributed an important role to environmental accountability and sustainability certification programs such as Leadership in Energy and Environmental Design (LEED-USGBC), and rating systems such as the Living Building Challenge (International Living Future Institute), and Building for Environmental and Economic Sustainability (BEES). However, the same AEC industry has failed to integrate questions of human health in how it conceptualizes and calculates a building's Life-Cycle.

The building industry lacks a holistic and integrated method for assessing the possible human health risks attendant to using materials that have been verified as toxic. In particular, it lacks an open-source, interactive interface for measuring the health risks associated with sourcing, manufacturing, selecting, installing, using, maintaining, and disposing of building-based polymers. Indeed, the ubiquitous use of plastics in architectural design and construction rarely acknowledges the serious environmental and health risks posed by all forms of polymer products derived from petroleum, coal, and natural gas. For well over fifty years, the majority of building products have been re-engineered to include polymers in order to achieve a range of advanced performance metrics. Even wood, the most 'natural' of materials, is widely manipulated using cold-cured synthetic resin glues for increasing its structural strength and moisture resistance. The current trend of building with cross-laminated timber (CLT) is only possible because of the availability of polymers such as melamine, Phenol-Resorcinal-Formaldehyde, and other resorcinol-based resins (woodproducts.fi). Moreover, polyvinyl chloride is yet another polymer used ubiquitously in plumbing supplies, exterior sheathing, interior surfaces, furniture, and landscaping. Indeed, nearly everything in our built environment is permeated by chemicals derived from fossil fuels. And yet, despite their unrestricted proliferation, very little data is freely disclosed about the potential health risks associated with using large quantities of plastics in the building industry.

Not only do most polymers exhibit elevated levels of embodied energy per material weight, but their high degree of chemical synthesis also makes them vastly more likely to be toxic than wood, glass, or concrete. Synthetic rubbers or elastomers, for example, have an embodied energy count of 110 Mj/Kg, whereas concrete's is but 1.9 Mj/Kg (Your Home Australia), and while it might be the case that a far greater amount of concrete is used in a building than rubber, close to fourteen billion tons of synthetic rubber are produced each year (Siemens, 2013); all of which, in one way or another, makes its way into the built environment, and all of which is typically sourced using natural gas, petroleum, coal, or hydrocarbons. The concern this poses for human health is significant.

Why, therefore, are architects, engineers, builders, clients, and the general public so poorly informed on the toxic accumulation of synthetic polymers that originate in the building industry and that pervade our air, water, and bodies? More precisely, why is this question not central to the very definition and practice of Life-Cycle Assessment? Surely, responsibility for answering these questions does lie exclusively with members of the AEC industry. Chemical companies, manufacturers, and governmental agencies have an important role to play. In Europe, significant strides have been made in what concerns regulatory and legislative requirements for the full material disclosure of industrial products. In 2006, the European Parliament ratified REACH (Registration, Evaluation, Authorisation, and Restriction of Chemicals), which is committed to "*the protection of human health and the environment through the better and earlier identification of the intrinsic properties of chemical substances. ... REACH Regulation places responsibility on industry to manage the risks from chemicals and to provide safety information on the substances. Manufacturers and importers are required to gather information on the properties of their chemical substances, which will allow their safe handling, and to register the information in a central database in the European Chemicals Agency (ECHA) in Helsinki* (European Commission, REACH)." This is not, however, the case in the United States and in developing countries, where there is no legislative requirement for the disclosure of harmful materials as comprehensive and holistic as that proposed by the European Union. Under these circumstances, translating the responsibility for data transparency, of the kind facilitated by REACH, to members of the building industry is neither an easy or obvious task.

In response, this paper discusses the first outlines of a process for developing a holistic and communicative method for itemizing, evaluating, and communicating the human health (HH) impacts of polymerized materials sourced from fossil fuels. It identifies policy priorities and value parameters for the AEC industry, including the benefits of increased access to data and material flow analyses (MFA), to building-related assessment tools that highlight hazardous materials, to the facilitation of material disclosures, and to expanding the role of education in changing the industry's position. Lastly, this paper identifies an intellectual infrastructure whose vocabulary, subject categories, and process protocols are needed when developing an industry-based Life-Cycle Index of Human Health in Building (LCI-HHB). After all, traditional LCAs are of little help to anyone not healthy enough to enjoy them.

Communities in Crisis

The reader might ask why is this important? What evidence is there to support dedicating so much time and effort to identifying the toxicity of materials in the built environment? In fact, the proof is staggering, and the pleas for action are loud and persistent. In May 2019, the *Guardian*'s investigative team devoted a series of focused articles on the high incidence of cancer-related deaths amongst the residents of Louisiana who lived along the banks of the Mississippi River, between New Orleans and Baton Rouge. Reporters Oliver Laughland and Jamies Lartey investigated the environmental, political, and corporate circumstances in this region of the United States known by the locals as "Cancer Alley" (Laughland & Lartey, 2019). For decades, the forgotten residents of Saint John the Baptist Parish have been victimized by substantially higher levels of chronic and life-threatening illnesses (fifty times the national average); the source of which is alleged to be the excessive amount of airborne pollutants (measured by the Environmental Protection Agency-EPA) discharged from a local chemical plant that produces Neoprene. This all-important rubber elastomer is used in a host of building products, including window seals, gaskets, bearing pads, washers, and conduits. It is highly resistant to chemical degradation and successfully wards off water infiltration. According to its inventor, "*Chloroprene rubber, is not attacked by hydrogen chloride, hydrogen fluoride, sulfur chloride, ozone, and many other chemicals. The high chlorine content of the chloroprene rubber also renders it very resistant to combustion* (Carothers et al., 1931)." What's not to like?

Maybe, the fact that it is deadly. Neoprene's chemical precursor is Chloroprene—a polymer invented in 1931 by engineer Wallace Carothers while working at Dupont's Experimental Station (Carothers et al., 1931). In the 1960s, Dupont decided to relocate its neoprene production along the winding Mississippi River in LaPlace, Louisiana, but a few kilometers from residential communities who, for decades, had called this part of the world, home. Today, the same plant is owned by the Japanese company Denka which produces the synthetic Denka Performance Elastomer using huge quantities of petrochemicals most likely sourced but a few miles from the Gulf Coast's petroleum reserves. Indeed, as noted by industry promoter Global Market Insights, the production of Neoprene "*involves tremendous technological and energy requirements which acts as a major entry barrier for new entrants in the market*" (Hedge, 2019). This is partly why the Pontchartrain plant operating in LaPlace is today the only producer of the product in the entire United States.

This is also the site of much unnecessary illness and death. In 2010, Chloroprene was classified by the EPA as a likely carcinogen (EPA, 2010). In a survey conducted by the University Network for Human Rights—a group of concerned academics and students operating out of Stanford University—it was reported that households living within a 1.5 and a 2.5-kilometer radius from the plant had substantially higher mortality rates and ailments—including headaches, chest pain, heart palpitations, eye irritations, fatigue, and skin rashes—than the national average for similar sample groups (UNHR, 2019). Not entirely surprising, the majority of inhabitants in the survey area (60%) of LaPlace are African American residents whose 'back yards' have been systematically targeted by chemical companies for locating their petrochemical plants; a practice highly consistent with other forms of environmental injustice inflicted on minorities with fewer financial and political resources (Elliot et al, 2004). Still today, the reported average concentration of Chloroprene in the air of Saint John Parish is magnitudes greater than the EPA allowable 0.2 micrograms per cubic meter (mcg/m3). Even after the Denka plant's stated improvements in manufacturing, and its reported 85% reduction in emissions, on-going monitoring confirms significantly higher levels of exposure to the carcinogen than is considered safe. In June 2019, the Attorney General of the State of Louisiana announced it would file a lawsuit against Denka Performance Elastomer on behalf of the Louisiana Department of Environmental Quality for a documented 50 violations of the United States Clean Air Act identified by the EPA in 2016: a courageous, albeit rare move.

Sadly, this is not the only instance of severe toxicity resulting from the chemical synthesis of polymers, nor is it the only instance in this region of Louisiana. The production of polyvinylchlorides, for example, is as toxic and as contested as their use is excessive in the building industry. According to Lena Moffit from the Sierra Club, the town of Mossville, Louisiana, has fourteen petrochemical plants, three of which produce vinyl chloride (Moffit, 2016). In this, Mossville has much to be concerned about, including the decision by South African petrochemical giant Sasol to develop yet another one-million-ton capacity polyethylene plant in Lake Charles, Louisiana. Following the announcement, Sasol began the all-too-expected 'voluntary' buy-out of properties, most of which were owned by descendants of freed black slaves who had voiced concerns with possible increases in the risk to their health. All too cunningly, by having paid for the relocation of these highly vocal residents, the company systematically disempowered those who were left behind. With air and water samples in Mossville already indicating higher than normal levels of Polychlorinated biphenyls (PCB) and Ethylene dichloride, residents left behind have much to be alarmed about (McDermott, 2013).

The impact which the chemical synthesis of building-based plastics has on human health, and the implications this has for social justice, continues to represent a serious blind spot in architectural design. So too does the ease with which the building industry assimilates new and untested materials with little to no knowledge of that which they contain. For these reasons and others, this paper investigates the practice of Life-Cycle Assessment and the role which it can play in answering some of the most pressing questions about the human health risks associated with the use of polymers in the built environment. It reviews and proposes a host of resources for empowering the AEC industry to reduce, if not eliminate, its use of known carcinogens. The key, of course, is our knowledge of such hazards.

PLASTICS AND TOXICITY

Matter and Chemical Synthesis

Plastics are ubiquitous because their properties are easily optimized to match an almost infinite variety of end-use performance criteria. Polymers in wall finishes, floors, sealants, pipes, paints, flashing, and furniture are almost always derived from fossil fuel feedstocks that are manipulated to become more or less transparent, porous, rigid, ductile, hypoallergenic, hydrophobic, or thermally conductive. By these standards, we should all be using plastics, all of the time. Indeed, all plastics share origins in chemical synthesis—in polymerization; the process by which molecules are combined and altered, typically from simple monomers (with few molecular bonds) into more complex polymers (with new covalent bonds). Molecular reorganization strengthens and reinforces otherwise weak links by creating longer molecular chains and three-dimensional structures (Addington & Schodek, 2005). This is what defines and conditions material performance in plastics. Surely, natural substances such as cellulose, wool, silk, and natural rubbers can be polymerized; however, the source of most original molecules used in the making of building industry polymers are petroleum products.

Typically, plastics in the building industry are one of three kinds: thermoplastics, thermosets, or elastomers. Thermoplastics consist of polymeric molecules that are not strongly bonded together. They are "*soft and ductile... [such that] external forces can cause chains to slide by one another relatively easily.*" (Addington & Schodek, 2005). This characteristic can be used to create plastics that are easily manipulated into a range of desired shapes, yet this is also what makes them prone to being less strong. And should thermoplastics re-liquefy, their performance may be further compromised. If properly deployed, however, precisely this ability to be cast, formed, and shaped is what makes thermoplastics interesting to designers. Typical thermoplastics include polycarbonates, ETFE (ethylene-tetraflouroethylene), PTFE (polytetraflouroethylene), polyamides such as Kevlar, polyethylenes, and polypropylenes (Fernandez, 2005).

Thermosets are polymers whose bonds are significantly stronger than thermoplastics. With polymeric molecules that are "*cross-linked or interconnected,*" these plastics are physically more durable and less resistant to degradation than thermoplastics (Addington & Schodek, 2005). Common thermosets include phenolics, epoxies, and polyesters (Fernandez, 2005). And while they too can be molded into a host of anticipated shapes, once they are set, they cannot easily be recast. In this sense, thermosets are often difficult, if not impossible, to recycle.

Lastly, plastics identified as elastomers are typically rubber-based compounds such as silicon, neoprene, and ethylene-propylene-diene-monomer (EPDM) (Fernandez, 2005). This category of polymers rarely enters into the architect's design vocabulary because elastomers are invisible. They are embedded in a building's seams, fastenings, sandwich panels, and composites (Kaltenbach, 2004), and this is precisely what makes them so ubiquitous and dangerous. Of the three polymer categories, thermoplastic elastomers are, in principle, recyclable; thermosets and elastomers are effectively not.

From a policy perspective, it is noteworthy that outside of the lab, these highly engineered materials are simply referred to by the general public as—plastics; less a denotation of what they are than a simple description of their most common mechanical property—the ability to bend and deform without breaking. Their larger material history, chemical definition, and associated health risks are so concealed to the layman that we continue to refer to this class of substances merely by how they look and feel, and not by what they are. Our general ignorance on the matter of plastics is all the more explicit by way of a comparison with another material; after all, we would no sooner refer to the richness and variety of the building material that is stone by the single moniker, *hard.*

More specifically, the vast majority of materials referred to as plastics are *organic polymers*: materials comprised of long chains of repeating molecular units which, albeit incredibly tough, in most cases are able to glide past each other contributing to the material's capacity for deformation. Much of plastic's versatility results from the highly tunable interactions between these chains. The modifier *organic* specifies that these chains are linked by carbon atoms. In some cases, such as with polyethylenes, the linking unit is simply one carbon atom. This *organic* distinction is important, as some polymers have repeating units of elements other than carbon, such as polysiloxanes (silicones) (Jones 2014). But as this study is concerned with the link between fossil fuels and the building industry, it is limited to plastics defined by their carbon footprint.

The Life-Cycle of Carbon-Based Plastics

Plastics have surprisingly carbon-intense Life-Cycles. The overwhelming majority of plastic resins come from petroleum, which requires extraction and distillation. Then the resins are formed into products and transported to market. All of these processes emit greenhouse gases, either directly or via the energy required to accomplish them. And the carbon footprint of plastics continues even after we've disposed of them.

Dumping, incinerating, recycling and composting (for certain plastics) all release carbon dioxide. All told, the emissions from plastics in 2015 were equivalent to nearly 1.8 billion metric tons of CO_2 (Tasoff, 2019).

According to the Center for International Environmental Law (CIEL), nearly "*all plastic produced today (more than 99 percent) is manufactured from fossil fuel feed stocks*," including natural gas liquids and by-products from crude oil refining (CIEL, 2019). Nearly 83% of all plastics produced by weight globally are sourced from just two industrial chemicals derived from fossil fuels: ethylene and propylene (CIEL, 2019). By 2050, the production of plastics will account for 20% of global fossil fuel consumption. It is not surprising that the largest oil and gas companies are also in the business of supplying raw materials for the making of plastics (CIEL, 2019). More critically, the construction industry is the second-largest consumer of plastics in the United States consuming between 16 and 20%, including more than 12,600 million pounds per year for applications such as windows, wall coverings, piping, roofing, and flooring (Plastics Market Watch 2016). The result of which is, sadly, that in 2015 plastics contributed 1.8 billion metric tons of CO_2, while in 2004, the entire building industry contributed but 2.236 billion metric tons of CO_2 (Tasoff 2019, USGBC). With such numbers, it is clear that whenever architects, engineers, and builders specify and install building-based plastics, they are directly implicated in the petrochemical to plastics cycle. And yet, they scarcely have resources to calculate and mitigate the risks attendant with their decisions.

Surely, our gluttonous use of building plastics recommends a rigorous review of their Life-Cycle impacts and costs. The building industry is increasingly aware of the energy embodied in materials and the carbon emitted in their production. Life-Cycle Assessment (LCA) quantifies both the embodied and operational energy of a product by analyzing the supply chain of its source materials, as well as the product's intended use and disposal (Bayer, 2010; Kohler, 2013; Menzies, 2007; Simonen, 2014). The building industry has embraced this expanded approach to material selection by incorporating LCAs in sustainability certification programs and in rating systems. However, it is with some irony that Life-Cycle Assessments often result in the promotion of all things plastic in the service of long-term energy savings. After all, compared to metals, plastics are significantly more thermally inert and less conductive. The total energy required to produce a PVC (polyvinyl chloride) window frame is about half that required for an aluminum alternative (Franklin Associates, 1991). And, the energy savings associated with the use of expanded polystyrene (EPS) insulation, when measured over a fifty-year period, returns two hundred times more energy savings than what was consumed in its production (American Chemical Council, 2009). It is clear that advocating for the phase-out of all building-based plastics solely due to their petrochemical origins is as reductive as categorically championing their use because, in some cases, they favor savings in operational energy.

What is needed is an integrated Life-Cycle index for communicating the type and sum of all risks to the quality of life and human health posed by building-based plastics (LCI-HHB). Existing LCAs poorly account for the larger ecological costs incurred when sourcing, manufacturing, selecting, installing, using, maintaining, and disposing of plastic (Zheng and Suh, 2019). According to a comprehensive review of existing LCAs published in 2014, which aimed to identify important gaps in how assessments categories are defined and deployed, human health represents precisely such a gap (Finkbeiner et al. 2014). The cumulative effect of plastics on human toxicity, contamination by particulate matter, and the likelihood of endocrine disruption are all subjects poorly considered in existing LCAs. So too are the effects of nanomaterials, microbiological pollution, and the impact of noise and odors on human health (Finkbeiner et al. 2014). Even LCAs more directly related to the building industry, such as Tally®, which is REVIT's plug-in supported by the U.S. Green Building Council, fails to acknowledge material toxicity amongst its many impact categories. And, while the BEES software developed by NIST (National Institute of Standards and Technology) integrates categories such as Ecological Toxicity, Indoor Air Quality, and Human Health, this data is only available for 230 material products associated with twenty-three building components.

A more holistic and transparent method for itemizing, evaluating, and communicating the human health (HH) impacts (and toxicology) of using polymerized materials sourced from fossil fuels is called for, and a good place to begin is with the risks associated with the industry's excessive use of polyvinyl chloride (PVC) (Plastics Europe, 2017).

Human Health Risks of Building with Polyvinyl Chloride

Polyvinyl chloride (PVC) is an incredibly versatile material used near ubiquitously in the construction industry. It is found in roof scuppers, floor tiles, water pipes, wallpaper, windows, and the exterior siding of many a single-family home in the United States. Vinyl products of all kinds are embedded in products touted for their water and chemical resistance. According to the Globe News Wire, Reports and Data, PVC is the single largest plastic used in the building industry with a total market share of 57.1% (Reports and Data 2019).

Critical for the question of human health, it originates in two chemical feedstocks, ethylene and chlorine, sourced from petrochemicals and saltwater, respectively (Perkins + Will 2019). And as with many other synthetic fabrication processes, PVC production involves precursors that pose risks to human and environmental health. PVC's monomer, VCM (vinyl chloride monomer), is a known carcinogen and must be manipulated in closed systems to protect workers from exposure (DOW 2019). Those who work in manufacturing PVC products are protected from monomer precursors through regulation and industry practices (OSHA 2018). However, PVC is often plasticized (softened) with chemical additives that are at risk of leaching into the environment (Lithner 2012). During the addition of phthalate plasticizers, which are 'loosely' connected to PVC molecules, the product acquires its toxicity. The vinyl industry has been under pressure to phase out the use of these plasticizers, which are proven endocrine disruptors and which pose risks to reproductive health, immune response, and embryonic development (NIEH, 2010). Additionally, children are at the highest risk of asthma, rhinitis, and eczema when exposed to phthalates in their environment via dust and when in contact with PVC surfaces and flooring (Vesterberg, 2005).

Moreover, most PVC products are substantially not recycled; at the end of life, they are typically incinerated or disposed of in landfills (Geyer, 2017). Even in Europe, where recycling is highly championed, only three percent of post-consumer PVC is diverted from the waste stream (Plinke, 2000), possibly because of the material's high chlorine and additive content (Baitz, 2004). And, if PVC incineration is not executed in accordance with environmental guidelines, the high chlorine content results in the formation of toxic, carcinogenic dioxin pollutants at concentrations over three to four orders of magnitude greater than other common plastics (Katami, 2002). In the United States, EPA regulations protect against dioxin emission from industrial incineration facilities (Venezia 2010); however, these restrictions do little to protect the ecosystems and well-being of rural communities that dispose of their waste through 'backyard burning.' Such community-level incineration is the largest source of domestic dioxin (NCEA, 2006), which is carcinogenic, and whose direct exposure to PVC gas results in skin lesions, eye irritation, and problems with circulation (Freiknel, 2011). Moreover, in the case of uncontrolled building fires, the release of dioxins poses serious risks for firefighters and others who are exposed; as noted in the case of the Binghamton State Office Building fire in 1981 and the World Trade Center fire in New York in 2011 (Fitzgerald et al., 1986; Dalgreen et al., 2007).

Just as tragically, plastics are with us forever. They may degrade, but they persist. Particulates and microplastics routinely leach and disperse their pollutants in water and air. As studied by Deonie Allen, of the École Nationale Supérieure Agronomique de Toulouse (ENSAT), trace plastics no bigger than dust were recorded as having been transported and contaminating sites as far afield as a hundred kilometers away from their source (S. Allen, D. Allen, et al., 2019). Landscapes register particulates in their environment, even where no manufacturer or use of the material is common. Such tenaciousness on the part of plastics is neither desirable nor life-sustaining. Whether through leaching, burning, or dispersal, plastics accumulate in the human body and contribute to increased risks for interference with the proper functioning of estrogen and testosterone, and with long term effects such as asthma, diabetes, obesity, infertility, and even attention deficit disorders.

These are but a few of the health risks associated with the production, use, and disposal of PVC, even as there are dozens of additional polymers sourced from fossil fuels that pose similar threats. In all cases, the deployment of these polymers in the building industry requires a fully integrated Life-Cycle Index of Human Health in Building (LCI-HHB). The remainder of this paper is dedicated to a discussion of data mining, existing assessment tools, material disclosures, and increased professional education as vehicles for precipitating industry change and for facilitating the first outlines of an Index for Human Health in Building.

POLICY PRIORITIES

Data Mining

Data mining is an essential first component in the creation of a Life Cycle Index for Human Health in Building. Naming, selecting, and organizing the data associated with the chemical synthesis fossil fuel-based polymers is the only way of coming to know of the risks associated with their fabrication and handling. However, in many a manufacturing industry, the lack of globally ratified regulations that require the material disclosure of commercial products protects suppliers and producers from having to communicate the full sourcing and raw material flows of that which they bring to market. The building industry is no different. It is a highly difficult, time-consuming, and resource-intensive process to follow the material supply chain of any product employed in a construction project. Notwithstanding efforts by organizations such as Verité which seeks to eliminate the use of forced labor in material procurement and does so by analyzing global supply chains for construction building materials as such as bamboo and bricks, as well as provisions taken by U.S. Customs and Border Protection to avoid the entry of all products into the United States produced using illegal labor, there are no state, federal, or globally sanctioned imperatives to disclose the supply chain of a building's materials. In this climate of professional secrecy, securing accurate data on the origins of material products is challenging for professionals and near impossible for the general public. Despite this, informed architects, researchers, material scientists, and policy experts have made significant strides in identifying, cataloging, and communicating measurable information about a host of material hazards that should be avoided in the design and construction of buildings.

For example, the Center for International Environmental Policy (CIEL), published in 2019 in collaboration with Earthworks and UPSTREAM, the report *Plastic & Health, The Hidden Costs of a Plastic Planet,* which identified the Life-Cycle phases of plastics in which humans are most at risk. These included the handling of raw chemicals, the use of plasticizers and chemical additives, and the larger ecological tragedy of micro-particulates, which are forever lost to air and water. The report detailed "*the physical impacts of ingesting, inhaling, and touching plastic, as well as the toxic chemicals associated with those plastic particles, whether chemical additives, processing agents, or byproducts of plastic*" (CIEL, 2019). Thankfully, data mining is of prime interest to many industry affiliated groups such as Ecoinvent (ecoinvent.org), GaBi (gabi-software.com), and USEtox® (usetox.org), amongst others who are singularly committed to creating and deploying transparent datasets for the production of ever more precise LCAs. Their respective repositories of rigorously collected information on the origins and impacts of all material types significantly assist all manner of industries, be they automotive, chemical, energy, electronic, or building and construction. Ecoinvent, for example, is a subscription-based inventory database, developed by the Swiss Centre for Life Cycle Inventories. With its claim to 10,000 data sets, it has serviced the construction sector in its work with LafargeHolcim, Saint-Gobain, and others, and it has done so by including the question of human and ecotoxicity potential within their modeling profiles. GaBi software, for example, is another industry-forward tool whose data profiles make life cycle assessments available to most product manufacturers. It also lists Human Toxicity as part of its impact potentials, identifies ecotoxicity and particulate matter within its Environmental Footprint reporting, and supports the use of Environmental Product Declarations. Lastly, USEtox® is a modeling interface for specifically and directly assessing the toxicity of chemicals to humans and the larger ecological environment. Seeking a globally deployable model, the United Nations Environment Program and the Society of Environmental Toxicology and Chemistry created the USEtox® model to unify disparate existing models and gain scientific consensus on the best way to identify the toxic risks associated with thousands of chemicals.

What remains unclear, however, is the extent to which any of these databases and interfaces are easily employable by construction industry members. Highly fractured, at times still artisanal, and far from given to conducting research, the building industry—contrary to medicine, agriculture, automotive, and defense—struggles to engage with datasets in any meaningful manner. For this reason and others, needed are easy to use assessment tools which offer design and construction professionals ways of evaluating said risks on a project to project basis.

Building Based Assessment Tools and Certifications

Increased availability of digitally based assessment tools is essential for the effective deployment of a Life Cycle Index for Human Health in Building. Thankfully, the design, engineering, and near commercialization of project-based interfaces that help one measure, evaluate, and benchmark the health hazards of material flows for all building-based polymers have been the priority of a number of key knowledge groups active in this space. The Healthy Building Network (HBN), for example, founded in 2000 by Bill Walsh, has developed a subscription-based on-line interface whose data sets can be queried via a comprehensive and detailed portal (healthybuilding.net). HBN releases many fact-finding reports, including reports on the use of chlorines in building products. It also makes available tools such as PHAROS whose search engine probes more than 150,000 chemicals and nearly 150 building products (pharosproject.net), and whose searchable database includes Hazards Lists (commons. healthymaterials.net) which deliver a GreenScreen Assessment Score on endpoints such as Neurotoxicity, Developmental Toxicity, and Skin-Eye Irritation/ Corrosivity amongst others. On the subject of Benzene, for example, Pharos communicates that the chemical triggers life-threatening or chronic human health risks at very low dosages and that it is found on 18 different Restricted Substances lists. HBN also connects to the Chemical Hazard Data Commons, which offers modeled information on chemical factors associated with Carcinogenicity, Reproductive Toxicity, and Acute Oral Toxicity, amongst others. PORTICO is yet another on-line software interface created by HBN in collaboration with GOOGLE, for evaluating a project's healthy material choices (portico.healthymaterials.net).

Additional assessment interfaces dedicated to the human health of materials include industry-based certification programs such as the Cradle to Cradle Product Standard™ (c2ccertified.org), the WELL Certification (wellcertified.com), and the all-important work of the International Living Future Institute with its Declare. Nutrition Label for Products (living-future.org). In the case of WELL, for example, managed by the International WELL Building Institute, Air, Water, and Materials are all subjects analyzed for building project health. In the context of Materials, the WELL certification is expressly designed to "reduce human exposure to hazardous building material ingredients through the restriction or elimination of compounds or products known to be toxic." (https://v2.wellcertified.com/v/en/materials). Points are available for abatement of hazardous materials, for proper waste management, for reducing the number of volatile compounds, and for having protocols that control the use of cleaning products.

In the end, dozens of repeat offender materials are brought to our attention via these assessment tools and certification processes. Their deployment easily corroborates information on the most obvious hazardous and toxic materials. And yet, few building professionals have yet to engage with such protocols as part of their typical offer of services.

Re-Educating AEC Professionals

In an environment controlled by market forces and consumer responses, there is little to no incentive for chemical distillers and product manufacturers to change their practices unless they hear from the market. Manufacturers are far less concerned with the issue of environmental health, even when it directly impacts their own workers. Absent regulatory mechanisms, required is a market-based approach deployed in conjunction with a successful education campaign to inform professional bodies who specify said materials and who control purchasing contracts. These are the architects, engineers, and builders who, on the front line of specifying said materials, are the individuals in the supply chain who are more easily convinced of the toxic dangers involved in the various lifecycle stages of material products. And these are the professionals involved in delivering building-based services to a general public who require re-education in their respective disciplines on the subject of material health.

This question is of great interest to design professionals. The architectural firm of Perkins+Will, for example, has invested much effort in the creation of a database supported website that educates architects and the larger public on the question of material disclosure in service to human health. Appropriately called Transparency (transparency.perkinswill.com), this resource is aimed at fellow professionals, educators, and members of the general public interested in the 'Precautionary List' of materials whose use should be avoided at all costs in the design and construction of built environments. The site identifies common risks associated with the use of chloroprene, formaldehydes, bisphenol A, flame retardants, all manner of phthalates, toxic metals, mercury, lead, cadmium, styrenes, and PVCs. It identifies their 'Health Hazards,' be they Carcinogenicity, Developmental Toxicity, Endocrine Disruption, and Neurotoxicity, amongst others; their 'Environmental Hazards,' be they Aquatic or Terrestrial Toxicity; and, it introduces the various 'Pathways of Exposure' that one may be subject to during Manufacturing, Installation, Use Phase, or Disposal of the material.

The American Institute of Architects (AIA) has also been an important player in this space. In spring 2018, it published in collaboration with engineering and architecture firm ARUP, *Prescription for Healthier Building Materials: A Design and Implementation Protocol,* which offered design-based strategies for making healthier building-based choices. The same year, the AIA updated its Code of Ethics such that architects are now required to design "*a built environment that equitably supports human health and well-being and is resistant to climate change*" (AIA, 2018). The AIA also offers its members a continuing education *Certificate Program on Materials*, whose third module is on "Health Impacts: Connecting Building Materials, Human Health, and the Environment."

University curricula, however, are not as enlightened. They too would benefit from the introduction of language and metrics associated with material toxicology, its delivery mechanisms, and possible human health impacts. At present, few architecture students in the United States are educated on the impact which materials have on water pollution, poor indoor air quality, and environmental toxicity. The National Architectural Accrediting Board (NAAB) makes no mention of human health issues resulting from poor architectural design decisions. NAAB lists among its "Student Performance Criteria (SPF)" knowledge of the environmental impact of materials, issues of social equity, and the code of professional ethics; however, there are no Student Performance Criteria that reference the ethical and social justice imperative of delivering toxic-free material environments. Changing this is an important policy priority.

At this time, leading the charge in schools of architecture is the Healthy Materials Lab at The New School Parsons (healthymaterialslab.org). Its directors are invested in a model of engagement and enlightenment, which offers the most constructive example of education-based leadership in the field. The Lab offers interested students and professionals a building-based clearinghouse for matters of 'human health' and a portal for accessing all manner of information on the subject. The website and physical library offer educational resources, online certification programs, and a host of links to seminal Tools + Guides. This one-stop Lab introduces its users to a repertoire of resources with which to initiate one's own research. And one can only hope that this is but the first of many such educational ventures devised to support the full material disclosure of building based risks to human health.

Material Disclosures

Surely, few would disagree with the premise that full material disclosure is essential to this success of a Life Cycle Index for Human Health in Building. And yet, it is unlikely that a majority of industry suppliers voluntarily disclose the origins, quantities, and risk factors associated with their materials. The building industry employs Material Safety Data Sheets (MSDS), whose goal is to collect and disclose data on the chemical origins, performance, and possible risks associated with all building-based materials; yet, these sheets are indeterminate in the area of human health risks. In principle, typical Material Safety Data Sheets call for information on Chemical products and Company Identification, on the Composition/Information of Ingredients, on Hazards, Fire Fighting Measures, Accidental Release Measures, Exposure Control, Stability and Reactivity, on Toxicology, and on Ecological and Regulatory Information. However, at present, there are no internationally shared protocols and regulations that compel a manufacturer to complete all sixteen MSDS categories. As a result, designers have little confidence they are fully informed on the true nature of the materials they come into contact with when designing, building, and renovating buildings, let alone the full toxicological profile of said materials.

Similar concerns might be noted in what concerns the Master Format® interface published by the Construction Specifications Institute and used by architects, engineers, and builders for effecting building based legal decisions that accompany building contracts. An initial review of the nomenclature used in the Master Format reveals a number of disquieting observations in what concerns the language surrounding polymers (See Figures 1 and 2). Only thirteen terms are commonly used to indicate polymer-based categories that require detailed specifications; these include the families of acrylic, epoxy, fiberglass, linoleum, polyethylene, polypropylene, polyurethane, polyvinyl, resilient, resin, and vinyl. Many of the most toxic and embedded elastomers are not discussed. Moreover, amongst the thirteen terms, two are descriptive of no materials at all: 'laminate' and 'plastic' are not material definitions. Laminate is a method for binding multiple thin material layers to each other and can reference any material, and plastic can refer to nearly 100 different polymers. Hence, anyone using the Master Format might not recognize the relevancy of these sections for their project delivery aims. In addition, while Specification categories such as Wood, Plastics, and Composites; Thermal and Moisture Protection; Openings; Finishes; and Utilities are the five which most referred to polymer-based materials, the degree to which each of these categories acknowledges equivalent performance metrics for each polymer is inconsistent. In fact, performance and sustainability are rarely concerns for the Master Format. Contractually delivering the project to the owner is its principal focus; the aftermath, maintenance, longevity, and human health impacts of the product are hardly issues. And hence, other than in the section dedicated to the risks posed by asbestos remediation, there are few health-centered contract guidelines for the writing of architectural specifications using the industry's Master Format.

More optimistically, legislative bodies in Europe have made significant advances in regulating the material disclosure of industrial products. This is clearly evident in the REACH program, administered by the European Parliament and as previously discussed. The regulatory approach requires, however, the following pre-condition—strong governmental support of the evidence and acceptance of the values subtended by those who argue for the avoidance of hazardous materials in the built environment. Where this exists, as in Europe, this offers a likely path for significant transformations in how materials are sourced, made, used, and discarded. However, where this does not exist—as in the United States and in most developing economies—there remains much skepticism about the impact of chemical contaminants in the air, water, and soil. In this political climate, to seek and hope for regulations at the federal or state level is an ineffective vehicle to action. Moreover, notwithstanding the availability of legislation that protects and renders transparent the information on chemical substances in material products, translating this information into usable content that building industry members can adopt in their professional activities remains outstanding.

Thankfully, material disclosure and transparency have become the priority of Environmental Product Declarations, and of the aggregation work being done by the Healthy Product Declaration Collaborative (HPD). These are the models that sustain this policy priority that insists that building material manufacturers be tasked with revealing the material flow origins and full chemical content of fossil fuel-based polymers used in their products so that building professionals, educators, medical professionals, and consumers can, more easily, access this information and gage its human health risks. It is proposed that product supply companies be legally responsible for fully disclosing all relevant information required when completing industry-specific MSDS. Subscribing to Environmental Product Declarations and participating in the Healthy Products Declaration Collaboration (HPD) are two ways of ensuring greater transparency. After all, material transparency in the building industry could, potentially, be equivalent to many city, state, and federal government mandates for full disclosure of the amount of energy consumed by commercial buildings. This reporting checklist could be instituted at the level of Building Code reviews, which take place with Building Permit issuance; no differently than is done with Code reviews that protect the safety of individuals against the propagation of fire. Toxic materials are no less dangerous, even if their effects are far less immediate and spectacular.

Conclusion: Life-Cycle Index of Human Health in Building (LCI-HHB)

The building industry continues to benefit from the leadership of those who champion transparency in material flows and sourcing. However, this work remains largely uncoordinated, independent, and not referenced to a set of integrated metrics. As such, the final policy priority which this paper proposes is the development of a holistic protocol for collating, visualizing, and communicating the human health risks associated with the use of fossil fuel-based polymer materials in the building industry. Once transparency in material flows is possible, and disclosure of all raw materials (especially synthetic chemicals) achieved, the health risks attendant to material production, use, and disposal could be diagnosed and diffused. Internationally accepted protocols across all three divisions of the AEC industry do not yet exist, yet attempts to devise such protocols should be discussed and deliberated. In anticipation of full material disclosure, however, beginning with available Hazards

Figure 1. Analysis of master format categories and polymer nomenclature. the occurrence of fossil fuel-based polymers in architectural specification documents
Source © Trubiano-Rinaldi

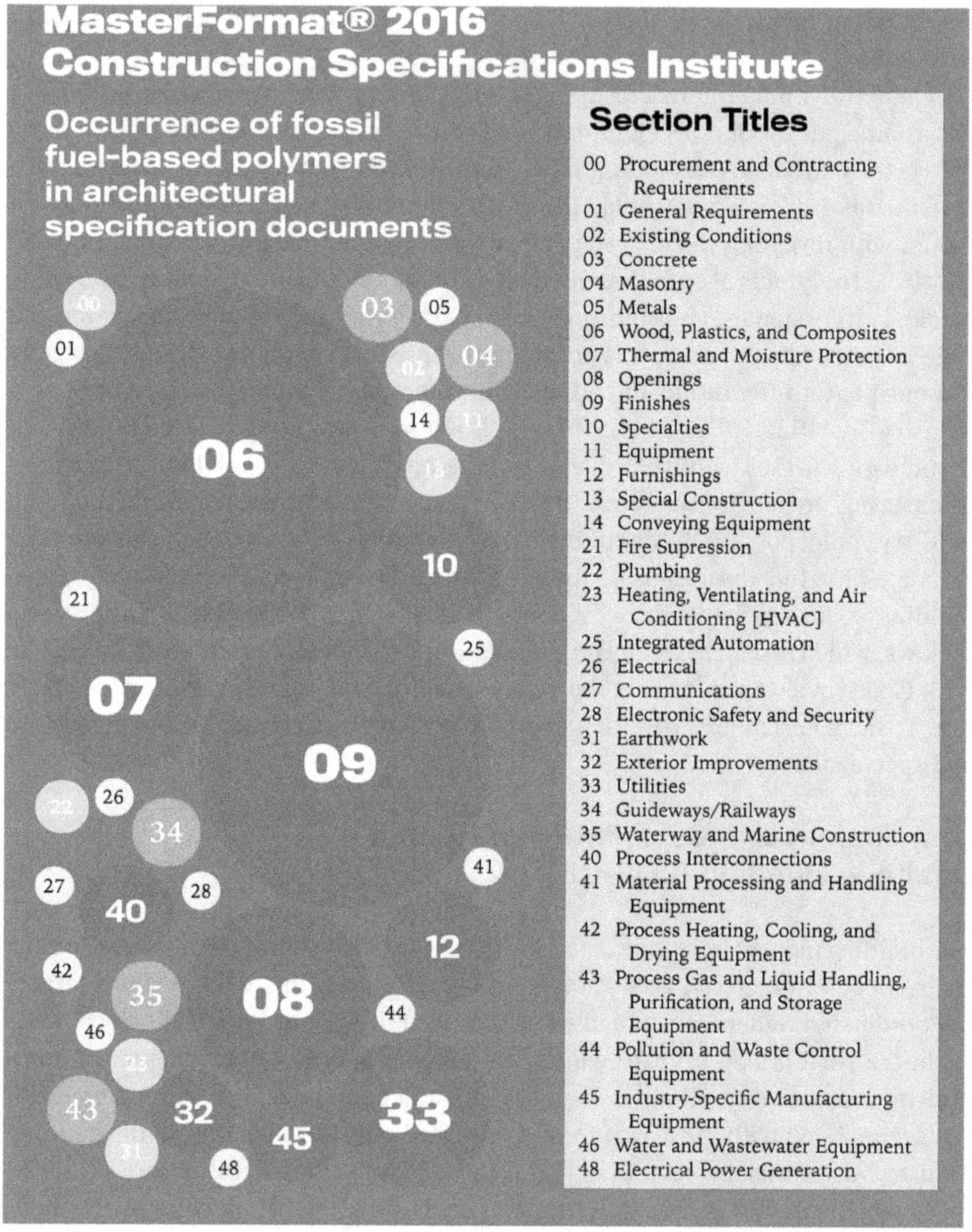

Figure 2. Analysis of master format categories and polymer nomenclature. Term search, 13 polymers in each of the 5 MasterFormat categories
Source © Trubiano-Rinaldi

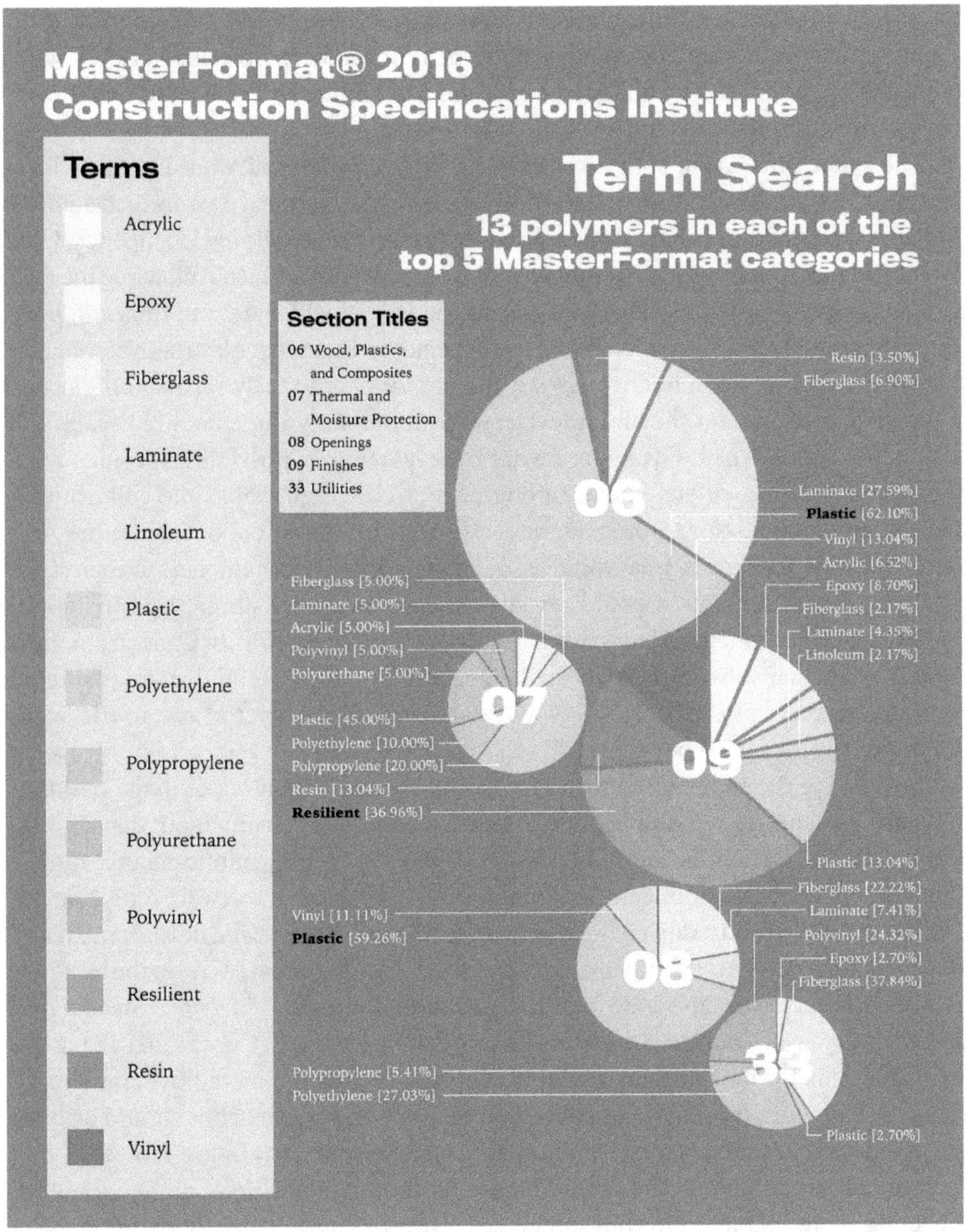

Lists described here above, an operable and easy to use Life-Cycle Index of Human Health in Building (LCI-HHB) can be developed for use by the AEC industry and its clients. This LCI-HHB could calculate and evaluate the health risks of using fossil fuel sourced building materials across a building's life and afterlife (See Figures 3 and 4). It could do so at each moment in a material product's Life-Cycle, requiring collaboration by all members of the AEC industry, medical toxicologists, environmental scientists, and material scientists.

The main Life-Cycle stages considered for each material would include Raw Material Extraction & Transportation, Refining and Chemical Synthesis, Material Manufacturing, Product Fabrication, Product Installation, Building Occupancy/Use, Waste Management, and Environmental Persistence of Materials. And while the first seven categories are consistent with the stages of typical LCAs, the final category 'Environmental Persistence of Materials' speaks to the particular risks associated with difficult to contain micro-plastics. It is impossible to fully recover all plastics produced, as they enter the air and waterways at humanly imperceptible scales.

Thereafter, each Life-Cycle stage would be evaluated for possible Exposure Paths, these being the particular vehicles through which one might come into contact with the contaminant, be this through Dermal Absorption, Ingestion, or Inhalation. And lastly, the Life-Cycle analysis would associate possible Health Impacts to each stage and Exposure Path. These could reveal ailments such as Asthma, Eye Irritation, Pulmonary, Reproductive Health Impacts, Endocrine Disruption, or Cancer, amongst others. And each sub-stage along the Life-Cycle timeline would declare (using an easy to interpret color wheel) its level of risk from Very Low, to Low, to Elevated, Medium, High, and To be Avoided.

Identifying the Index of Human Health for any material product in the building industry is, surely, a complex undertaking. Moreover, using and distributing the results from such an Index necessitates an easy to use, public-facing, digital interface. Architects, engineers, and builders respond best to software tools that are computationally robust during the early stages of a project's design development and detailed material selection processes. A recent review of typical LCAs in the building industry characterizes the state of current methodologies as "*fairly fragmented and spread over several national and international publications*" (Cabeza 2014). On the one hand, this level of fragmentation is expected to give the incredible complexity of accounting for all possible impacts, and case-by-case approaches should be both expected and encouraged over standardizing a singular impact Index. However, this lack of coherence leaves design/build teams without holistic comparative tools for selecting materials and building processes. Developing new methods of analysis and supply chain inventories for each project is capital intensive and improbable (Jensen 1998). If entire papers are published in order to report on the LCA of two different insulation materials (Papadopoulos 2006), how might we expect to achieve

Figure 3. (Draft) Proposed graphic design for the life-cycle index of human health in buildings, sample for PVC
Source © Trubiano-Onbargi

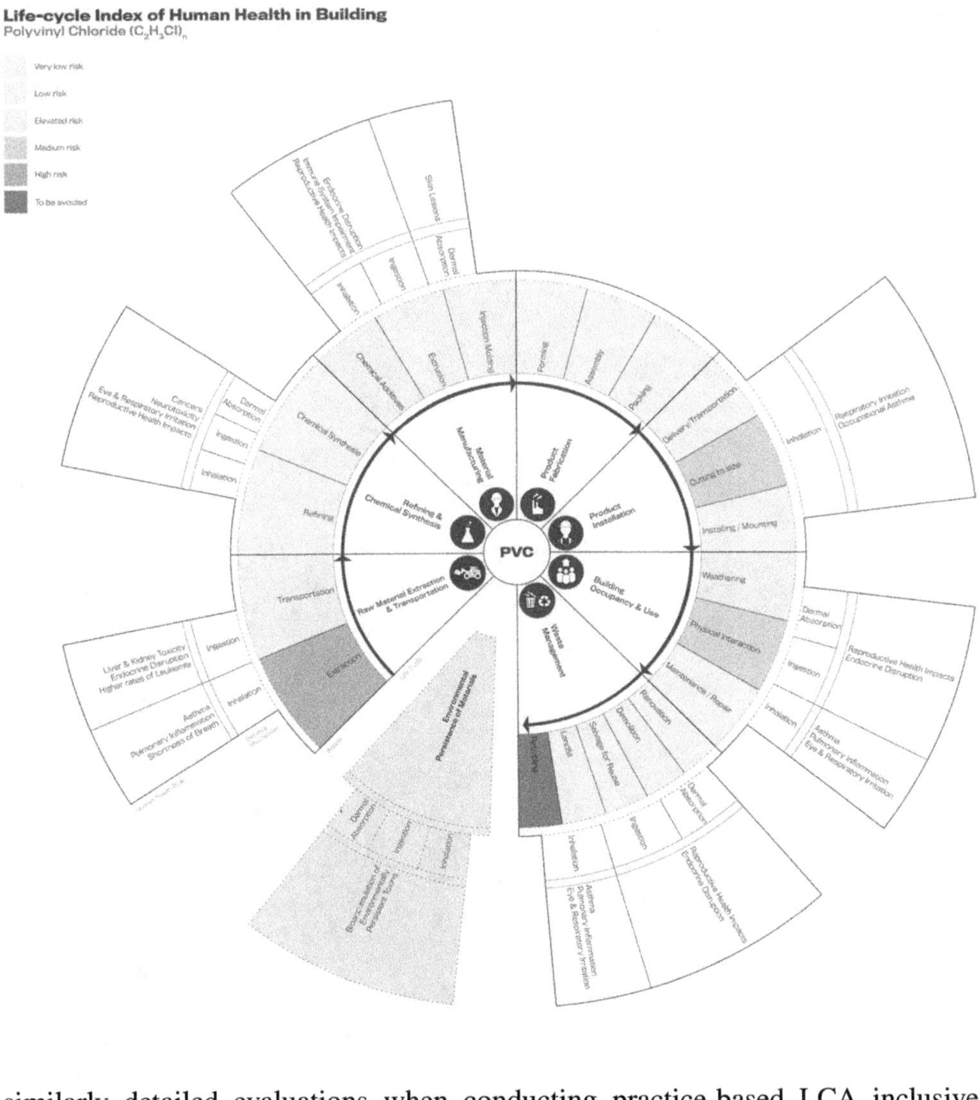

similarly detailed evaluations when conducting practice-based LCA inclusive of human health factors on an actual building project with several hundreds of different material products? While a number of LCA tools show early signs of digital integration in the design and construction BIM space, so too could human health factors be introduced in an intuitive and integrated software that computes the full risk of material toxicity in building. A robust digital interface for calculating the **LCI-HHB** can achieve this.

Figure 4. (Draft) Detail of the proposed graphic design for the life-cycle index of human health in buildings, sample for PVC
Source © Trubiano-Onbargi

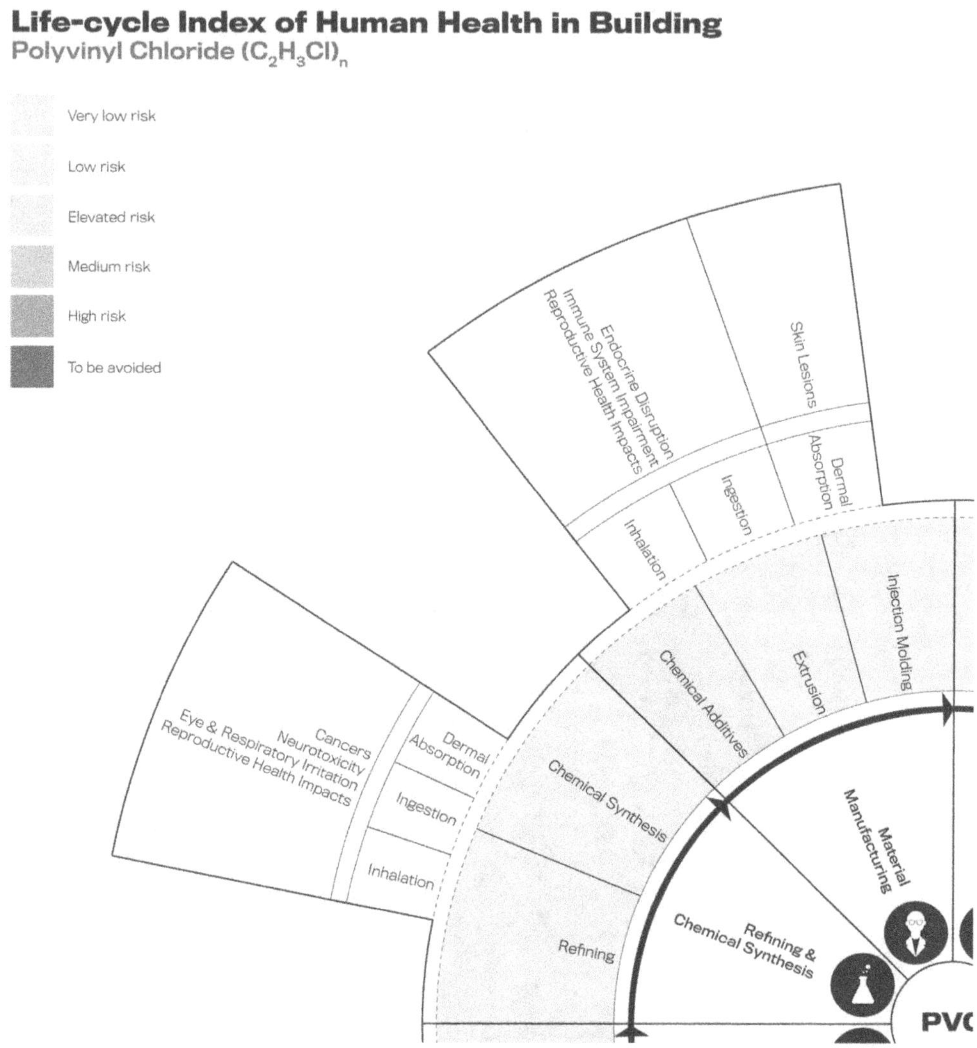

The policy priorities of interest to this research call for continued AEC engagement in lobbying for and achieving Material Data Disclosures for all polymers introduced in the building supply chain; changes to the industry's educational priorities in order to facilitate ethical decisions based on knowledge of the human health risks of materials; and finally, the creation and deployment of a Life-Cycle Index of Human Health in Building (LCI-HHB).

ACKNOWLEDGMENT

The author acknowledges funding support by The Kleinman Center for Energy Policy at the Weitzman School of Design of the University of Pennsylvania. Parts of this text have been previously published in a Kleinman Center digest from September 11, 2019, " Fossil Fuels, the Building Industry, and Human Health, Evaluating Toxicity in Architectural Plastics." The author also recognizes the important research assistance provided by undergraduate students at the University of Pennsylvania: Chloe Onbargi, Antonio Rinaldi, and Zachary Whitlock.

REFERENCES

Addington, M., & Schodek, D. (2005). Smart Materials and New Technologies for the architecture and design profession. Oxford, UK: Architectural Press.

Allen, S., Allen, D., Phoenix, V., Le Roux, G., Durántez Jiménez, P., Simonneau, A., ... Galop, D. (2019, May). Atmospheric transport and deposition of microplastics in a remote mountain catchment. *Nature Geoscience*, *12*(5), 339–344. doi:10.103841561-019-0335-5

American Chemistry Council. (2009). *Plastics, Major Markets: Building and Construction*. Retrieved from https://plastics.americanchemistry.com/B-C/

American Institute of Architects (AIA). (2018). *Ethics from the Office of General Counsel - 2018 Code of Ethics and Professional Conduct*. Retrieved from https://aianova.org/pdf/codeofethics.pdf

American Institute of Architects (AIA), & Arup. (2018). *Prescription for Healthier Building Materials: A Design and Implementation Protocol*. Retrieved from http://content.aia.org/sites/default/files/2018-06/2018-06-07-SAI18_Materials_Protocol_Handbook_v04_FINAL.pdf

Baitz, M., Kreißig, J., Byrne, E., Makishi, C., & Kupfer, T. (2004). *Life-Cycle Assessment of PVC and of principal competing materials*. European Commission. Retrieved from https://ec.europa.eu/environment/waste/studies/pdf/pvc-final_report_lca.pdf

Bayer, C., Gamble, M., Gentry, R., & Joshi, S. (2010). *AIA Guide to Building Life-Cycle Assessment in Practice*. Retrieved from https://www.brikbase.org/sites/default/files/aiab082942.pdf

Building for Environmental and Economic Sustainability (BEES). Retrieved from https://www.nist.gov/services-resources/software/bees and https://ws680.nist.gov/bees/(A(s4j522g11QEkAAAAM TZhMjA2ZDgtYmZkYy00OWJjLTgwYjAtMGUyNjk1NzNkNGRl 92eu0lpz1iKrFVoBsvKGUeA4fms1))/default.aspx

Cabeza, L. F., Rincón, L., Vilariño, V., Pérez, G., & Castell, A. (2014). Life-Cycle assessment (LCA) and Life-Cycle energy analysis (LCEA) of buildings and the building sector: A review. *Renewable & Sustainable Energy Reviews*, *29*, 394–416. doi:10.1016/j.rser.2013.08.037

Carothers, W. H., Williams, I., Collins, A. M., & Kirby, J. E. (1931, November). Acetylene Polymers and Their Derivatives. II. A New Synthetic Rubber: Chloroprene and Its Polymer. *Journal of the American Chemical Society*, *53*(11), 4203–4225. doi:10.1021/ja01362a042

Center for International Environmental Policy (CIEL). (2019). *Plastics & Health: The Hidden Costs of a Plastic Planet*. Retrieved from https://www.ciel.org/wp-content/uploads/2019/02/Plastic-and-Health-The-Hidden-Costs-of-a-Plastic-Planet-February-2019.pdf

Dahlgren, J., Cecchini, M., Takhar, H., & Paepke, O. (2007). Persistent organic pollutants in 9/11 world trade center rescue workers: Reduction following detoxification. *Chemosphere*, *69*(8), 1320–1325. doi:10.1016/j.chemosphere.2006.05.127 PMID:17234251

Dow Chemical Company. (2019). *Vinyl Chloride Monomer, Product Safety Assessment*. Retrieved from http://msdssearch.dow.com/PublishedLiteratureDOWCOM/dh_08d7/0901b803808d77d0.pdf

Elliott, M. R., Wang, Y., Lowe, R. A., & Kleindorfer, P. R. (2004). Environmental justice: Frequency and severity of US chemical industry accidents and the socioeconomic status of surrounding communities. *Journal of Epidemiology and Community Health*, *58*(1), 24–30. doi:10.1136/jech.58.1.24 PMID:14684723

Environmental Protection Agency (EPA). (2010). *Toxicological Review of Chloroprene, (CAS No. 126-99-8) In Support of Summary Information on the Integrated Risk Information System (IRIS),* Retrieved from https://cfpub.epa.gov/ncea/iris/iris_documents/documents/toxreviews/1021tr.pdf

European Commission. (n.d.). *REACH*. Retrieved from https://ec.europa.eu/environment/chemicals/reach/reach_en.htm

Fernandez, J. (2006). Material Architecture, emergent materials for innovative buildings and ecological construction. London, UK: Routledge.

Finkbeiner, M., Ackermann, R., & Bach, V. (2014). Challenges in Life Cycle Assessment: An Overview of Current Gaps and Research Needs. In *Background and Future Prospects in Life Cycle Assessment.* Springer Science+Business Media.

Fitzgerald, F. E., Standfast, S. J., Youngblood, L. G., Melius, J. M., & Janerich, D. T. (1986, November/December). Assessing the Health Effects of Potential Exposure to PCBs, Dioxins, and Furans from Electrical Transformer Fires: The Binghamton State Office Building Medical Surveillance Program. *Archives of Environmental Health, 41*(6), 368–376. doi:10.1080/00039896.1986.9935781 PMID:3113345

Franklin Associates. (1991). *Comparative Energy Evaluation of Plastic Products and Their Alternatives for the Building and Construction and Transportation Industries.* Retrieved from https://www.greenbuildingsolutions.org/wp-content/uploads/2016/05/Comparative-Energy-Evaluation-of-Plastic-Products-1991.pdf

Freinkel, S. (2011). *Plastic: A Toxic Love Story.* Boston: Houghton Mifflin Harcourt.

Geyer, R., Jambeck, J. R., & Lavender Law, K. (2017). Production, Use, and Fate of All Plastics Ever Made. *Science Advances, 3*(7), 1–5. doi:10.1126ciadv.1700782 PMID:28776036

Hedge, A. (2019), *Global Neoprene Market size is set to achieve 2.3% CAGR over 2019 to 2025.* Retrieved from https://www.reuters.com/brandfeatures/venture-capital/article?id=101765

International Living Future Institute. (n.d.). Retrieved from https://living-future.org/education/Life-Cycle-assessment-environmental-product-declarations-need-know/

Jensen, A. A., Hoffman, L., Møller, B. T., & Schmidt, A. (1998). Life-Cycle assessment (LCA): a guide to approaches, experiences and information sources. *Environmental Issues Series.* Retrieved from https://www.eea.europa.eu/publications/GH-07-97-595-EN-C/Issue-report-No-6.pdf/view

Jones, R. A. L. (2014). *Soft Condensed Matter.* Oxford, UK: Oxford University Press.

Kaltenbach, F. (2004). Plastics -- translucent semi-finished sheet products. In F. Kaltenbach (Ed.), Translucent Materials - Glass, Plastic, Metals. Birkhauser, Edition Detail, 41.

Katami, T., Yasuhara, A., Okuda, T., & Shibamoto, T. (2002). Formation of PCDDs, PCDFs, and Coplanar PCBs from Polyvinyl Chloride during Combustion in an Incinerator. *Environmental Science & Technology, 36*(6), 1320–1324. doi:10.1021/es0109904 PMID:11944687

Kohler, N., König, H., Kreissig, J., & Lützkendorf, T. (2013). *A Life-Cycle Approach to Buildings: Principles, Calculations, Design Tools*. DETAIL, Green Books.

Laughland, O., & Lartey, J. (2019a). Residents of US Cancer Town diagnosed at 'highly unusual' rates, study says. *The Guardian*. Retrieved from https://www.theguardian.com/us-news/2019/jul/24/reserve-louisiana-cancer-highly-unusual-rates-study

Laughland, O., & Lartey, J. (2019b). First slavery, then a chemical plant and cancer deaths: one town's brutal history. *The Guardian*. Retrieved from https://www.theguardian.com/us-news/2019/may/06/cancertown-louisiana-reserve-history-slavery

Lithner, D., Nordensvan, I., & Göran, D. (2012). Comparative acute toxicity of leachates from plastic products made of polypropylene, polyethylene, PVC, acrylonitrile-butadiene-styrene, and epoxy to *Daphina manga. Environmental Science and Pollution Research International, 19*(5), 1763–1772. doi:10.100711356-011-0663-5 PMID:22183785

McDermott, S. (2013). *Sasol Heeds Community's Call for Environmental Justice in Mossville, LA*. Trillium Asset Management. Retrieved from https://trilliuminvest.com/sasol-heeds-communitys-call-for-environmental-justice-in-mossville-la/

Menzies, F., & Banfill, P. F. G. (2007). Lifecycle assessment and embodied energy: a review. *Proceedings of the Institution of Civil Engineers: Construction Materials,* (160), 135-143. doi:10.1680/coma2007.160.4.135

Milne, G. (2013). *Australian Government "Your Home, Australia's guide to environmentally sustainable homes*. Retrieved from https://www.yourhome.gov.au/materials/embodied-energy

Moffitt, L. (2016). *A Toxic Tour Through the Gulf South, Compass,* Retrieved from https://content.sierraclub.org/coal/posts/toxic-tour-through-gulf-south

National Center for Environmental Assessment (NCEA). (2006). *An Inventory of Sources and Environmental Releases of Dioxin-Like Compounds in the United States for the Years 1987, 1995, and 2000*. Retrieved from https://cfpub.epa.gov/ncea/risk/recordisplay.cfm?deid=159286

National Institute of Environmental Health Science (NIEH). (2010). *Endocrine Disruptors*. Washington, DC: National Institutes of Health. Retrieved from https://www.niehs.nih.gov/health/materials/endocrine_disruptors_508.pdf

Occupational Safety and Health Administration (OSHA). (2018). *1910.1017 - Vinyl chloride*. Washington, DC. United States Department of Labor. Retrieved from https://www.osha.gov/laws-regs/regulations/standardnumber/1910/1910.1017

Papadopoulos, A., & Giama, E. (2007). Environmental performance evaluation of thermal insulation materials and its impact on the building. *Building and Environment*, *42*(5), 2178–2187. doi:10.1016/j.buildenv.2006.04.012

Perkins + Will. (2015). *Healthy Environments: What's New (and What's Not) With PVC*. Retrieved from https://perkinswill.com/site/default/filesPerkinsWll_PVC_2015_Whitepaper_2.pdf

Plastics Market Watch. (2016). *Plastics – Building for a more sustainable future*. The Plastics Industry Trade Association. Retrieved from www.plasticsmarketwatch.org

PlasticsEurope. (2017). *Plastics - The Facts 2017: An analysis of European plastics production, demand and waste data*. Retrieved from https://www.plasticseurope.org/application/files/5715/1717/4180/Plastics_the_facts_2017_FINAL_for_website_one_page.pdf

Plinke, E., Wenk, N., Wolff, G., Castiglione, D., & Palmark, M. (2000). *Mechanical Recycling of PVC Wastes, Study for DG XI of the European Commission*. Retrieved from https://ec.europa.eu/environment/waste/studies/pvc/mech_recylce.pdf

Reports and Data. (2019). *Construction Plastics Market To Reach USD 128.12 Billion By 2026*. Retrieved from https://www.globenewswire.com/news-release/2019/06/03/1863469/0/en/Construction-Plastics-Market-To-Reach-USD-128-12-Billion-By-2026-Reports-And-Data.html

Siemens, A. G. (2013). *Production of Synthetic Rubber, Process Gas Chromatograph MAXUM edition II controls and optimizes Butyl Rubber production*. Retrieved from https://w3.siemens.com/mcms/sensor-systems/CaseStudies/CS_Butyl_Rubber_2013-01_en_Web.pdf, 2.

Simonen, K. (2014). *Life-Cycle assessment*. London: Routledge. doi:10.4324/9781315778730

Tasoff, H. (2019). Plastic's Carbon Footprint, Researchers conduct first global assessment of the lifecycle of greenhouse gas emissions from plastics. *The CURRENT*. Retrieved from https://www.news.ucsb.edu/2019/019424/plastic-s-carbon-footprint?fbclid=IwAR3JqmOqIWj5EePTCYOmaLeMMfJSI0qYhSx__hR911b90r0emO4ICIxw3gk

University Network for Human Rights (UNHR). (2019). *Waiting to Die, Toxic Emissions and Disease Near the Louisiana Denka / DuPont Plant*. Retrieved from https://www.humanrightsnetwork.org/waiting-to-die

US Green Building Council (USGBC). (2019). *LEED v4 for Building Design and Construction, Point Breakdown Spreadsheet.* Retrieved from https://www.usgbc.org/resources/leed-v4-building-design-and-construction-checklist

US Green Building Council (USGBC). (n.d.). *Buildings and Climate Change.* Retrieved from https://www.eesi.org/files/climate.pdf

Venezia, D., Joshi, S., & Szurgot, A. & Van Brunt, M. (2010). Energy-from-waste and dioxin emission control: is there a role for PVC separation. *Proceedings of the 18th Annual North American Waste-to-Energy Conference*, 1-8.

Vesterberg, A., Hendermark, M., & Vass, A. M. (2005). *PVC in Medical Devices: An inventory of PVC and phthalates containing devices used in health care*. Retrieved from https://noharm.org/sites/default/files/lib/downloads/pvc/PVC_Med_Devices_Karolinska.pdf

WOODPRODUCTSfi. (n.d.). *Adhesives in wood construction*. Retrieved from https://www.woodproducts.fi/content/adhesives-wood-construction

Zheng, J., & Suh, S. (2019). Strategies to reduce the global carbon footprint of plastics. *Nature Climate Change*, *9*(5), 374–378. doi:10.103841558-019-0459-z

KEY TERMS AND DEFINITIONS

Building Life Cycle: The various stages through which all materials and systems go through in the birth to waste cycle of a building including, Raw Material Extraction and Transportation, Refining and Chemical Synthesis, Material Manufacturing, Product Fabrication, Product Installation, Building Occupancy/Use, Waste Management, and Environmental Persistence of Materials.

Environmental Product Declarations: Forms of visual and numerical accounting for all of the content, risks, and possible impact of material found in architectural building products.

Fossil-Fuels: Ancient forms of energy stored in the earth resulting from decomposition of carbon-based organisms, such as natural gas, coal, and petroleum.

Human Health: The register of life-affirming processes in the biological, the physical, and the emotional register of a population.

Material Toxicity: The amount and polemical definition of a material substance that is recognized scientifically to have a harmful effect on human health and the environment, in general. Effects of material toxicity include endocrine disruption to reproductive health, immune response, and embryonic development.

Polymers: Molecules and units of molecules found in natural and synthetic (fossil fuel) sources that are the building blocks of architectural plastics.

Polyvinyl Chloride: A synthetic polymer that originates from vinyl chloride and used in the making of pipes, cables, architectural flooring, and clothing. Vinyl chloride and its phthalate plasticizers are known carcinogens.

Chapter 5
Green Charcoal:
Developing Biodegradable Construction Materials for a Circular Economy

Shreyas More
ISDI School of Design and Innovation, India

Meenal Sutaria
ISDI School of Design and Innovation, India

ABSTRACT

The two main challenges that future cities will face are the unavailability of material resources and the waste generated as a result of resource consumption. The chapter exhibits applied research into green charcoal that addresses the crisis of the fourth industrial revolution through the development of a biomaterial consisting of luffa, charcoal, and soil. It justifies that building materiality must be intentionally designed to transform over time and support an ecosystem of plants, insects, and birds to create self-sustaining natural habitats for all lifeforms. The approach to building materiality and building systems is performance-based, circular, and net positive, thus representing a departure from conventional architectural practices. It provides a framework for high-growth countries like India to reverse the resource crisis and achieve a competitive advantage over mature economies through such initiatives.

INTRODUCTION

Rampant urbanization in growing economies has created some of the principal issues faced by cities, like the scarcity of all-natural and human-made resources and the unchecked generation of waste. The prevalent air pollution in urban settings caused

DOI: 10.4018/978-1-7998-2426-8.ch005

by carbon dioxide emissions is one such silent urban emergency. To address these conditions, designers of the built environment have been compelled to investigate radical and innovative approaches in all areas of the built environment. One such area of design concerns building materiality, where design cannot be limited to preservation and aesthetics and needs to include considerations of performance, circularity, and regeneration. Green charcoal, an ongoing research endeavor, is one such approach that could transform these crises into opportunities within high-growth markets.

This chapter investigates the meaning of the term materiality in the building industry and the potential role of architecture to inform ecological and economic sustainability. It is composed of three parts. The first part demonstrates the existing trends in urbanization and its impacts on Southeast Asian as well as global levels. The second part emphasizes the new material economy for India, followed by a biomaterial research investigation on green charcoal. The final part discusses scaling-up performative architecture using biomaterials to bring rural knowledge into urban developments and provide access to emerging technologies in rural settlements for decentralized, organized, and advanced material innovations across India.

BACKGROUND

Towards a Culture of Materiality

The Indian Context

In the West, religion stands in opposition to science. However, in India it is believed that the material world and spiritual world coexist. According to "India's changing consumer Economy: A cultural perspective" by Alladi Venkatesh at the University of California, Irvine (1994), Indians believe that materials have symbolic meaning at three levels: aesthetic, functional, and spiritual. These principles have been ingrained in India's cultural systems and economy for thousands of years. Less than a century ago, Mahatma Gandhi advocated sufficiency in consumption for a circular future. As mentioned by Shankar Venkateswaran, traditionally, the Indian economy has been one where reusing, repurposing, and recycling has been second nature. In a world that is increasingly running out of natural resources, this thinking is an asset that must be leveraged by businesses, policymakers, and citizens in an organized manner and expanded to include other elements to make the economy truly circular.

Figure 1.

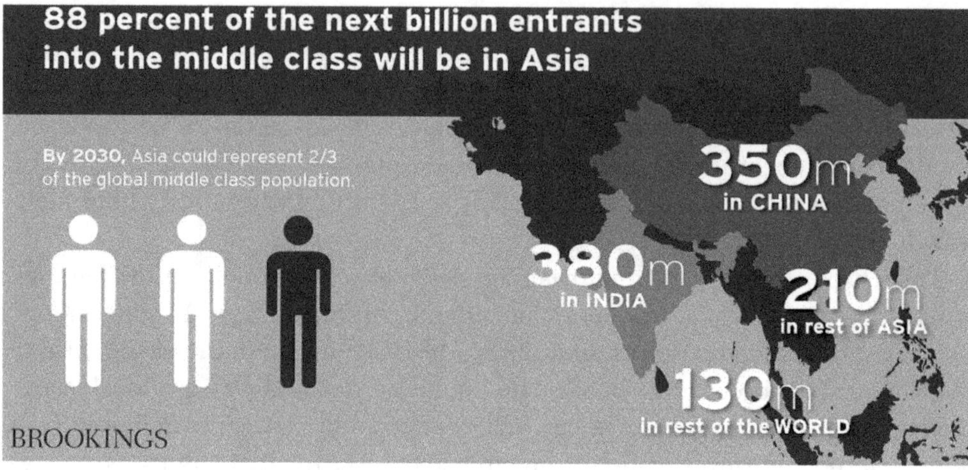

Transitioning from a Linear Economy to a Circular Economy

The current linear models of economic growth have lifted millions out of poverty. The size of the global middle class will increase from 1.8 billion in 2009 to 3.2 billion by 2020 and 4.9 billion by 2030 (Pezzini, 2012). According to Homi Kharas (2017), the bulk of this growth will come from Asia. By 2030, Asia will represent 66% of the global middle-class population and 59% of middle-class consumption, compared to 28% and 23%, respectively in 2009, as indicated in Figure 1. While business leaders and governments across the world are reevaluating these linear models, many are looking to India and Southeast Asian countries that are facing the challenges of unprecedented consumer demands, leading to the rapid depletion of resources and irreversible damage to the planet. The need here is to rethink the economic model altogether and prosper within the regenerative capacity of the Earth.

Design for Degradation: Towards a Future Economy

For manufacturers to repair or remanufacture products in the circular economy and recover material components, products will need to be designed for disassembly from the outset. This approach requires a radical overhaul of the design process, with consideration paid to how components come apart, how the user will upgrade products (if desirable and possible), and what the component pieces can subsequently become (Ellen MacArthur, 2013). While designing for disassembly becomes inevitable in a circular economy, it also essential to integrate material life cycle parameters like aging and transformation. Critical questions include: Should buildings materials be designed to last forever? Can the aging of materials be perceived positively? Can building materials and their function change with time? Can buildings, like humans, be allowed to age with time and intentionally provide changing experiences for the user over time? Can buildings be seen as material banks where the essential resources stored and energy produced by one building can be supplied to other buildings and people?

Material Innovations in the Concrete Industry Using Bio-Reinforcement and Biochar

The four interdependent domains of the new material culture are also reflected in the agriculture sector, which is one of the primary industries in the Indian subcontinent. Traditionally, agro-waste has been used in the generation of heat, biogas, or compost across rural areas in India, on a small informal scale, with an effort to utilize resources efficiently. Today, with increased waste streams, new research points to the redirection of agro-waste in the development of materials for the building industry. Previously, researchers have attempted to utilize agricultural waste materials such as oil palm shells, palm oil clinker, and coconut shells, as lightweight aggregates to produce structural grade concrete. These and other research efforts have measured the effects of using agro-waste as a partial or full replacement of sand—or using agro-waste as a reinforcing material—to increase the compressive strength of concrete (Hung, el at, 2014). In India, coconut husk is used in the manufacture of building boards, roofing sheets, insulation boards, building panels, lightweight aggregate, coir fiber-reinforced composites, cement boards, geo-textiles, and rubberized coir (Verma, Gope, & Maheshwari, 2012).

Hemp is a plant fiber also used in the development of natural building materials that, in recent years, has also been increasingly popular in several European countries. Concrete made with hemp fiber-reinforcement, called Hempcrete, is used in non-bearing walls, finishing plasters, and floor and roof insulation. Recently, the environmental performance of a non-load-bearing wall made of Hempcrete blocks were evaluated via Life Cycle Assessment (LCA) (Gupta, Kua, & Low, 2018)

Furthermore, there is research on amplifying the performance of cement mortar with the use of biochar as a carbon-sequestering additive. It is found that the addition of fresh biochar offers significantly higher mechanical strength and improved permeability compared to biochar saturated with carbon dioxide. These results suggest that biochar has the potential to be deployed successfully as a carbon-sequestering admixture in concrete that also provides a method for recycling waste (Arrigoni et al., 2017).

These research endeavors suggest opportunities for the use of bio-reinforcement to enhance the performance of concrete. Green charcoal investigates the possibilities of using similar bio-reinforcements and charcoal together to explore a co-existing material for higher environmental performance. It also looks at a selection of fibers that are "pre-engineered" by nature in a particular arrangement that enhances the structural integrity and permeability of concrete.

GREEN CHARCOAL

Green charcoal is an ongoing, practice-based research effort that addresses the issue of rising pollution and temperatures by creating healthy building materials. The research consists of the development of new construction materials made of charcoal, fibers of the luffa cylindrica plant, soil, and air. The resulting building system is biodegradable, lightweight, and allows the growth of living plant and insect ecosystems on its surface. The characteristics, industries, and ingredients used in the green charcoal material are discussed in the following section.

Key Material Components

Luffa Cylindrica

Luffa cylindrica (LC) is a tropical plant belonging to the family of cucurbitaceous, with a fruit possessing netting like a fibrous vascular system, as shown in Figure 2. The LC strut is characterized by microcellular architecture with a continuous hollow micro-channel that forms vascular bundles and yields a multimodal, hierarchical pore system (Mohanta, 2016). Natural elements like bones, skin, shells, and coral, which can withstand extreme compressive and tensile forces, are porous or fibrous in their microscopic structure. The composition observed in luffa consists of interlinked, branching fibers that make it extremely flexible, with the capacity to resist high tensile forces. Across its longitudinal and transverse sections, the luffa has different forms of interlocking fibers. These fibers are arranged in multiple directions, and are used in green charcoal to control the composite density, structural integrity, porosity, and flexibility of the mixture.

In addition to the structure, the fibers themselves can absorb and retain moisture. Studies conducted by the Forest and Garden College at Anhui Agricultural University on the matured dried fruit of luffa cylindrica, as shown in Figure 2, highlight the mechanical and water-absorbing capacities of the plant due to its interwoven fibrous network. The results demonstrate that, in comparison with other plant fiber-reinforced composites, the luffa sponge-reinforced composite material has a greater moisture absorption and a lower mechanical strength (Chen, Su, et al, 2017). This particular micro-structure has a direct impact on the performance of the material in terms of

Figure 2.

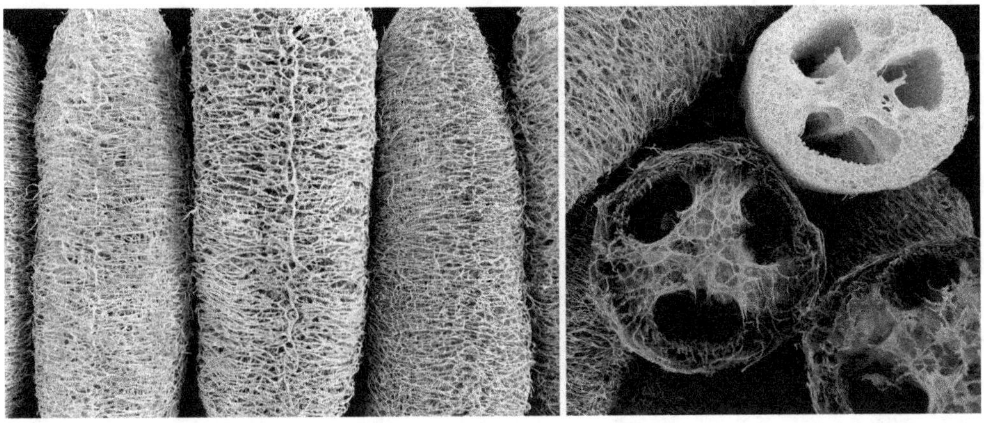

water retention, flexibility, and thermal capacity. On a microscopic scale, the luffa has two types of fibers: high-density fiber bundles and low-density fiber bundles. As shown in Figure 3, each luffa sponge column is composed of outer, inter, middle, and inner layers. These four types of fiber bundles are used in various material constructions, depending on the flexibility and the strength and size of the pores.

Charcoal and Coal

In 400 BCE, the Ancient Hindus and Phoenicians used charcoal to purify water because of its antiseptic properties. A refined form of charcoal called activated carbon is used in a variety of industries today, including corn- and cane sugar-refining, gas adsorption, dry-cleaning, pharmaceuticals, and fat and oil removal. 40 million years ago, a primitive moss that absorbs carbon from the air created topsoil for the first vascular plants, and increased atmospheric oxygen to present levels (Rodger, 2014). Mosses act as an indicator of soil and water pH levels, retain water up to 30 times their weight, and take up huge amounts of atmospheric CO_2 and NO_2 (Radford, 2016). This diversification of biological properties, which evolved over billions of years, holds significant importance from a design perspective. It is imperative to construct the world in alignment with such natural systems and materials.

Globally, coal resources have been estimated at over 861 billion metric tons, and India accounts for 286 billion metric tons of coal resources (as of 31 March 2011). The coal industry contributes to more than 1.5% of the GDP of India's economy. Despite having the fifth largest reserves of coal in the world, India still faces challenges in production capacity, leading to the import of coal from other countries. The unavailability of coal will have a significant impact on power generation in the

Figure 3.

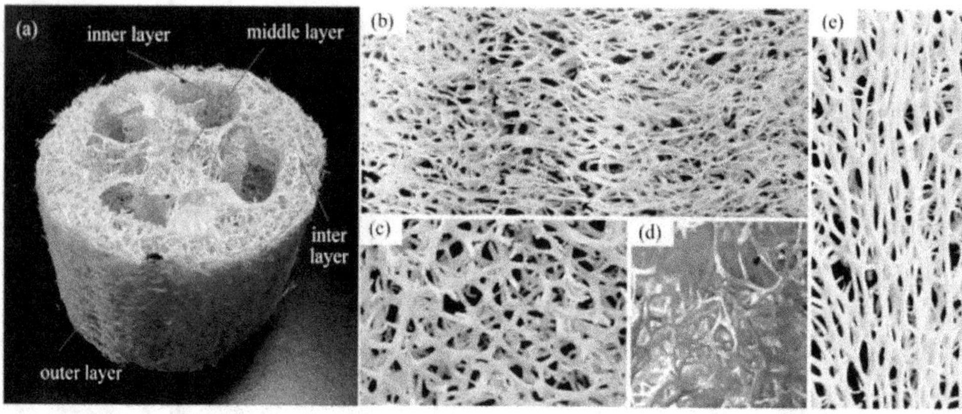

Figure 3

The geometrical features of luffa sponge column: (**a**) different regions; (**b**) orientation of outer layer; (**c**) orientation of inter layer; (**d**) orientation of middle layer; and (**e**) orientation of inner layer.

country, which in turn will impact proposed projects in the manufacturing and cement sector and retard overall economic growth (Indian Chamber of Commerce, 2017).

While the majority of coal is used for power generation, it is also employed in the generation of cement, which itself is a polluting industry. India is the second-largest producer of cement in the world. A large amount of coal-based energy is utilized during the production of cement. With around 450g of coal consumed in the production of 900g of cement (Indian Chamber of Commerce, 2017), coal is responsible for 50% of the carbon emissions caused in the process of manufacturing cement. An agenda to reduce the embodied energy of cement includes the consideration of alternative materials that reduce the use of coal in cement production. One process treats materials such as charcoal as a carbon bank. Once charcoal is transformed into activated charcoal, due to its large surface and adsorptive nature, it can continue to adsorb a broad range of pollutants with varying dimensions based on its broad pore distribution of micro- and mesopores (Reimerink & Kleut, 1999).

Defining Green Charcoal

Green refers to the natural reinforcement by luffa cylindrica, and charcoal refers to an adsorptive material. These two key components are used in the initial phase of developing a novel biomaterial for the circular economy. In employing these naturally sourced key components, this research questions the permanence of natural building materials, since biomaterials are subject to more rapid decay than traditional building materials. In many climates, mosses, creepers, and other plants will grow on facades as buildings age. Rainfall, humidity, wind, and heat will alter the material, neutralizing the pH value sufficiently to support new life. This process commonly occurs when buildings are dilapidated and materials decay in the presence of moisture and can, therefore, support the growth of biological organisms. However, this phenomenon is not an outcome of an informed process, but the result of uncontrolled environmental forces acting upon a building to transform its materiality. If allowed to continue, the process can result in the ultimate destruction of a building. For example, Ross Island, located 800 miles away from the coast of mainland India, consists of abandoned settlements built in the 18th century that are now consumed by forest growth, as indicated in Figure 4 (Nessy, 2015). Green charcoal challenges the notion that buildings should require decades or centuries of age to be aligned well with the ecosphere. It argues that building materiality can be intentionally designed to decay and allow the growth of natural life. The research indicates an opportunity to develop an entire ecosystem of bryophytes, insects, birds, plants, trees, and man-made buildings that coexist in habitats to sustain all planetary life.

Green Charcoal Applications

A composite mixture of soil, luffa, charcoal, aggregate, and cement is referred to as soilcrete because of the high volume of soil and the use of concrete as a binding agent. This research emphasizes the need for the development of building materials that can actively absorb pollutants from the air, thereby increasing the local air quality. Surface plant growth further increases carbon absorption from the atmosphere, thereby regulating micro-temperatures and transforming buildings in urban areas into vertical carbon sinks. With the use of organic reinforcement like luffa fibers, which are prone to eventual decay and weathering, the research questions the permanence of architecture. It capitalizes on the transformation of natural fibrous materials when mixed with other components, as well as the plant root systems that grow within the material. The research also looks into the procurement of raw materials from local waste to reduce the effective carbon footprint of the mixture. For instance, the soil removed during excavation to construct building foundations can be directly used in the making of new materials. Green charcoal also integrates the use of computational modeling and digital fabrication as tools to contextualize the design of building blocks or cladding systems made using the bio-mineral mixture. It provides a direction for the development of new materials that can be manipulated through composition and form-oriented design, in accordance with local resources, climate, site, program, and user-experience. Green charcoal aims to demonstrate how architecture can be redefined, particularly in developing countries, using the lenses

Figure 4.

of materials, design, and the circular economy to positively impact the environment, resources, gross domestic product (GDP), and human health.

Material Mixture and Design

Composite Development: Concrete Luffa Pods (Material Experiment Series 1)

Dry whole luffa pods were soaked in water overnight to soften the fibers. After draining excess water, the luffa pods were soaked in a charcoal-concrete mixture, as shown in Figure 5. The experiment aimed to evaluate the capacity of luffa fibers to absorb and bond with the concrete mix and measure the achieved porosity. Increasing amounts of charcoal were added to the concrete mix to observe the impurity-filtration capacity of the charcoal. This property was tested by pouring impure water through the concrete luffa pods to collect filtered residual water.

Concrete Luffa Block (Material Experiment Series 2)

In this series of experiments, luffa pods were cut along their lengths to create a densely interwoven sheet of fiber bundles. The central core was separated to be used later in the mixture. These luffa sheets were immersed in carbon concrete and layered one above another within a brick formwork, with the outer fibers of the luffa always facing up. The core of the luffa was laid at the bottom of the brick

Figure 5.

Figure 6.

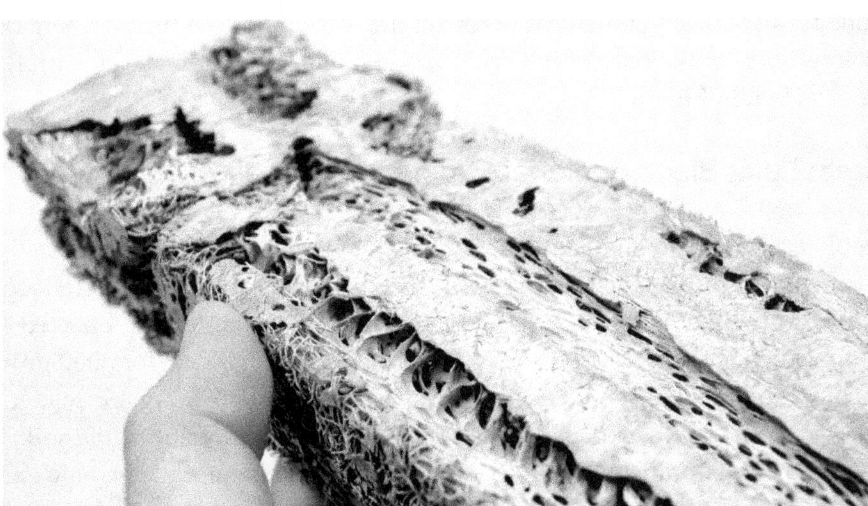

Figure 7.

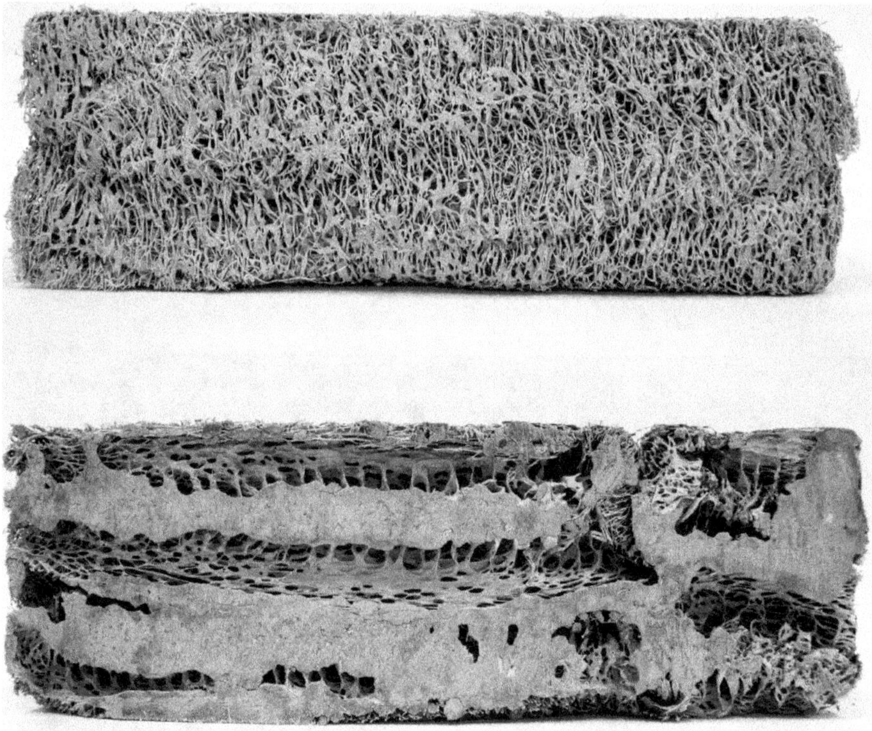

formwork. Because of the complexity of the fibrous network, these layers fused together with the concrete to create a lightweight, porous luffa-charcoal brick as shown in Figure 6. Due to the stacking of different types of fibers, the outer layer developed a significant amount of pores on the top side of the brick, whereas the stiffer core fibers at the base of the brick resulted in a stronger material with fewer pores, as shown in Figure 7.

Soilcrete Luffa Block (Material Experiment Series 3)

Cement is an extremely alkaline material, with a pH value ranging between 9 and 11. Most species of plants cannot grow in such alkaline conditions. To reduce the pH value of the base mixture, the aggregate from the concrete was replaced with local soil, which has a pH value ranging between 5.5 and 7—ideal for plant growth. The local topsoil typically contains microbes, bacteria, and other organic matter like dried leaves, roots, and wood chips that enhance the nutrient value of the soil. This soil used in the development of soilcrete was intended to enhance the organic quality of the green charcoal mixture. Unlike coarse aggregate typically employed in cement

Figure 8.

mixes, topsoil has a much higher capacity to bind with itself under compression. Hence the soilcrete mixture binds and sets well with lower quality cement. The mixture of soil, luffa, cement, and charcoal was cast in cuboidal swatches to test the material's ability to remain intact and sustain plant growth, as shown in Figure 8. After positive results, the green charcoal soilcrete was cast in smaller prototypes of organic brick forms. Two variants, with and without luffa fibers, were cast to test the integrity of the mix and the effectiveness of the luffa reinforcement, as shown in Figure 9.

Figure 9.

Figure 10.

Material Interrelationships

The luffa fibrous network acts as reinforcing, increasing compressive strength and flexibility while ensuring a high porosity to provide a mixture with optimal anchorage for plant roots. Due to the water retention capacity of the luffa, these pores function like tiny water tanks to provide adequate moisture, meanwhile reducing the temperature of the material. As a key component of the biodegradable mixture, charcoal was used in small portions in the mix. This combination created a codependent system where the charcoal, being hyper-porous, adsorbed particulate impurities like nitrates and sulphates from the air (Chyka, 2001). These, in turn, served as plant nutrients, as indicated in Figure 10.

Digital Design and Fabrication

The process of designing a building block prototype using the soilcrete mixture was inspired by principles of modularity and growth. In the ecosphere, elements are created by the organization of billions of cells. These cells are structured in certain patterns for the development of modules. Multiple modules together create a larger apparatus. This large apparatus, in turn, is constantly dependent on the nature of a single cell. This phenomenon exhibits not a simple multiplication but the relative growth of individual parts, which is different than a mere modular increase (An &

Figure 11.

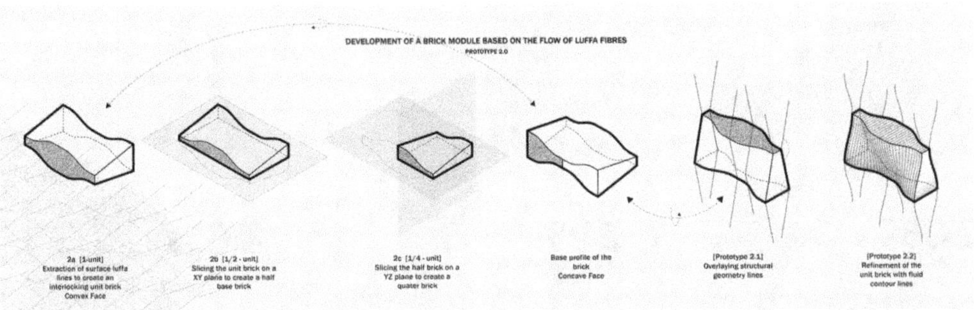

Figure 12.

Fan, 2016). Cellular meta-morphogenesis is designed by a process through which cell-growth geometry is captured, layered, and connected. The singularity of the cell and the functional organism are united. This approach to design from micro to macro was adapted in the development of the brick modules. The interlocking fibrous system of luffa was interpreted as a model to generate interlocking brick prototypes and surface-treatments.

Brick Prototype

The evolution of the prototypes was based on two principles: 1. modularity—providing the ease of assembly of multiple blocks to make them appear as seamless natural patterns instead of mere parts put together to create a whole, and 2. growth—providing a continuous dynamic surface with anchor points for the plant roots to grow vertically and horizontally on the exterior surface.

The brick prototype represents an evolution of interlocking forms that assemble themselves in various three-dimensional compositions. The geometry has two

Figure 13.

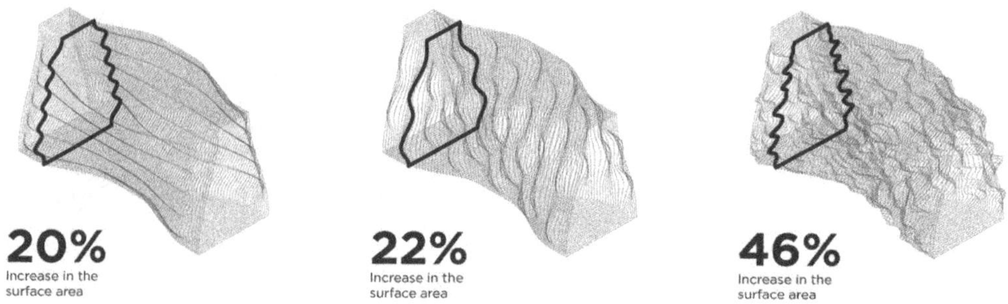

Figure 14.

opposing curves, as indicated in Figure 11, that interlock with one other when organized horizontally, and create a fluid surface when layered vertically. Three types of surface undulations were explored using CNC-milling to create formwork to increase the brick's adsorption surface area. These horizontal linear projections consisted of defined outgrowths on a diagonal grid and organic outgrowths spread across the surface, creating continuous valleys as indicated in Figure 12 and Figure 13. These organic undulating surfaces were intended to facilitate better anchorage for plant roots and also to guide the roots to grow in a certain fashion. When exposed to direct sunlight, these also created a micro-shading environment for small plants and insects. The organic visual patterns also provided a comforting and stimulating material connection with nature, as shown in Figure 14.

Design for Flexible Applications

By changing the ratios of the material's composition, there is the potential to create diverse applications for built environments. This material could be developed for road curbs, dividers, plastering, and cladding materials or biobricks for construction. The phase 1 research culminated in a self-shading brick prototype using the green charcoal mixture, as indicated in Figure 15 and Figure 16, to test the anchorage and growth of plants, material processing and casting, curing of the mix, and physical changes over time. When adapted for various built elements, this mixture is intended to purify the air, regulate ambient temperature, and provide natural habitats for multiple species of organisms.

Chapter Results

The research adopts a cradle-to-cradle approach in which construction materials are designed to degrade safely into minerals and organic matter that provide nutrition for other life cycles. The green charcoal phase 1.0 material composite has a 90% reduction in the use of coarse aggregate, a 4% reduction in cement and fine aggregate, and a 21% increase in entrained air compared with conventional concrete. It exhibits a 54% increase in organic matter from local materials compared with a standard concrete block. In the process of designing such circular biomaterials, the research presents the following four critical challenges to consider and overcome in the upcoming phase: material extraction, increased performance, adaptability for digital fabrication, and overall effectiveness. Further aims include life-size testing and mass-manufacture to measure environmental impact at a building scale.

Material extraction further considers the carbon footprint of the raw materials used and the energy consumed in manufacturing. This challenge includes the possibility of using high-volume materials like soil from local construction excavation sites to minimize the pressure on virgin resources.

- **Increased Performance:** With the support of scientific partners, new materials must be developed and tested for their responsiveness to various environmental conditions so they can provide optimum performance in real urban environments.
- **Adaptability for Digital Fabrication:** Cost- and resource-intensive formwork can be replaced with additive manufacturing techniques like 3D printing for the production of accurate, non-rigid, climate-adaptive, and dynamic forms.
- **Overall Effectiveness:** A common challenge faced by material manufacturers is that biomaterials often exhibit an abbreviated lifespan compared with synthetic products. However, the trade-offs between material performance

Figure 15.

Figure 16.

and longevity need to be considered carefully regarding the overall material effectiveness from an ecological-accounting standpoint. Although the composite mixture is predominantly safe for the environment and contains organic ingredients, it still uses cement as a binder to stabilize the mix. A further goal of green charcoal is to utilize organic methods of binding the composite in order to be 100% biodegradable. The aspiration is that urban elements can be made completely free of traditional concrete.

- **Material Testing for End of Life (EoL):** One of the green charcoal soilcrete bricks was crushed to see if new plant life would grow at the end of the module's life cycle. The germination of the seeds in the green charcoal mixture, as indicated in Figure 17, represents a cradle-to-cradle material life cycle. However, more time is required to evaluate the transformation of the material over a significant duration, such as ten years, to determine the nutritional value and support of plant growth after such a period. Statistical reports of material testing in various contexts, in collaboration with scientific partners, will enable further material refinement.

Figure 17.

Compressive and Water Absorption Test Results

Studies on stabilized mud block developed at the Mar Athanasius College of Engineering, Kothamangalam, India are compared with the results of the green charcoal blocks. The investigation carried out in the Department of Geology at the above-mentioned institute aimed to find the suitable proportion of locally available materials such as soil, coir, and straw with cement for improving the strength of locally available mud blocks. The compressive strength and water absorption test results of these materials and the green charcoal (TGC) are provided in Table 1.

In comparison to Hempcrete, which has a compressive strength of 1 N/mm^2 (Energy4farms, 2009), the green charcoal mixture has over twice the compressive strength achieved in the first phase of this research.

Table 1. Compressive strength test result of green charcoal in comparison to bio-reinforced mud blocks developed at the Mar Athanasius College of Engineering, Kothamangalam, India

Sr.	Description	Material Mixture	Compressive strength (N/mm^2)	Water absorption (%)
1	S	Sand Only	1.06	23.61
2	L5	5% lime	1.09	20.94
3	C5	5% cement	1.33	19.83
4	C5C	5% cement + 3% coir	2.03	16.44
5	C5S	5% cement + 3% straw	1.99	19.56
6	C10C	10% cement + 3% coir	3.20	13.25
7	C10S	10% cement + 3% straw	2.53	18.13
8	TGC	The mix	2.1	16.6

FUTURE RESEARCH DIRECTIONS

The Indian Economic Binary

India's recent economic growth has been an impressive 7.2% GDP, making it one of the fastest-growing economies globally, and government initiatives such as Digital India and Make in India are generating massive opportunity (Robins, 2018). On September 25, 2014, Make in India announced the primary goal of establishing India as a global manufacturing hub by encouraging both multinational as well as domestic companies to manufacture their products within the country. While India's economy is healthy, the development of advanced tools is imperative for successful growth. Meanwhile, the rural population in India has the ancestral knowledge, wisdom, and skills in material crafts and construction that should be preserved. The ancient knowledge should be maintained and applied contextually for sustainable building. The green charcoal research thus provides a framework for integrating these two trajectories that define today's Indian material economy. The contrasting elements of this framework are as follows:

1. **Biology and Technology:** Seeing nature as a source of advanced, complex, and refined technologies for construction
2. **Traditional and Modern:** Integrating material practices with new manufacturing processes, balancing manual labor and intelligent labor to design with a human touch

3. **Low-Tech and High-Tech**: For a constantly evolving human- and machine-based feedback approach in the design process
4. **Physical and Digital**: Exploring the potentials of both virtual and real-world analysis for physical performance and large scale material and form simulations
5. **Micro and Macro**: Using technology to facilitate better tools for scaling production to a variety of scales
6. **Made and Grown**: Material structures in nature possess high levels of precision and the seamless integration of functional components. A key distinguishing trait of nature's designs is the capability to generate complex structures of multifunctional composites such as shells, pearls, corals, teeth, wool, silk, horn, collagen, and muscle fibers (Benyus, 1997). If two or more such materials from different environments are brought together, they create a new relationship, and a new performance that can be used towards a specific function. In green charcoal, the relationship between various ingredients creates a new complex material that combines what is made and what is grown.

Material Technology in the Rural and Urban Indian Context

As a result of the scientific and technological advances of the last few centuries, material performance is achieved through energy-intensive chemical processes and applications. While humans make by subtracting bulk material, nature assembles layer by layer, molecule by molecule. In nature, material performance is attained via structure and the optimal allocation of resources. Nature employs a small subset of elements from the periodic table; five polymer systems (keratin, collagen, chitin, etc.) are used to create intricate structures. Meanwhile, humans use over 350 polymers, one for every function, thus making recycling difficult. Bacteria started the "maker revolution" of life 3.8 billion years ago, while humans have been at it for only 250 years. Over this period of time, nature has chosen life-friendly chemistry and biological processes that are consistent and abundant.

Material Technology in a Rural Context

Today, cities rely on rural settings for their survival. With rampant migration to cities, rural economies are collapsing and shifting to adapt to urban economic needs, thereby falling behind in advancement and development—especially agricultural sectors. For any circular economy to function, its bio-economy needs to be efficient and sustainable, as indicated in Figure 18. The goal should be to add jobs in the rural sector that are profitable and offer long-term sustenance, thereby safeguarding resource efficiency and security using naturally renewable resources. Small centers of resource self-sufficiency where materials are manufactured, consumed, and

Figure 18.

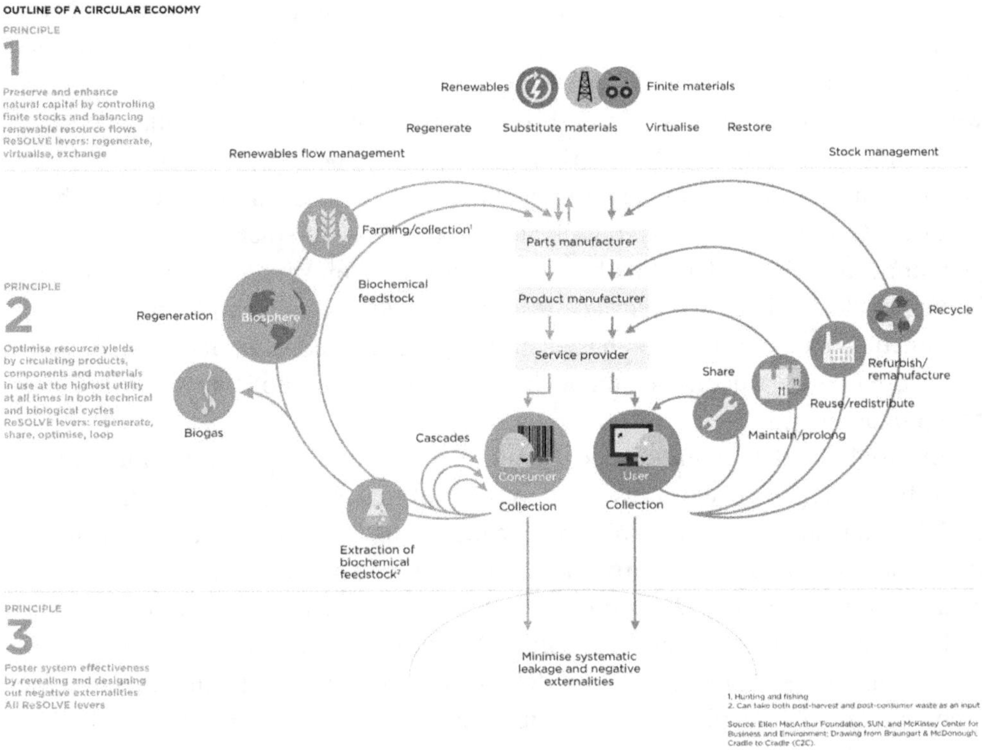

recycled, should be encouraged. An ingenious system is one where human labor becomes a renewable resource, thereby maintaining natural capital and allowing for sustained future ecosystem flows. It remains to be seen whether human society will develop urban village centers within smart cities or smart urban centers within self-sufficient village economies.

Material Technology in an Urban Context

India is experiencing a rapid urban transformation. The country's urban population reached 420 million, or 33% of its total population, in 2015. This growth is expected to almost double to 800 million by 2050, with close to 400 million more people living in towns and cities by 2050—one in every two Indians. By 2031, 75% of India's national income is expected to come from cities, and the majority of new jobs will be created in urban areas (Tewari et al., 2016). It is estimated that, as of 2007, at least 45% of India's total emissions had urban origins. The fact that one-third of the country's population lives in dense urban settings is a direct indicator of high amounts of carbon emissions, the loss of biomes, and an extreme influx of resources for infrastructure, food, water, and industry within a relatively small area.

There will inevitably be a growing number of new cities emerging in India, and the rural population will continue to migrate to urban centers where there are new work opportunities, resources, and global connectivity. The urban village concept rethinks how we design, build, and finance future neighborhoods and cities. We have an opportunity to design built environments that can provide long-term sustainability to residents. One approach is to look at the materiality of smart cities on an urban planning scale. Green charcoal offers an example to rethink interactions between the human-made and nature-made. The research provides a framework for developing sustainable- and biomaterials-based building components for facades, finishes, walls, urban infrastructural elements like sidewalks, and public spaces for healthy social intervention.

Inspired by the green charcoal initiative for energy-positive biomaterials, the urban village concept aims to adopt natural materials for the construction of buildings and infrastructure. This practice brings the values of circular biomaterials for construction and performative design into cities. The research suggests ways to counter the existing damage caused to the natural environment, air quality, biodiversity, and human quality of life in dense cities. The urban village provides a framework for new urban developments, using the lens of smart materials, that can redefine architectural relationships between nature and people. To transform existing cities incrementally, urban surgery could be performed via the redevelopment of existing architecture by superimposing or replacing it with new biomaterials. The green charcoal model provides principles for self-sufficiency in which building materials produce and store water, food, energy and resources, while being affordable and livable for future generations.

CONCLUSION

The Future of Performative and Circular Materials: India in 2030

Since the evolution of materials science and engineering in the latter part of the 20th century as an interdisciplinary combination of physics, chemistry, and several engineering disciplines, materials have been developed with increasing multi-functionality and the ability to survive in, and respond to, complex and challenging environments (Silberglitt, 2019). In the last decade, designers have taken an elevated interest in developing advanced materials that are driven by a process of creative thinking. These materials are either developed methodically with a specific purpose or with the goal of experimentation. Both methods are independent of the standard industrial framework of developing commercial materials, and hence allow material designers to pursue disruptive innovations.

Being future-proof today requires materials that are healthy for the environment and humans, adapt to their surroundings, become valuable resources at the end of their lives, and recover quickly from challenges (Material Driven). The knowledge of materials in architecture is quickly expanding. Physical matter is merging with computers and biological interfaces. The materiality of architecture must transform from a static to a dynamic state such that buildings can provide instant physical and digital feedback to occupants. The World Economic Forum (2019) forecasts that The Fourth Industrial Revolution will focus on solving the world's most pressing environmental challenges of resource scarcity, climate change, and waste-generation. Materiality in architecture can provide solutions to address these challenges. For the amount of damage done to the Earth in the last five decades, a net-zero goal is not enough. The approach to both building materials and systems must be performative, circular, and net-positive, thus representing a radical departure from current manufacturing and construction practices.

The two largest sectors of employment within India are agriculture and construction. More than 60% of the average household expenditure is spent on food, housing and mobility, while these three sectors also account for 80% of the resource use. The Ellen MacArthur Foundation (2016) found that a circular economy development path in India could create an annual value of ₹14 lakh crore (US$ 218 billion) in 2030 and ₹40 lakh crore (US$ 624 billion) in 2050, compared with the current development scenario. This goal would give impetus to the circular design process and new material exploration within the country, generating material savings, new resources, and increased profits. With its large population, India's material supply chains can become more sustainable while delivering cheaper products. Countries with high growth markets like India could reverse their resource crisis and achieve a competitive advantage over mature economies. The Indian construction sector is poised to be the third largest in the world, with over 70% of its building stock yet to be built. Transforming building design and rethinking the resources used in construction—as indicated by the green charcoal project—could contribute to the creation of resilient cities decoupled from the consumption of virgin, non-renewable materials (Baltic Cluster, 2017).

ACKNOWLEDGMENT

We thank Siddharth Shahani, the co-founder and executive director at ISDI School of Design and Innovation, who encouraged the development of faculty-led research in the institution. We are thankful to Professor Mookesh Patel, former Dean at

ISDI (2015-2018), who supported the research initiative under his mentorship. We also thank Meena Krishna, Senior Associate Dean at ISDI, for her endless support throughout the research development, and the ISDI School of Design and Innovation for funding the research and providing infrastructural lab facilities.

The research would not have been complete without the support of the students from the interior design sophomore, junior, and senior year at ISDI and the fabrication workshop lab team at ISDI who helped with the making of formwork and casting of the material blocks with utmost sincerity and enthusiasm.

Finally, we would like to thank Professor Blaine Brownell, the editor of the book, for providing us with an opportunity to write a chapter on the research development.

REFERENCES

An, X., & Fan, H. (2016, September 15). Hybrid design and energy absorption of luffa-sponge-like hierarchical cellular structures. *Materials & Design*, *106*, 247–257. doi:10.1016/j.matdes.2016.05.110

Arrigoni, A., & Pelosato, R. (2017). Life cycle assessment of natural building materials: the role of carbonation, mixture components and transport in the environmental impacts of hempcrete blocks. *Journal of Cleaner Production, 149.* doi:10.1016/j.jclepro.2017.02.161

Baltic Cluster Forum. (2017, June 28) *Switching to the circular economy in India.* Retrieved from http://circularconstruction.eu/

Benyus, J. M. (1997). *Biomimicry: innovation inspired by nature*. New York: Quill.

Chen, Y., & Su, N. (2017). In-Depth Analysis of the Structure and Properties of Two Varieties of Natural Luffa Sponge Fibers. *Materials Basel, 10*(5), 1-25. doi:10.3390/ma10050479

Chyka, P. A., Butler, A. Y., & Herman, M. I. (2001). Ferrous sulfate adsorption by activated charcoal. *Veterinary and Human Toxicology*, *43*(1), 11–13. PMID:11205069

Ellen MacArthur Foundation. (2013). The art of design for disassembly. *Engineering the circular economy a field manual for re-designing a regenerative economy.* Retrieved from https://www.ellenmacarthurfoundation.org/assets/downloads/news/EMF_Engineering-the-Circular-Economy_300913.pdf

Ellen MacArthur Foundation. (2016). The case for a circular economy in India. *Circular economy in India: rethinking growth for long-term prosperity, 11*. Retrieved from https://www.ellenmacarthurfoundation.org/

Energy4farms. (2009). *Hempcrete. American Lime Technology*. Retrieved from https://energy4farms.eu/hempcrete/

Gupta, S., Kua, H. W., & Low, C. Y. (2018) Use of biochar as carbon sequestering additive in cement mortar. *Journal of Cement and Concrete Composites, 149.* doi:10.1016/j.cemconcomp.2017.12.009

Homi, K. (2017). The Unprecedented Expansion of Global Middle Class, An Update; Brookings Instituition. *Global Economy & Development.*

Hung, M. K., Johnson, A. U., & Zamin, J. (2014). A Review on the Use of Agriculture Waste Material as Lightweight Aggregate for Reinforced Concrete Structural Members. *Advances in Materials Science and Engineering*. doi:10.1155/2014/365197

Indian Chamber of Commerce. (2017). Outlook. *The Indian coal sector: Challenges and future outlook. Impact of coal shortate. 14.* Retrieved from https://www.pwc.in/assets/pdfs/industries/power-mining/icc-coal-report.pdf

Indian Chamber of Commerce. (2017). Overview. *The Indian coal sector: Challenges and future outlook. Impact of coal shortate. 6.* Retrieved from https://www.pwc.in/assets/pdfs/industries/power-mining/icc-coal-report.pdf

Make in India. (2016). Retrieved from http://www.makeinindia.com/article/-/v/make-in-india-reason-vision-for-the-initiative

Mohanta, N., & Acharya, S. K. (2016). Investigation of mechanical properties of luffa cylindrica fiber reinforced epoxy hybrid composite. *International Journal of Engineering Science and Technology*, *7*(1), 1. doi:10.4314/ijest.v7i1.1

Nessy, M. (2015). *Island Buried by Nature: The Dark Colonial Past of Jungle Book Island*. Retrieved from https://www.messynessychic.com

Pezzini, M. (2012). *An emerging middle class.* Retrieved from http://oecdobserver.org/news/fullstory.php/aid/3681/An_emerging_middle_class.html

Radford, T. (2016, August 15). All hail the humble moss, bringer of oxygen and life to Earth. *The Guardian*. Retrieved from https://www.theguardian.com/international

Reimerink, W. & Kleut., D. (1999). Air pollution control by adsorption. *Air Pollution Control by Adsorption*, *120*(Part B), 807-819. doi:10.1016/S0167-2991(99)80380-2

Robins, C. (2018). *How technology is opening up the gates of bigger opportunities for India*. Retrieved from https://economictimes.indiatimes.com/

Rodger, L. (2014, February 9). *The History of Activated Carbon*. Retrieved from https://www.jurassiccarbon.com/blogs/news/12186281-the-history-of-activated-carbon

Silberglitt, R. (2019). *New and Critical Materials: Identifying Potential Dual-Use Areas*. doi:10.7249/CT513

Tewari, M., & Godfrey, N. (2016). *Better Cities, Better Growth: India's Urban Opportunity*. Retrieved from http://newclimateeconomy.report/workingpapers

Verma, D., Gope, P., & Maheshwari, M. K. (2012). Coir Fiber Reinforcement and Application in Polymer Composites: A Review. *Journal of Materials and Environmental Science.*, *4*, 263–276.

World Economic Forum. (2019). *Fourth Industrial Revolution: Beacons of Technology and Innovation in Manufacturing*. Retrieved from http://www3.weforum.org/docs/WEF_4IR_Beacons_of_Technology_and_Innovation_in_Manufacturing_report_2019.pdf

ADDITIONAL READING

Browning, W., Ryan, C., & Pegasus, C. J. (2014). *14 patterns of biophilic design - improving* health & well-being in the built environment. Retrieved from https://www.terrapinbrightgreen.com/

Council of Europe. (2017). *Materials innovation for the global circular economy and sustainable society*. Retrieved from http://www.esf.org/fileadmin/user_upload/esf/2018_WMS_online.pdf

Dickson, Dd., Aldred, T., & Carbeck, J. (2013) *Driving innovation: Advanced Materials Systems*. A report by the Deloitte Touche Tohmatsu Limited Global Manufacturing Industry group.

Ellen MacArthur Foundation. (2013). *Towards the circular economy*. Retrieved from https://www.ellenmacarthurfoundation.org/publications

Ellen MacArthur Foundation. (2016). *Urban biocycles*. Retrieved from https://www.ellenmacarthurfoundation.org/assets/downloads/publications/Urban-Biocycles_Ellen-MacArthurFoundation_30-3-2017.pdf

Ellen MacArthur Foundation. (2019). *Completing the Picture: How the Circular Economy Tackles Climate Change*. Retrieved from https://www.ellenmacarthurfoundation.org/publications

Kretzer, M. (2017). The Ever-Changing Nature of Materiality and the Meaning of Materials in Architecture and Construction. In M. Kretzer (Ed.), *Information Materials*. PA: Springer International Publishing. doi:10.1007/978-3-319-35150-6_2

KEY TERMS AND DEFINITIONS

Biochar: A charcoal-like substance made by burning organic material from agricultural and forestry wastes (also called biomass) in a controlled process called pyrolysis.

Bioeconomy: The production of renewable biological resources and the conversion of these resources and waste streams into value added products such as food, feed, bio-based products, and bioenergy.

Carbon Sequestration: Described as the long-term storage of carbon dioxide or other forms of carbon to either mitigate or defer global warming and avoid dangerous climate change.

Circular Economy: A systemic approach to economic development designed to benefit businesses, society, and the environment, where the finite resources are used in a constant loop of regeneration without creating waste.

Cucurbitaceous: A family of chiefly herbaceous tendril-bearing vines that bare fleshy edible fruits.

End of Life (EOL): Refers to the final stages of a product or material's phase of use.

Life Cycle Assessment (LCA): A systematic analysis of the environmental impact of products during their entire life cycle.

Microchannel: The natural structure of the luffa sponge fibers, which are arranged in parallel with diameters ranging from 4 to 10 μm and wall thicknesses of 0.3 to 1 μm.

Chapter 6
Fluid Matters:
The Water Footprint of Building Materials

Aletheia Aida
https://orcid.org/0000-0003-4389-0309
University of Arizona, USA

ABSTRACT

Water interactions with building materials are addressed for major material groups including natural materials, non-technical ceramics, technical ceramics, metals, polymers, elastomers, and foams. Water quantities and qualities are identified across the life-cycle stages of building materials from sourcing and extraction, manufacturing, construction installation, operation and maintenance, and recyclability. With background information on the water cycle and physiochemistry properties, chemical interactions of building materials are highlighted to demonstrate the range of environmental impacts that building materials have upon water resources. Water consumption metrics are also correlated to the energy footprints of building material production and manufacturing processes. Various water impact calculation methods are referenced, and an overall assessment theorem is introduced for calculating the embodied water footprint of building materials. Example sum totals are indicated for each major material group in a comparative sourcing-to-operation framework.

INTRODUCTION

The environmental ecology of materials in building design and construction practices encompasses complex layers of information. Practitioners are enabled to partially assess material selection with the expanding toolsets and methods for life cycle analysis (LCA) (Hunt, Franklin, & Hunt, 1996), ecological footprint (EF) (Wackernagel,

DOI: 10.4018/978-1-7998-2426-8.ch006

1996), and carbon emission impacts. However, the data required for inputs in such calculation tools and methods is not always readily available or may require detail-oriented research that is non-billable in standard Architecture, Engineering, and Construction (AEC) practice. Professional organizations, non-profits, product manufacturers, and research practices are beginning to accumulate numerous modes for qualifying the accountability of building materials in terms of environmental impact characteristics through labeling mechanisms and certification practices. Yet, there is still an intuitive challenge in the design process to comprehend the intersection between the accumulation index of total building composition environmental impact with the ever-expanding and vast field of products and building materials offered on the market. While the challenges to addressing the environmental accountability of building materials are handled in part with labeling, (providing information on chemistry and embodied energy), the full comprehension of environmental impact is less known.

In particular, the water quantity and water quality impacts of building materials are not readily documented with accessible data sets, although they can be (Simonen, 2014). Furthermore, the holistic assessment of water impacts through interactions with building materials during operation and maintenance as a result of the building design characteristics cannot currently be traced directly into environmental contexts. Though we have basic calculation methods for obtaining water consumption impacts associated with building operations (i.e., due to source energy production, water use patterns, and wastewater treatment energy), we do not yet have integrated calculation methods for attributing the water consumption and quality impacts resulting from material sourcing and extraction, manufacturing, construction, and operational design standpoints. These aspects may be attainable in isolation, or readily available in high consumption industries such as food, fashion, and electronics, but are not yet apparent in a clear pathway format for AEC practitioners. The Water Footprint Assessment (WFA) method has evolved in other sectors over the past fifteen years or so (Hoekstra, 2017), but is not yet fully translated into the building material industry.

Beginning with the fundamentals of water physio-chemistry and the built environment, this chapter addresses the primary material groups (natural materials, non-technical ceramics, technical ceramics, metals, and polymers, elastomers and foams) through both water quantity and water quality interaction measures. With a specific lens of water impacts, the scope covers aspects of material sourcing and extraction, manufacturing and production, construction and installation, maintenance and operation, and end-of-life deconstruction and recycling. The results indicate a holistic water footprint metric for building materials (WF_{BM}). While there are numerous gaps in aspects of embodied water impacts for specific building materials, the research to date indicates a scale of impact for metals on the high end, to technical ceramics (glass), non-technical ceramics (concrete, masonry, stone), and polymers at the low end, with natural materials in the mid-range. Composites vary depending upon the material compositions and fabrication processes. Furthermore, a general correlation between energy footprint and water footprint can be made, indicating a proportional relation. A distinction is also made for water stress measures between *water withdrawal* from surface or groundwater and *water consumed* in the building material production process. The latter is no longer available for the replenishment of water resources.

Overall, the chapter intends to provide architects and engineers a basis for water stress aspects that may be considered important in the building design and material specification process. There is no fundamental measure that can completely satisfy the values and balance of considerations involved in the process of building design. However, the constructed environment has direct impacts on all other living systems, including human life, and should provide healthy ecosystems to the greatest extent possible. As a result, the material potentials for building design may be re-envisioned in conjunction with entomologists, ornithologists, biologists, and ecologists to ensure healthy water and soils. Might we establish new modes of material production and material chemistry to provide nutrients, probiotics, and environmental benefits through both the processing and water interactions? To address such a challenge, both the water cycle and the physiochemistry of water are presented as fundamental background information so that future designers might think more critically about the scientific basis of their material choices. General water impact information for each major material group is also presented to help inform designer decisions.

BACKGROUND

Water Cycle and Physiochemistry

Water is considered the oldest substance on earth to support living cells that originated in the depths of oceans over 4-billion years ago (Strang, 2015). More recently, astronomers have observed a hydrosphere (water vapor cloud), estimated to be 140 trillion times the mass of water in earth's oceans, surrounding a black hole that is more than 10 billion light-years away (Strang, 2015). The scientific realizations for the water molecule composition and properties did not occur until the 18th century. The founder of the atomic theory, John Dalton (physicist, chemist, and meteorologist), provided the first documented vision of water in molecular form (Knickerbocker, 1927). However, it was the Swedish chemist Jons Jakob Berzelius who proved that the ratio of hydrogen to oxygen in water molecules is 2:1, thus defining water as H_2O (Szabadvary, 1966). Later in the 1700s, Swedish astronomer Andres Celsius defined the melting and boiling point properties of water (Knickerbocker, 1927).

Figure 1. Water molecule in relation to particulate matter standard sizes

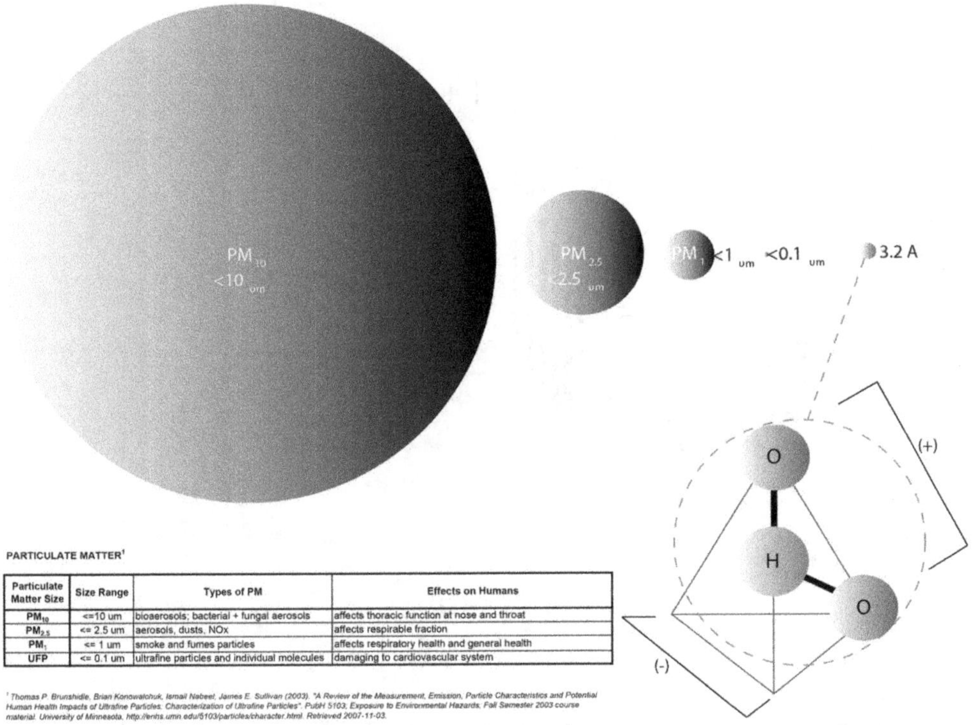

PARTICULATE MATTER[1]

Particulate Matter Size	Size Range	Types of PM	Effects on Humans
PM_{10}	<=10 um	bioaerosols; bacterial + fungal aerosols	affects thoracic function at nose and throat
$PM_{2.5}$	<= 2.5 um	aerosols, dusts, NOx	affects respirable fraction
PM_1	<= 1 um	smoke and fumes particles	affects respiratory health and general health
UFP	<= 0.1 um	ultrafine particles and individual molecules	damaging to cardiovascular system

[1] *Thomas P. Brunshidle, Brian Konowalchuk, Ismail Nabeel, James E. Sullivan (2003). "A Review of the Measurement, Emission, Particle Characteristics and Potential Human Health Impacts of Ultrafine Particles: Characterization of Ultrafine Particles". PubH 5103: Exposure to Environmental Hazards; Fall Semester 2003 course material. University of Minnesota. http://enhs.umn.edu/5103/particles/character.html. Retrieved 2007-11-03.*

The water molecule, at 3.2 Angstrom, is comprised of two hydrogen atoms and one oxygen, resulting in a dipole or electrically charged condition as shown in Figure 1. Because of the negative charge on one end and a positive charge on the other end, water has a unique ability to bond to itself as well as form a range of more complex molecules with other substances. Water is also a universal solvent, so it can easily dissociate and release substances that it has bonded with (Franks, 1984). This ability to separate and recombine with other molecules makes water both a carrier for depositing substances (such as nutrients and oxygen) for living systems as well as a filter for adsorbing substances (such as waste and toxins) out from organisms and hydrologic systems. The reversibility, though, infers that water can also be contaminated and carry such problematic chemicals into other entities and deposit them there.

Water also has a unique ability to undergo phase transitions and transform from liquid to gas and solid states interchangeably. This ability pertains to all scales, micro and macro. In the environment, water is in a constant state of motion. Water movement is influenced by solar energy and gravity at the macro-scale, and by sorption kinetics and capillary action at the micro-scale.

Figure 2. Interscalar water thermophysical properties, phases, residence timeframes, and chemistry

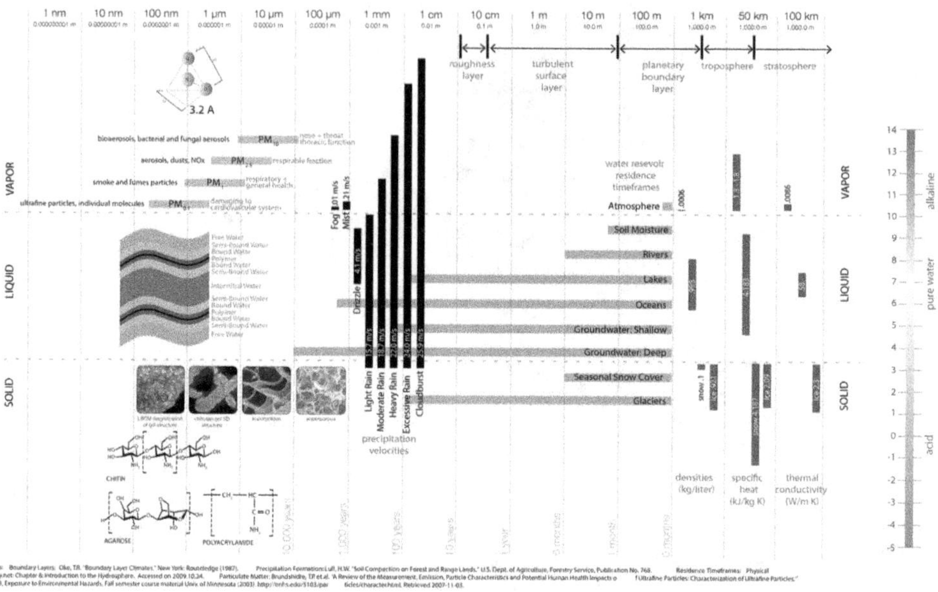

An inter-scalar matrix of water phases and phenomena is visualized in Figure 2 to correlate the interconnected role of water medium and the environment across multiple scales. The monolithic nature of the air-water entity intersects at the scale of the micron with solid materials. The water molecule is concomitant with particulate matter (PM) at the micron-scale in the air, and carries the PM effects at a much larger scale within the atmospheric boundary layer conditions of the environment. This matrix also depicts the water residence timeframes among varying reservoirs and indicates that water in the atmosphere has the shortest residence timeframe of only about one week. The brief atmospheric residence timeframe of water vapor is indicative of a fairly rapid cyclical transition of vapor into water condensate. Also included within this matrix is a depiction of the varying densities, specific heat, and thermal conductivity of water in its various phases (solid, liquid, gas). The density and specific heat are greatest in the liquid phase, while thermal conductivity is greatest in the solid phase. Water in the vapor phase has minimal density, low specific heat, and minimal thermal conductivity value. Each of these properties plays a role in the energy performance of building materials that experience water interactions (sorption of vapor or rainwater, thermal and emissivity impacts of snow and ice).

Freshwater is a global resource, yet it is limited in availability based on space and time since the water cycle and water reservoirs each have spatial and temporal variables associated with them. As land development and population growth continue to impact water resources in many locations globally, the surrounding soil resources are also impacted. Soil and water maintain an interconnected chemistry in the built and natural environments. Soil is the host for water and nutrients provided to plants and agriculture. The soil systems also directly enable natural resources utilized for building materials, such as fibers and wood, as well as ceramics and stone. Because soil often serves as a receptacle for waste materials, it processes and filters the contaminants and toxins that are deposited there. Water moves these substances through the soil systems in different modes, including infiltration of rainwater, filtration, synthesis and decay, precipitation weathering, ion exchange, adsorption partitioning, and desorption (Essington, 2003). At the surface of soils around building sites, matter can enter soil via rainfall, atmospheric particulates, fertilizers and other anthropogenic inputs, diffusion of gases, and deposition.

One particular challenge with building materials and soil chemistry impact through water is due to the biocides that are added to the surface chemistry of exterior cladding products such as wood and composites. The biocides ultimately leach off from the building materials during rain-driven processes and find their way into surrounding soil environments (Bollmann et al., 2017). Because metal cations and anions are easily hydrated (surrounded by a sphere of water molecules), potentially toxic metals also leach from metal building materials into soils via water-driven processes. While aluminum (Al) is the most common and highest in concentration in urban soils, other common toxic metals include lead (Pb), cadmium (Cd), mercury (Hg), arsenic (As), chromium (Cr) (Essington, 2003). Numerous chemical reactions occur in aqueous soils, and it is water that enables such interactions that have an effect on the environment, including hydration-hydrolysis, acid-base, oxidation-reduction, and complexation (Essington, 2003). Hydrolysis, in particular, can significantly impact the chemistry of a metal in an aqueous environment. As such, the chemical form of the metal may change, thus enabling chemical reactivity with other soluble constituents.

FLUID MEASURES

Water Types and Processes

The water cycle in the natural environment consists of the movement of freshwater sources through different states and bodies, including deep groundwater, surface runoff, oceans, lakes, reservoirs, rivers, snow-melt, evaporation, and precipitation amidst others. In the built environment, the movement and quality of water are affected by building materials, infrastructure, anthropogenic activities, and industrial

Figure 3. Water types by color, mode (quantity or quality), and process

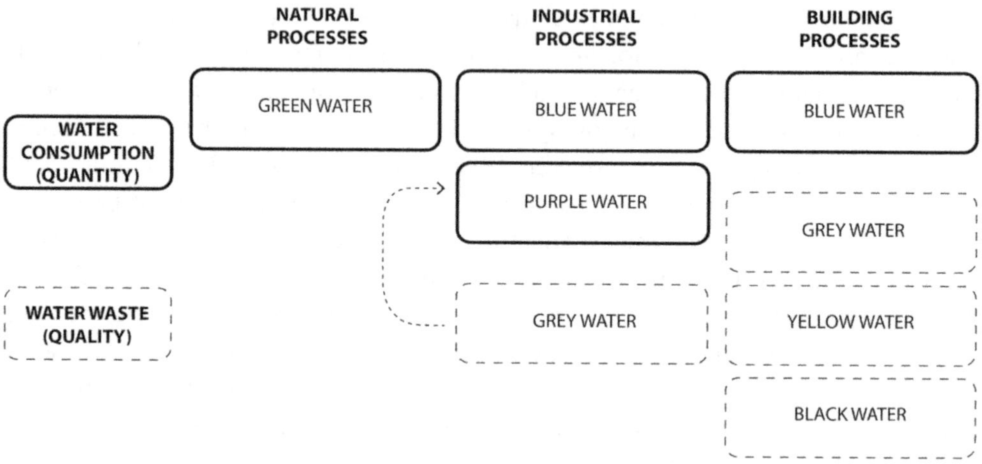

processes. While freshwater (from ground and surface water) is identified as blue water, Figure 3 depicts the other standardized 'colors' of water based on quality types include: green water, greywater, black water, and purple water. Green water is rainwater that is absorbed by plants and soil systems, which carries nutrients for agriculture and natural systems. Greywater in the Water Footprint Assessment (WFA) method established by Hoekstra is the polluted water output, which requires filtration and dilution to achieve existing ambient water quality standards (G.P. Zhang, Hoekstra, & Mathews, 2013). However, greywater in the building industry is identified as mild wastewater from building plumbing fixtures other than toilets. Both forms of greywater are slightly contaminated and require dilution and filtration for treatment prior to potential re-use purposes. Black water is wastewater with urine and fecal matter from toilets. Purple water is reclaimed water after particular treatment of greywater resulting from industrial chemical processes. The sixth type of water in the built environment that is not clearly identified in prior literature is the contaminated surface runoff from building and urban infrastructure materials, which will be identified here as yellow water. In essence, yellow water is another form of greywater that combines into stormwater systems and often remains untreated and feeds back into freshwater bodies and soils carrying problematic chemical constituents.

Along the production sequence for building materials, water is both consumed and polluted in various processes. In the built environment water cycle, some of the water types are consumed in the process of making and maintaining building materials (i.e., blue, green, purple) while some of the water types are byproducts of making and maintaining building materials (i.e., grey, purple, yellow). Within the building processes, both greywater and black water are based on human water consumption patterns in buildings and will not be included in the water footprint impact of building materials.

Another facet of water consumption that should be taken into account is the freshwater use for source power production as a function of energy use for building materials. In general, the aggregate total water consumption (unrecoverable losses) for power plant generation in the United States (U.S.) is found to be 2.0 gal/kWh (7.6 L/kWh) (Torcellini, Long, & Judkoff, 2004). There are both regional variances as well as power plant type differences. Hydroelectric power plants lose almost 40-times more water than thermoelectric plants to irrecoverable evaporative processes. The thermoelectric power production is found to be around 0.47 gal (1.8 L) of freshwater evaporated per kWh of end-use electricity in the U.S. (Torcellini et al., 2004). Hydroelectric power plants evaporate approximately 18 gal (68 L) of freshwater per kWh consumed by the end-user (Torcellini et al., 2004).

While agricultural production consumes the greatest quantity of water resources for human activity (Zhang et al., 2013), the indirect use of water in the building material production supply chain is not insignificant. Steel, cement, and glass construction materials alone are produced in millions of tons globally per year (Gerbens-Leenes, Hoekstra, & Bosman, 2018). The quantities of blue water consumed in the material sourcing and production processes are much higher than the construction and building operation phases, and are often invisible to the AEC disciplines who are responsible for specifying the materials.

Water Consumption from Building Materials

Water consumption for building material sourcing is not readily documented, but there are some baseline metrics for extraction activities such as mining and forestry. In addition, there is limited information available on the water-consumption for building material manufacturing, but some extrapolations from comparable industries can be made for certain materials. Some prior studies on the life cycle of common building materials depict the input of water resources and output of wastewater at different junctures along each manufacturing flow-scheme (Müezzinoğlu & Toprak, 1995). Based on the available information, water quantities required for material sourcing and extraction are defined in ranges for the general material categories. In some cases, the water consumption for a material group's bio-geochemical and/or recyclability phase is also identified.

Natural Materials

Natural materials, such as wood and bamboo, have a series of supply chain processes before implementation in building construction. Natural materials require solar energy and nutrients supplied through water uptake during the growth phase. The global water consumption (blue and green) attributed to roundwood forestry production for lumber is 961 x 109 m^3/year for 2001-2010 (Schyns, Booij, & Hoekstra, 2017). Standardizing sawn lumber into dimensional sizes requires specialized steam heating chambers that control the moisture content of the wood, both through induced evaporation and addition of vapor (Shmulsky & Jones, 2011). Research shows that recycling lumber products from building demolition and reusing as downcycled products, such as wood flooring materials, significantly reduces the water footprint of lumber building materials (Falk & McKeever, 2004).

Non-Technical Ceramics

Non-technical ceramics, including stone, concrete, and masonry, are comprised of various soil compositions. Many stone materials that are quarried from deep ground sources are a result of long-term sedimentation processes facilitated through water-interactions that took place hundreds of thousands of years ago. The quarrying for stone consists of large-scale machinery and equipment capable of digging and cutting large slabs of solid earth, which requires water flushing for dust control and cooling. Soils utilized for masonry products are filtered through water to extract the soil constituents desired for the mixes. This is a basic technique that is used even in the earliest of masonry building methods, such as soil-water filtration measures for making adobe blocks. Water is consumed for cooling processes in masonry kiln and curing stages. For concrete, the main constituents of Portland cement and aggregates both require water consumption during sourcing and extraction. The amount of water consumed in the mixing and curing of concrete is typically in the range of 0.25 – 0.3 L/lb, but much more water is required in the manufacturing of Portland cement. Water reporting from the Cement Sustainability Initiative excludes the supply chain water as it solely focuses on the carbon emissions impacts (Cook & Ponssard, 2011).

Technical Ceramics

Technical ceramics, particularly glass, require water for various production tasks in the manufacturing process. There is a lot of heat energy required for vitrifying aluminosilicates into glazing materials; thus the equivalent water consumption per power plant energy source is relatively high. Glass requires a controlled cooling process so that fractures and impurities in the amorphous structure during curing do not result. For the production of float glass, the typical building glazing product for windows, there are a series of stages in the manufacturing. The five basic stages include: 1) melting and refining; 2) float bath; 3) coatings; 4) annealing; and 5) inspection and cutting. The float bath entails a controlled cooling process, which utilizes a large quantity of water as a cooling medium (Dakwala, Mohanty, & Bhargava, 2014). Float glass industry is generally under pressure to optimize the conservation of both energy and freshwater consumption, including minimizing wastewater (Dakwala et al., 2014). The design interactions between the heat recovery processes and water reuse is also an important consideration being addressed in glass production.

Metals

Metals require a large amount of water during the deep earth mining activities. The geothermal high heat conditions in underground mines, and the extraction of earthen metal ore, together constitute a need for water to both cool the extraction materials and filter the substances to separate the desired metals. Alongside both forestry and stone quarrying, mining requires the highest amounts of water consumption during sourcing activities. Historically, the water utilized during mining processes would result in highly contaminated greywater with numerous fine particles of various metal byproducts, which were included in the water source to enable the extraction and filtration process. Today, the water treatment methods have advanced to allow for mining wastewater to be readily filtered for reuse. Recyclability of metals is one of the great benefits of this building material. While there is a high embodied energy of the material in first use, the recycling process is technically advanced and readily available. Utilizing recycled metal for building products can greatly reduce the water footprint, since the majority of water consumption is due to the originating mining activities. Water is used at many junctures along the production sequence for steel and aluminum. Water is consumed for dust emission control, sorting material, cleaning, cooling, and gas treatment processes during steel production (Gerbens-Leenes et al., 2018). Compared to other building materials, steel uses 10-times more water than Portland cement production and 4-times more water than glass production (Gerbens-Leenes et al., 2018). The typical amount of water consumed to produce steel that might be used as building materials ranges from 70,000 – 80,000 ga/ton.

Polymers, Elastomers and Foams

Polymers, elastomers and foams, when synthetically produced, require solvents for the chemical bonding of polymer chains to occur. Often the solvent will be a pure water source, such as deionized water. The typical amount of water consumed to produce plastics and foams that might be used as building materials ranges from 18 – 22 ga/lb. Many building insulation materials are composed of polymer foams, such as polyisocyanurate and expanded polystyrene (EPS). Insulation materials are considered crucial to providing the heat resistance values needed in building enclosure systems to reduce the building heating and cooling loads. Some insulation materials are recycled from other material streams (such as blue-jean pulp from fashion floors or EPS from packaging industries), and thus the water footprint is greatly reduced because the supply chain attribute for water consumption was in an alternate industry altogether.

In some cases, polymers are used as a primary building skin, such as ethylene tetrafluoroethylene (ETFE), which is a fluorine-based plastic. Wastewater from plastics manufacturing will typically contain substances and potentially toxic chemicals. The wastewater is generally to be treated on-site at the manufacturing plant (C.C.R.O.T., 2019). Water quality for textiles, a large sector of interior building finish materials comprised of polymers, is one of the most challenging to treat. The wastewater from textile manufacturing is highly contaminated and must be transported directly to specialized treatment plants (Rott, 2003). Textiles that utilize dyes for coloration result in additional dye chemicals in wastewater streams that must be removed by specialized treatment. Natural plant dyes are an environmentally friendly alternative to synthetic dyes.

While some general information regarding basic water quantity and quality for building material sourcing and manufacturing is available, there is a vast gap of data across the building material and product industry. In addition, the Environmental Product Declaration (EPD), as established by the International Office of Standardization (ISO) 14025:2016 Life Cycle Assessment method, does not yet account for water consumption or quality impacts (I.S.O., 2019). However, a handful of web-based resources and initiatives are beginning to make the tools and data-tracking for water footprint assessments accessible to industry. These include the Alliance for Water Stewardship (a4ws.org, 2019), the Water Footprint Network's Virtual Water project (W.F.N., 2019), and the CEO Water Mandate (U.N., 2019).

Material Measures

Building materials are composed of a range of material chemistries and properties from each of the major material groups depicted in Figure 4. The primary material groups include natural materials, non-technical ceramics, technical ceramics, metals, and polymers, elastomers, and foams (Ashby, 1980). Composites are not specifically referenced in this study since they are a function of the combined characteristics from the fundamental material groups. The primary material groups are defined through relative characteristic relations with water in terms of hydrophobicity, hydrophilicity, porosity, permeability, corrosivity, leaching ability, absorptivity, moisture diffusivity, and flow friction ratios. The specificity of these material aspects lends to the design and engineering potentials for building integration and water impact improvements.

Figure 4. Material groups and their relative water footprints (volume water/volume material)

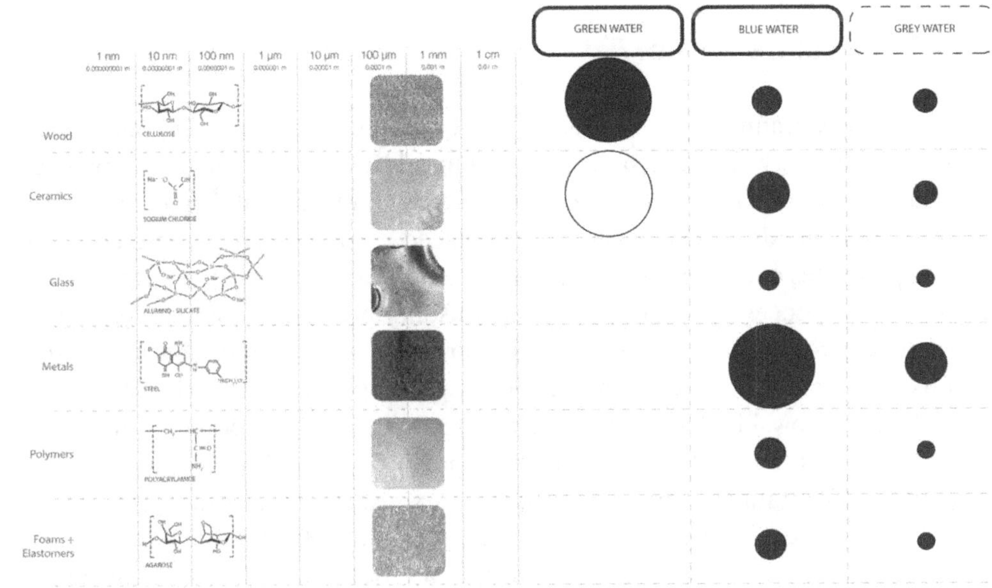

Natural Materials

Natural Materials consist of cellulose molecules typically formed into bundles of macro- and micro-fibrils amidst a lignin matrix. Natural materials have anisotropic structural properties due to the concurrent outward and upward growth patterns. Due to the fibril bundles in natural materials, numerous interstitial pores result (in some cases, such as in bamboo, the pores are hexagonal). Because natural materials are generally hygroscopic, the pores are prone to water sorption, causing expansion of the natural materials and also susceptibility to mold and mildew development (Shmulsky & Jones, 2011). Natural building materials are often treated with additives such as biocides or other chemicals to prolong the life of the material and reduce exposure degradation effects with moisture and water, including within the adhesives for natural laminates and composites (C. Zhang et al., 2018). Moisture infiltration into natural building materials also impacts the thermal performance if used within the exterior building enclosure system. Hygrothermal behaviors of bio-based building materials will affect the thermal insulation value of an enclosure system by reducing thermal resistance when saturated with moisture (Amziane & Arnaud, 2013). Inevitably, as moisture infiltrates exterior natural building materials (such as natural insulation), the energy load for the building may be impacted, resulting in higher water consumption at the power plant source.

Non-Technical Ceramics

Non-technical Ceramics span a range of soil-based minerals, including numerous forms of silicates, as well as gypsum, lime, sodium, feldspar, quartz, calcium carbonate, and others. The bonding process and formation of non-technical ceramic building materials is typically a result of magma induced or precipitation-sedimentation induced processes over thousands of years (in the case of raw stone) or water-mixed and heat-cured processes for manufactured masonry products. Concrete formation is a chemically-induced exothermic curing process, which requires water in the wet mix to initiate the bonding between cement and aggregates (Hewlett, Liska, & Lea, 2019; Kurdowski, 2014). As many non-technical ceramic building materials maintain various sizes of air-pockets, the porosity makes them prone to water sorption and infiltration of moisture (Ludwig, Rosina, & Sansonetti, 2018; Pavlík, Žumár, Medved, & Černý, 2012; Togkalidou, Karoglou, Bakolas, Giakoumaki, & Moropoulou, 2013). In these cases, the moisture can affect internal reinforcing metals (in reinforced concrete) causing corrosion and decay (Freiesleben Hansen, 2009), result in efflorescence on the material surfaces (N.E.H.A., 2003), and otherwise cause premature degradation and material failure (Gómez-Laserna et al., 2013).

Technical Ceramics

Technical Ceramics are generally glass materials comprised of silicon-oxides, alumina-silicates, or soda-lime. The vitrification process for glass formation occurs at extremely high temperatures and results in an amorphous molecular bonding pattern that is typically very solid with minimal air pockets when executed under controlled conditions. The resultant structure of glass is cohesive, brittle, and impermeable. The silicon-dioxides of glass are polar, and therefore hydrophilic. The surface chemistry of glazing materials can be modified with coating procedures to ensure hydrophobic as well as self-cleaning attributes.

Metals

Metals present dynamic crystalline covalent bonding patterns with seas of free electron sharing on the outer valence bands of the molecules' atoms. The resulting structure is very dense but also malleable and ductile under higher temperatures. The densities of metals generally prevent moisture transport through the material. However, the surface chemistry of non-anodized metals makes them prone to corrosion through reduction-oxidation processed. In some cases, the corrosion results in patina or modification in the coloration of the metal. In other cases, the corrosion of a metal building material is the cause of structural failure (Kermani & Harrop, 2019). Leaching of metal particulates can occur during corrosion processes, impacting surrounding soil and water bodies (Vollpracht & Brameshuber, 2010).

Polymers, Elastomers, and Foams

Polymers, elastomers, and foams are comprised of long-chain molecules bound together either covalently (strong chemical bonds) or physically (weak cross-links). The structure of foams is based on polymer solids that have an extremely porous interface throughout the material. Elastomers have a physical elongation-contraction reversibility due to the polymer chain cross-link strength (covalent) and large mesh-pore sizes (the area between cross-links) that allows the deformation to take place. Many of the polymer-based building materials are utilized on interior finishes, but a handful are incorporated in exterior envelope systems (vapor barriers, insulation, gaskets) and integral to many composites. There is quite a range of water-related properties for polymers and foams, many of which are controllable during the synthesis and manufacturing design. Polymer sheets, such as vapor barriers, are impermeable to water and hydrophobic. Polymer-based insulation materials are foam-structures that are porous and prone to moisture infiltration (Bedane, Eić, Farmahini-Farahani, & Xiao, 2016). Polymer materials used in building construction are typically synthetic (covalent bonds), but many studies and interests are developing around natural polymer building materials, which can also be designed and engineered with chemical synthesis (physical bonds) (Habibi, Lucia, & Wiley, 2012; Krieg & Rybtchinski, 2011).

The challenges for water-related environmental impacts of building materials is complex and cannot be reduced to a basic water consumption metric. The factors to consider encompass: 1) water consumption and wastewater due to material sourcing and production; 2) water related infiltration of exterior building materials affecting thermal insulation values; 3) premature material degradation resulting from water-borne effects constituting material replacement prior to full lifetime; 4) water consumption required for building façade maintenance and cleaning; and 5) chemical runoff from water-induced interactions with exterior building materials due to biocides and finishing treatments.

Figure 5. Diagram of building-integrated materials and water cycle interactions and potentials

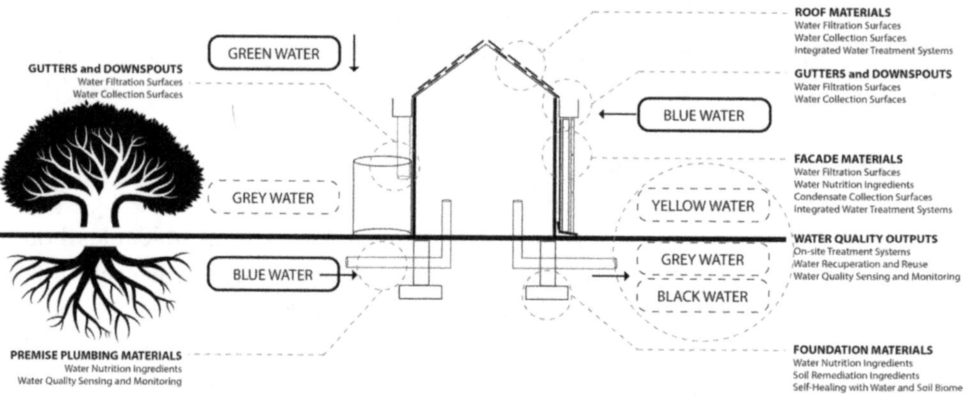

BUILDING INTEGRATION MEASURES

The water cycle provides comprehension of what phases and forms water moves through the natural environment at multiple scales and rates. At the same time, a complete building water cycle is needed to inform potential quality impacts with materials and quantity impacts with spatio-material design configurations. By lensing our understanding of building materials' design and construction practice through a water quantity and water quality accounting method, the resulting material typologies and compositions might be fully re-informed and emerge on an alternate trajectory of built ecology futures.

The building-integrated water cycle is presented in Figure 5, depicting the water types in contact with typical materials. Similar to the natural water cycle, the building water cycle can be addressed at multiple scales of water interactions and states as well as different water types. This includes water in the solid phase (ice or snow), liquid phase in different compositions (blue, grey, black, green, and yellow water) and forms (rainwater, flooding, municipal water, etc.), and vapor (humidity). The different water types are defined, and typical building material interactions are described. The building material systems that interact with water flows are categorized by a building hierarchy scheme by sub-grade or foundation systems, exterior envelope systems, roof systems, and interior plumbing systems. The building material typologies described in detail include elements such as gutters, roofing materials, cisterns, plumbing pipes (copper, PVC, galvanized steel, etc.), façade rainscreens, vapor barriers, waterproofing, etc. Building-integrated material sensing and water monitoring tools and techniques are also introduced to provide holistic methods for ongoing water impact data collection.

Building-integrated materials interact with water flows at different scales and locations. Water tends to follow gravity, capillary action, or pressure differences. There are particular directional flows in which water will engage with different building materials. Roofing materials are exposed to rainfall, snow, and ice. Façade materials are exposed to wind-driven rain, humidity, and vapor sorption. Plumbing materials are exposed to various chemistries and particles of wastewater streams, while potable water flows and stands at different rates dependent on building occupant water use patterns. Foundation materials are exposed to soil moisture, hydrostatic pressure, and freeze-thaw conditions. Gutters and downspouts are exposed to directed water flows during rainfall, which might include pollutants and particulate matter. Water cisterns retain various amounts of standing water depending on the rates of collection and use.

The water from rainfall in urban environments may be acidic or otherwise heavily polluted and accelerate building material degradation (Baboian, 1985; Sanjurjo-Sánchez & Alves, 2012). Other issues when water interacts with building materials may result in microbial communities and mycotoxins that may have negative human health impacts (Lichtfouse, Schwarzbauer, & Robert, 2015; Tuomi et al., 2000). Architectural finishes applied to exterior building materials may provide some barriers and delay in these processes but require ongoing maintenance and also have their own water impacts in the production supply chain (Abd El-Hameed, Mansour, & Faggal, 2017).

Furthermore, the leaching of material chemicals into the environment through water interactions is another challenge with building material design and specification. In addition, standard plumbing materials have proven problematic with low-flow or standing water at the premise-plumbing connections between municipal supply and building use. In particular, the copper, lead, and PVC typically used for building pipe materials affect the resulting water chemistry with varying respective concentrations of Cu (copper), Pb (lead), and Zn (zinc) after stagnation (Pan, Jeffrey, Marc, & Amy, 2015). The building plumbing pipes are smaller in diameter relative to main water distribution lines; therefore, the water's surface area exposure to the pipe material is higher.

These issues and others are cause for an overall questioning of why and how we use particular building materials where there are potentially harmful water impacts. The future challenge will be for designers to re-envision both the typology of materials that will interact with water and the form of applications on the building site. In some cases, there might be emerging zoning and wastewater regulations that require building sites to incorporate capture basins for all yellow water runoff and treat on-site before the water can be absorbed into the soils and ground water. Perhaps our building enclosure materials and systems will become water filters to purify and enhance the yellow water runoff. Such material filtration properties

Figure 6. System boundaries for primary, secondary, and tertiary phases of the building material production, installation, and operation life-cycle for water footprint impact assessment – water consumption inputs (top) and water quality outputs (bottom)

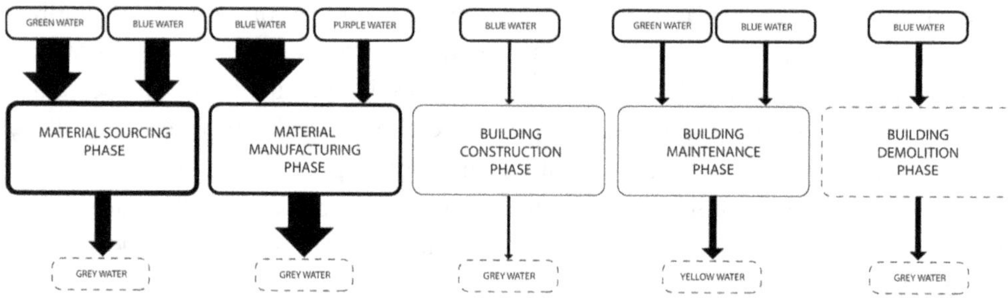

might provide positive benefits with nutrients and probiotics to support local soil and water systems. Our long outdated modes of employing nineteenth-century building material and construction practices are in need of a holistic environmental overhaul. The direct and indirect impacts of building materials on the water ecology should be a primary reconsideration in the calibration towards environmentally beneficial building design. Deeper collaborations between microbiologists, material scientists, biotechnologists, ecologists, entomologists, and others are crucial for an optimistic outlook of building material futures.

Water Impact Calculation Methods

An important aspect of water footprint assessments for building materials is to establish clear system boundary parameters for the stages of the material life cycle. The other important aspect is to establish clear water quality and water consumption types in order to maintain direct comparisons across building material groups. In this study, the material phases are identified as: 1) material sourcing, 2) material manufacturing, 3) building construction, 4) building maintenance, and 5) building demolition. The relative inputs of water consumed for each phase are depicted by the size of arrows along the top of Figure 6. The relative wastewater outputs for each phase are depicted by the size of arrows along the bottom of Figure 6. The relative estimates of quantities for water inputs and outputs at various phases are based on available literature review data, including select studies specifically identifying water footprints of building materials. One particular study for buildings in China assessed a cradle-to-gate (sourcing and manufacturing phases) of twelve typical building construction materials including steel, cement, concrete, wood, brick, sand, gravel, lime, glass, paint, ceramic tile, and ceramic ware (Chang, Huang, Ries, & Masanet, 2016).

The upstream phases of material sourcing and manufacturing constitute the highest quantities of water consumption and wastewater output for building materials. The building maintenance phase does not include occupant water use here, as the main goal is to identify material-related impacts with water. The construction and demolition phases, bookending a building's operational lifetime, have minimal water footprints relative to the other phases. While water is utilized in various aspects of standard building construction practices, such as dust control during site grading and general building material and debris cleaning activities, the quantities are found to be negligible in comparison with material production manufacturing (X. Zhang et al., 2011). Some in situ building material construction, such as concrete pouring, requires larger amounts of water, but for purposes of this study is accounted for upstream in the supply chain. In some locations, additional construction site energy is required to pump groundwater for use during the construction activities (Husain & Prakash, 2019).

An overall assessment theorem is provided for calculating the embodied water footprint of building materials (WF_{BM}). The variables identified for the water footprint of building materials relate to the phases of the material supply chain, the time for building operation life, and the mode of water input and output. The development of the theorem is founded in prior water footprint calculation methods (Hoekstra, 2015, 2016; Wang, Ding, & Wu, 2014) in combination with the particular considerations addressed for building materials and water. According to Solis-Guzman, "the water footprint of a company is defined as the total volume of freshwater that is used for production of goods and services consumed by the company" and is expressed in units of volume (Solis-Guzman, 2015). In Vickers' Handbook of water use and conservation, supply-chain water-stress measures are defined as water withdrawal and water consumption (Vickers, 2001). Water withdrawal is the water extracted from ground or surface water sources, causing impacts on the surrounding water ecology. Water consumption is the water that is no longer available after it is withdrawn from its source. Other more detailed models for impacts to freshwater sources are characterized in prior life cycle studies (Milà i Canals et al., 2009).

For a holistic building material water footprint (WF_{BM}) calculation, the following theorem is proposed:

$$WF_{BM} = (GW_1 + BWF_1 - WWF_1) + (BWF_2 - WWF_2) + ((GW_4 + BWF_4)YW)/t$$

In this assessment method, phases 1, 2, and 4 represent the material sourcing, manufacturing, and building maintenance phases, respectively. Phase 3 and 5 represent construction and demolition stages respectively, and are excluded from the proposed building material water footprint assessment due to negligible relevant impact for building material water use. The green water (GW) value for phase 1 material sources is one of the most complex to calculate as it is dependent on bio-geochemical physics, long-term histories, regional climate conditions, ecological system types, etc. The GW variable in phase 1 should be assessed by experts in the respective fields (i.e., forestry, agriculture, geology, mining, etc.). The green water in phase 4 is specific to local rainfall and snow during a building's operational life, which can be calculated based on climate data records or site-specific precipitation data.

The blue water footprint (BWF) at each phase takes into account both the direct blue water used for material processes (BW_{BM}) and the indirect blue water used for power plant source energy (BW_{PE}) required for the material processes:

$$BWF = BW_{BM} + BW_{PE}$$

$$BW_{PE} = E_{BM} \times WF_{PE}$$

The blue water footprint of phase 4 (BWF4) is the direct blue water used for building material maintenance (i.e., façade washing), and takes into account the indirect blue water used for power plant source energy required during the material maintenance processes. The wastewater footprint (WWF) for phase 1 and 2 is calculated as the greywater output from material sourcing and manufacturing activities (WW) in addition to the indirect blue water used for power plant source energy required as a function of the volume of greywater treatment:

$$WWF = WW + (WW(E_{WT} \times WF_{PE}))$$

The yellow water (YW) in phase 4 is the by-product water from interactions with building materials, either as a result of natural green water precipitation sources or as a result of blue water sources for material maintenance. The yellow water is factored into the assessment as a qualitative impact on a scale of 1.1 – 1.9. The time (t) in phase 4 is the number of years for the building material lifespan.

FUTURE RESEARCH DIRECTIONS

Numerous knowledge gaps for the water impacts from building materials are identified, especially in the detailed data inventories for water resources required during sourcing and manufacturing. Regional assessments are needed for water valuation mechanisms in relation to water poverty and stress indices and various zoning and policy regulations. Regional assessments of building design and construction practice in terms of water footprint summations are also needed to provide guidance on challenge areas related to building material supply chains, climate contexts, and construction methods. Ercin and Hoekstra also differentiate the attributes of economically-driven development vs. socially- and environmentally-driven development needs and weigh this in the impact of water footprint scenarios leading to water stress and water management tactics (Ercin & Hoekstra, 2014). An extremely important aspect of regional considerations will be the policies and zoning codes around site development and building integrated water processes, including building codes for new material standards with water impact assessment. On-site water treatment and re-use will be one of the major challenges in policies to address. Building owners and operators may have to become pseudo water utilities in the way that water quality is managed. Distributed sensing and monitoring technologies and techniques will become more necessary and prevalent in future water quality management.

Furthermore, a deeper historical and cultural analysis of the building technology practices for materials, methods, and water relations could be informative. At the same time, numerous advancements in material science and biotechnology are being adapted and translated into building materials and construction practices (Palko, 2014). There are a handful of bio-metabolism approaches to in situ building construction taking place, including the remediation of wood and water foundations by scientist Rachel Armstrong, or the site-specific tree-growth structural forming with the aid of polylactic-acid biodegradable formwork. Numerous mycelium or fungi-based growth mechanisms in biotechnology are informing new building insulation materials with minimal water and energy footprints. The biopolymeric bubbles of Zbigniew Oksiuta, and other conceptual works by the Living Architecture Systems Group led by Philip Beasley, demonstrate a new architectural relationship between material systems and water phenomena interactions or production techniques. A handful of material innovations, such as bio-bricks, utilize wastewater to induce curing through microbial interactions.

Water-based nanotechnology and new generations of nano-materials and coatings for building materials are enabling responsive mechanisms to functionalize integrated treatment or control of water interaction with materials. Some examples include switchable hydrophobic-hydrophilic aluminum surfaces (Ryu et al., 2016) as well as exploratory research with doping water molecules for applied molecular biology materials (Shoseyov & Levy, 2008). Developments with graphene are enabling removal of contaminants from water solutions through photocatalysis and filtration of toxic metal ions, bisphenols, pesticides, nitrates, and phosphates (fertilizers), control of bromide formation, or removal of arsenic, etc. (Naushad, 2018). Emerging building material systems for integrated water treatment processes include examples by Ann Dyson and the Center for Ecosystems and Architecture at Yale, including the Solar Enclosure for Water Reuse (SEWR).

CONCLUSION

The study demonstrates that there is a direct correlation between energy consumption and water footprint for building material production due to the power plant source energy water use. Reductions in water footprint could be realized when building material manufacturing plants convert to on-site renewable energy sources that do not require water-driven steam processes. The study also shows that the steel and metal building materials constitute the highest water footprint based on direct and indirect blue water consumption. The impact of water stress is variable by region and dependent on available and accessible water resources as well as population and water demands in sectors other than the building material industry (i.e., in agriculture, the highest water use sector). Material sourcing and extraction water impacts are more readily known through translation from related product industries. Future work is required to engage both industry and governments to take responsibility for tracking water impacts and transparency in information sharing (Ercin, 2012).

As our efforts continuously focus on energy denominators because of direct alignment with economic value, we are failing to realize the true impact on the ability for all of life to exist without shifting to a water valuation basis of environmental accounting. For all water that is contaminated or greywater by-product from material sourcing and manufacturing, additional energy is required to treat and purify the greywater source. So, in some sense, the energy dilemma is compounded by the water resource dilemma. Humans were once capable of survival without modern forms of energy production, but never without water. Ultimately, the framework provided for the water impacts from building materials sets out an agenda for building material and system design to engage with water ecologies for more beneficial outcomes in terms of the chemical environment, building metabolism, longevity, and the effective water stress.

REFERENCES

a4ws.org. (2019). *Alliance for Water Stewardship*. Retrieved from https://a4ws.org/

Abd El-Hameed, A. K., Mansour, Y. M., & Faggal, A. A. (2017). Benchmarking water efficiency of architectural finishing materials based on a "cradle-to-gate" approach. *Journal of Building Engineering*, *14*, 73–80. doi:10.1016/j.jobe.2017.10.001

Amziane, S., & Arnaud, L. (2013). *Bio-aggregate-based building materials applications to hemp concretes*. Hoboken, NJ: Wiley. doi:10.1002/9781118576809

Ashby, M. F. (1980). Engineering materials: an introduction to their properties and applications. New York, NY: Pergamon Press.

Baboian, R. (1985). *Materials degradation caused by acid rain*. Paper presented at the 20th State-of-the-Art Symposium of the American Chemical Society, Arlington, VA.

Bedane, A., Eić, M., Farmahini-Farahani, M., & Xiao, H. (2016). Theoretical modeling of water vapor transport in cellulose-based materials. *Cellulose (London, England)*, *23*(3), 1537–1552. doi:10.100710570-016-0917-y

Bollmann, U. E., Fernández-Calviño, D., Brandt, K. K., Storgaard, M. S., Sanderson, H., & Bester, K. (2017). Biocide Runoff from Building Facades: Degradation Kinetics in Soil. *Environmental Science & Technology*, *51*(7), 3694–3702. doi:10.1021/acs.est.6b05512 PMID:28287716

C.C.R.O.T. (2019). *Polymer Databse*. Retrieved from http://polymerdatabase.com/

Chang, Y., Huang, Z., Ries, R. J., & Masanet, E. (2016). The embodied air pollutant emissions and water footprints of buildings in China: A quantification using disaggregated input–output life cycle inventory model. *Journal of Cleaner Production*, *113*, 274–284. doi:10.1016/j.jclepro.2015.11.014

Cook, G., & Ponssard, J.-P. (2011). A proposal for the renewal of sectoral approaches building on the Cement Sustainability Initiative. *Climate Policy*, *11*(5), 1246–1256. doi:10.1080/14693062.2011.602552

Dakwala, M., Mohanty, B., & Bhargava, R. (2014). Simultaneous water and energy conservation through graphical and mathematical programming: A case study for float glass industry. *Journal of Cleaner Production*, *78*, 78. doi:10.1016/j.jclepro.2014.04.059

Ercin, A. E., & Hoekstra, A. Y. (2014). Water footprint scenarios for 2050: A global analysis. *Environment International*, *64*, 71–82. doi:10.1016/j.envint.2013.11.019 PMID:24374780

Ercin, E. (2012). *Governance of globalized water resources: the application of the water footprint to inform corporate strategy and government policy* (Unpublished doctoral dissertation). University of Twente, Ankara, Turkey.

Essington, M. E. (2003). Soil and Water Chemistry an Integrative Approach: An Integrative Approach. Baton Rouge, LA: CRC Press LLC.

Falk, R., & McKeever, D. (2004). *Recovering wood for reuse and recycling: a United States perspective.* Paper presented at the European COST E31 Conference.

Franks, F. (1984). Water (Rev. 1st ed.). London: Royal Society of Chemistry.

Freiesleben Hansen, P. (2009). *The Science of Construction Materials.* Berlin: Springer. doi:10.1007/978-3-540-70898-8

Gerbens-Leenes, P. W., Hoekstra, A. Y., & Bosman, R. (2018). The blue and greywater footprint of construction materials: Steel, cement and glass. *Water Resources and Industry, 19*, 1–12. doi:10.1016/j.wri.2017.11.002

Gómez-Laserna, O., Olazabal, M. Á., Morillas, H., Prieto-Taboada, N., Martinez-Arkarazo, I., Arana, G., & Madariaga, J. M. (2013). In-situ spectroscopic assessment of the conservation state of building materials from a Palace house affected by infiltration water. *Journal of Raman Spectroscopy: JRS, 44*(9), 1277–1284. doi:10.1002/jrs.4359

Habibi, Y., Lucia, L. A., & Wiley, I. (2012). *Polysaccharide building blocks a sustainable approach to renewable materials.* Hoboken, NJ: John Wiley & Sons, Inc. doi:10.1002/9781118229484

Hewlett, P. C., Liska, M., & Lea, F. M. (2019). Lea's Chemistry of Cement and Concrete (5th ed.). Elsevier Science.

Hoekstra, A. (2015). The water footprint of industry. In Assessing and Measuring Environmental Impact and Sustainability (pp. 221-254). Academic Press.

Hoekstra, A. (2017). Water Footprint Assessment: Evolvement of a New Research Field. *An International Journal - Published for the European Water Resources Association, 31*(10), 3061-3081.

Hoekstra, A. Y. (2016). A critique on the water-scarcity weighted water footprint in LCA. *Ecological Indicators, 66*(C), 564–573. doi:10.1016/j.ecolind.2016.02.026

Hunt, R., Franklin, W., & Hunt, R. (1996). LCA— How it came about. *The International Journal of Life Cycle Assessment, 1*(1), 4–7. doi:10.1007/BF02978624

Husain, D., & Prakash, R. (2019). Life Cycle Ecological Footprint Assessment of an Academic Building. *Civil, Architectural. Environmental and Agricultural Engineering, 100*(1), 97–110.

I.S.O. (2019). ISO 14025:2006. In Environmental labels and declarations - Type III environmental declarations - Principles and procedures. ISO Standards.

Kermani, B., & Harrop, D. (2019). *Water Chemistry.* Chichester, UK: John Wiley & Sons, Ltd.

Knickerbocker, W. S. (1927). *Classics of modern science (Copernicus to Pasteur)*. New York, New York: A.A. Knopf.

Krieg, E., & Rybtchinski, B. (2011). Noncovalent Water-Based Materials: Robust yet Adaptive. *Chemistry (Weinheim an der Bergstrasse, Germany)*, *17*(33), 9016–9026. doi:10.1002/chem.201100809 PMID:21726009

Kurdowski, W. (2014). *Cement and concrete chemistry*. Dordrecht: Springer. doi:10.1007/978-94-007-7945-7

Lichtfouse, E., Schwarzbauer, J., & Robert, D. (2015). Pollutants in Buildings, Water and Living Organisms: Vol. 7. *Environmental Chemistry for a Sustainable World*. Cham: Springer International Publishing.

Ludwig, N., Rosina, E., & Sansonetti, A. (2018). Evaluation and monitoring of water diffusion into stone porous materials by means of innovative IR thermography techniques. *Measurement*, *118*, 348–353. doi:10.1016/j.measurement.2017.09.002

Milà i Canals, L., Chenoweth, J., Chapagain, A., Orr, S., Antón, A., & Clift, R. (2009). Assessing freshwater use impacts in LCA: Part I—inventory modelling and characterisation factors for the main impact pathways. *The International Journal of Life Cycle Assessment*, *14*(1), 28–42. doi:10.100711367-008-0030-z

Müezzinoğlu, A., & Toprak, H. (1995). *A Life Cycle Study for Common Construction Materials*. Paper presented at the Cleaner Technologies and Cleaner Products for Sustainable Development, Berlin, Germany.

National Environmental Health Association. (2003). Efflorescence. (Technical Briefs). *Journal of Environmental Health*, *65*(10), 35. PMID:12800819

Naushad, M. (2018). *A New Generation Material Graphene: Applications in Water Technology*. Cham: Springer.

Palko, M. (2014). Advanced Building Construction and Materials II. Durnten: Trans Tech Publications, Limited.

Pan, J., Jeffrey, P., Marc, A. E., & Amy, P. (2015). Impact of Water Chemistry, Pipe Material and Stagnation on the Building Plumbing Microbiome. *PLoS One*, *10*(10), e0141087. doi:10.1371/journal.pone.0141087 PMID:26495985

Pavlík, Z., Žumár, J., Medved, I., & Černý, R. (2012). Water Vapor Adsorption in Porous Building Materials: Experimental Measurement and Theoretical Analysis. *Transport in Porous Media*, *91*(3), 939–954. doi:10.100711242-011-9884-9

Rott, U. (2003). Multiple Use of Water in Industry—The Textile Industry Case. *Journal of Environmental Science and Health. Part A, Toxic/Hazardous Substances & Environmental Engineering*, *38*(8), 1629–1639. doi:10.1081/ESE-120021485 PMID:12929813

Ryu, H.-Y., Yoon, S. H., Han, D.-H., Hafeez, H., Paluvai, N. R., Lee, C. S., & Park, J.-G. (2016). Fabrication of hydrophobic/hydrophilic switchable aluminum surface using poly(N-isopropylacrylamide). *Progress in Organic Coatings*, *99*, 295–301. doi:10.1016/j.porgcoat.2016.06.008

Sanjurjo-Sánchez, J., & Alves, C. (2012). *Pollutant-Induced Decay of Building Materials*. Dordrecht: Springer Netherlands. doi:10.1007/978-94-007-2439-6_2

Schyns, J. F., Booij, M. J., & Hoekstra, A. Y. (2017). The water footprint of wood for lumber, pulp, paper, fuel and firewood. *Advances in Water Resources*, *107*, 490–501. doi:10.1016/j.advwatres.2017.05.013

Shmulsky, R., & Jones, P. D. (2011). *Wood and Water*. Academic Press.

Shoseyov, O., & Levy, I. (2008). *NanoBioTechnology bioinspired devices and materials of the future*. Totowa, NJ: Humana Press.

Simonen, K. (2014). Life Cycle Assessment. London: Routledge. doi:10.4324/9781315778730

Solis-Guzman, J. (2015). *Ecological footprint assessment of building construction*. Sharjah: Bentham Science Publishers, Limited. doi:10.2174/97816810809871150101

Strang, V. (2015). *Water: Nature and Culture*. London: Reaktion Books, Limited.

Szabadvary, F. (1966). *The Period of Berzelius*. Elsevier Ltd. doi:10.1016/B978-0-08-010980-0.50009-2

Togkalidou, T., Karoglou, M., Bakolas, A., Giakoumaki, A., & Moropoulou, A. (2013). Correlation of Water Vapor Permeability with Microstructure Characteristics of Building Materials Using Robust Chemometrics. *Transport in Porous Media*, *99*(2), 273–295. doi:10.100711242-013-0184-4

Torcellini, P., Long, N., & Judkoff, R. (2004). Consumptive Water Use for U.S. Power Production. *ASHRAE Transactions*, *110*, 96–100.

Tuomi, T., Reijula, K., Johnsson, T., Hemminki, K., Hintikka, E.-L., Lindroos, O., ... Haahtela, T. (2000). Mycotoxins in Crude Building Materials from Water-Damaged Buildings. *Applied and Environmental Microbiology*, *66*(5), 1899–1904. doi:10.1128/AEM.66.5.1899-1904.2000 PMID:10788357

UN Global Impact. (2019). *CEO Water Mandate.* Retrieved from https://ceowatermandate.org/

Vickers, A. (2001). Handbook of water use and conservation. Choice Reviews Online, 39(3), 39-1594-1539-1594.

Vollpracht, A., & Brameshuber, W. (2010). Investigations on the leaching behaviour of irrigated construction elements. (Report). *Environmental Science and Pollution Research International, 17*(5), 1177–1182. doi:10.100711356-009-0264-8 PMID:19937278

Wackernagel, M. (1996). *Our ecological footprint: reducing human impact on the earth.* Philadelphia, PA: New Society Publishers.

Wang, L., Ding, X., & Wu, X. (2014). Careful considerations when reporting and evaluating the greywater footprint of products. *Ecological Indicators, 41*, 131–132. doi:10.1016/j.ecolind.2014.01.039

W.F.N. (2019). *Water Footprint Network.* Retrieved from https://waterfootprint.org/en/

Zhang, C., Yu, L., Ferdosian, F., Vijayaraghavan, S., Mesnager, J., Jollet, V., & Zhao, B. (2018). Behavior of Water/pMDI Emulsion Adhesive on Bonding Wood Substrates with Varied Surface Properties. *Industrial & Engineering Chemistry Research, 57*(48), 16318–16326. doi:10.1021/acs.iecr.8b03485

Zhang, G. P., Hoekstra, A. Y., & Mathews, R. E. (2013). Water Footprint Assessment (WFA) for better water governance and sustainable development. *Water Resources and Industry, 1-2*(C), 1–6. doi:10.1016/j.wri.2013.06.004

Zhang, X., Pint, C. L., Lee, M. H., Schubert, B. E., Jamshidi, A., Takei, K., & Javey, A. (2011). Optically- and thermally-responsive programmable materials based on carbon nanotube-hydrogel polymer composites. *Nano Letters, 11*(8), 3239–3244. doi:10.1021/nl201503e PMID:21736337

KEY TERMS AND DEFINITIONS

Dipole: In chemistry, a molecule that has a concentration of negative electric charge separated from a positive charge (i.e., water).

Dope: To intentionally introduce impurities to a substance (i.e., silicon, glass fibers) in order to modify the electrical, optical, and physical structure properties.

LCA: Life cycle analysis and also life cycle assessment.

Nanotechnology: Technological study and developments that focus on materials at the nano-scale.

Protonation/Deprotonation: The addition or removal of a proton from an atom, molecule, or ion.

Solvent: Typically, a liquid substance able to dissolve or breakdown other chemical substances into a solution.

Universal Solvent: Typically, a liquid substance able to reversibly attract or release other chemical substances (i.e., water).

Water Footprint: A volumetric measure that defines the freshwater resources consumed as a result of an industrial, agricultural, or human endeavor or process.

Section 3
Expanded Frameworks

These chapters navigate territories that exist outside of conventional material performance methods, ranging from incremental transformations to disruptive paradigms. Authors consider the future of life cycle assessment, the need for increased emphasis on social performance, the adaptation of interior climates, and "post-normal" approaches to building design and construction. The common theme concerns the potential for new contributions in environmental accounting to emerge by questioning limitations, exploring interdisciplinary connections, or subverting entrenched models.

Chapter 7
Life-Cycle Assessment-Based Environmental Performance Targets for Buildings:
What Is Next?

Getachew Assefa
University of Calgary, Canada

ABSTRACT

The role of targets in delivering meaningful performance improvements for designing new buildings and retrofitting existing building stocks is important. A piecemeal approach of incomprehensive assessments around insignificant changes falls short of achieving deep cuts in impacts. Most of the current assessments are not based on well-defined performance targets. The chapter is centered around exploring the utility of the concept of planetary boundaries for setting well-grounded benchmarking systems in guiding the transformation of the built environment that significantly contributes to the overall environmental impact of the economy. It discusses the role of life cycle assessment, environmental product declarations and product category rules, and how these and relevant standards and guides can be used in tandem with tools and processes used in design offices such as building information modeling. It concludes by charting the need for research on taking concepts such as planetary boundaries to building level benchmarking systems that support better design practices.

DOI: 10.4018/978-1-7998-2426-8.ch007

INTRODUCTION

Commonly used building environmental performance assessments are based on the notion of less is better, rather than on well-defined performance targets grounded in first-order principles. Performance benchmarking systems currently used in building assessments have little, if any, connection with planetary boundaries or equivalent constraints. This chapter argues for the need to develop science-based absolute performance benchmark values derived from environmental limits such as planetary boundaries. It starts with a background review of works undertaken on life cycle assessment-related environmental performance targets of buildings; argues for more meaningful impact targets; highlights the potential of including performance targets in communication tools; sets the context for early design intervention; highlights the utility of industry standards, guidance, and software tools; and ends with highlighting areas for future research that informs better building design practice.

Climate change, resource depletion, biodiversity, and toxicity are among the many global challenges humanity is increasingly facing. Cities with their buildings play a critical role as entry points for developing and implementing solutions that aim at addressing these challenges (Seto et al., 2014; Kennedy et al., 2015; Solecki et al., 2018; Swilling et al., 2018). Buildings globally are responsible for 40% of global energy and resource use, 33% of greenhouse gas emissions, and 25% water use (UNEP, 2013). For the world to meet the ambition of limiting global temperature increase by 2100 to 1.5°C based on the Paris Agreement, and thereby avoiding unprecedented climate change driven catastrophes, greenhouse gas emissions from the building sector need to be reduced by 80–90% by 2050 according to Kuramochi et al. (2018) as cited in IPCC (2018). The current global building stock at 223 billion m^2 is expected to increase by 230 billion m^2 in the next four decades (Bionova, 2018). The magnitude of the new additions is equated to adding to the building stock an entire city of New York every month for the next forty years. The 1.5°C goal requires addressing both new construction and retrofitting existing building stock.

Life cycle assessment (LCA) is commonly used in assessing the environmental impacts of buildings at different levels with varying emphasis on different aspects. Diverse studies focusing at building-level environmental performance have explored the circular economy for buildings (Hossain and Ng, 2018) and building elements (Eberhardt et al., 2019a); construction waste optimization (Jalaei et al., 2019); integrated design (Leoto and Lizarraldea, 2019); parametric design and building optimization (Szalay and Kiss, 2019); building rehabilitation, refurbishment, and repurposing (Thibodeau et al., 2019; Vilches et al., 2017; Assefa and Ambler, 2017); embodied environmental cost (Roh et al., 2019); design for disassembly (Eberhardt et al., 2019b); datasets and tools (Emami et al., 2019); streamlined data collection (Tecchio et al., 2019b); uncertainty and surrogate data (Tecchio et al., 2019a); and result visualization (Kiss and Szalay, 2019). Other studies have looked for solutions and insights at scales larger than buildings (Fouguet et al., 2017; Nault et al., 2018; Assefa, 2019).

Addressing environmental impacts of buildings at the early design stage provides the right timing for making major decisions regarding the material and operational aspects of the buildings. There is a growing body of literature focusing on exploring and improving the integration of LCA in the design process of buildings (e.g., Shadram et al., 2016, Soust-Verdaguer et al., 2017). The credibility of the LCA results for use in communicating performance or informing and influencing design decisions will depend on a number of factors such as data quality and uncertainty, assessment methodology, and assumptions made.

The use of targets and benchmarks in measuring relative performance improvements for new buildings and existing building stock is common (Lützkendorf et al., 2012; Hollberg et al., 2019). Benchmarks can represent typical (average, ambitious, or poor) values or levels of some form of building performance indicator, such as energy use (Häkkinen et al., 2012). The approaches used to develop these targets and what they aspire to achieve are critical in determining their effectiveness. A piecemeal approach of assessments around insignificant improvements and trivial reductions will fall short of achieving deep cuts in impact and meaningful reductions in emissions and extractions associated with the different life cycle stages of buildings. There is a concern that the performance targets commonly used in building LCAs are far from making meaningful improvements in reducing the environmental impacts of the building sector. There is a need for advances in different areas such as ensuring consistent use of a comprehensive system boundary, developing a solid performance target system, and establishing the right incentives. Table 1 depicts the structure of a comprehensive system boundary of building LCAs, organized under different levels of life cycle stages and modules as stipulated in relevant standards such as ISO 21930 (ISO, 2017), EN 15804 (CEN, 2012) and EN 15942 (CEN, 2011a).

BACKGROUND

LCA-based impact target setting approaches applied in different building environmental impact studies are diverse, pertaining to the different ways of benchmarking used. Hollberg et al. (2019) provide a good overview of the building-related benchmarking approaches in the literature. They differentiate between top-down and bottom-up benchmarks, between absolute and relative benchmark values, between whole life cycle and life cycle stage based benchmarks, and between whole building and building element benchmarks. They also discuss the utility of a performance corridor delimited by a limit value, reference value, best practice

Table 1. Life cycle stages in a building LCA (adapted from Spirinckx et al., 2018)

<table>
<tr><th>Life cycle stages</th><th>Module</th><th colspan="2">Description</th></tr>
<tr><td rowspan="3">Production Stages</td><td>A1</td><td colspan="2">Pre-processing and acquisition of raw materials and packaging of raw materials</td></tr>
<tr><td>A2</td><td colspan="2">Transport of raw (processed) materials to production site</td></tr>
<tr><td>A3</td><td colspan="2">Manufacturing of building products and related packaging</td></tr>
<tr><td rowspan="2">Construction stages</td><td>A4</td><td colspan="2">Transport packaged building products to the construction site</td></tr>
<tr><td>A5</td><td colspan="2">Construction processes including all ancillary materials, end-of-Life of packaging material discarded, and any losses during construction</td></tr>
<tr><td rowspan="7">Use stages</td><td>B1</td><td>Use stage</td><td rowspan="5">Building fabric related</td></tr>
<tr><td>B2</td><td>Maintenance</td></tr>
<tr><td>B3</td><td>Repair</td></tr>
<tr><td>B4</td><td>Replacement</td></tr>
<tr><td>B5</td><td>Refurbishment</td></tr>
<tr><td>B6</td><td>Operational energy use</td><td rowspan="2">Building operation related</td></tr>
<tr><td>B7</td><td>Operational water use</td></tr>
<tr><td rowspan="3">End of life (EoL) stages</td><td>C1</td><td colspan="2">Dismantling</td></tr>
<tr><td>C2</td><td colspan="2">Transport to EoL</td></tr>
<tr><td>C3/C4</td><td colspan="2">Disposal at EoL (sorting, recycling, incineration, and landfill of all materials at the EoL of the building)</td></tr>
</table>

value, and target benchmark value as used, for example, in Häkkinen et al. (2012). Limit value refers to the lowest acceptable value sourced from a legal or prescriptive minimum. Best practice value is experimental or demonstration project value representing the best value or upper quartile. Reference value indicates the median value of the current state of the art. Target values may refer to an upper limit of the scale based on political targets or technical or economic optimums. All four values are based on relative performance compared to individual buildings or a group of buildings without any connection to global absolute performance targets, which is the core argument of this chapter.

Within the context of green building rating systems, Ganassali et al. (2016) recognize the existence of external and internal benchmarks. Benchmark values are based on a statistical analysis of a building stock or performance of a building element relative to alternative elements within the jurisdiction of interest. The statistical analysis of a building stock approach is used, for example, in Germany, while the relative performance of alternative building elements is used in the BREEAM rating system. These systems differ in whether and how the use phase is included. In the BREEAM system, for example, all the modules under the production stage (modules

A1-A3), construction stage (modules A4 and A5), and end of life cycle stage (modules C1-C4) are included, but not any of the use stage modules (see Table 1). The other rating and certification systems covered in Ganassali et al. (2016) account for the three stages included in the BREEAM system and part of the use stage modules.

One strand of internal benchmarks refers to relative performance values based on a comparison between different iterations of the same building: the baseline and proposed building. In the case of LEED v4, the baseline and the proposed building need to be comparable in function, size, shape, orientation, and energy performance. The benchmark value or reference values are the LCA result of the baseline building covering production stages (A1-A3), transportation of building products to the site (A4), all modules of the use stages related to the building fabric (B1-B5), and all modules of the end of life stages (C1-C4). The LEED v4 LCA credit requires the proposed design to show a minimum of 10% reduction compared to the baseline design in three out of the six listed environmental impact categories, including global warming potential, with none of the remaining impact categories showing an increase of more than 5%.

Most of the current building environmental performance assessments are based on the notion of less is better, not on well-defined performance targets grounded in first-order principles. Most impact performance targets focus on limited aspects such as energy and greenhouse gas targets. On the energy portion, the dominant focus is on operating energy, not much on embodied energy and other parts of the life cycle of buildings. The target setting or benchmarking is often done based on a statistical analysis of limited samples that usually focus on reference cases such as buildings considered as best practices defined based on different criteria. Performance targets need to go beyond a narrow focus on energy and greenhouse gases to achieve more comprehensive performance, such as net positive overall environmental impact targets for new buildings and existing stock.

There are also studies that developed their benchmarking using a top-down approach of taking national targets of energy and greenhouse gas goals to sector-specific targets. One example of such benchmarking concerns energy and greenhouse gas targets calculated for different types of buildings in Switzerland, based on the Swiss 2000-Watt society target for 2050. The 2000-Watt society, first developed by the Swiss Federal Institute of Technologies and later adopted by the government, aspires to limit Switzerland's annual per capita energy consumption and greenhouse gas emissions to 2000 Watts and a 1-ton carbon dioxide equivalent, respectively, by 2150 (Riera Perez et al., 2014). Building on this society-wide vision, the Swiss Society of Engineers and Architects has set sectoral energy and greenhouse gas targets for 2050 for new buildings and renovations of 440 MJ/m^2 and 16.5 kg CO_2-eq/m^2 that cover embodied energy, operating energy and building-related mobility.

Table 2. Greenhouse targets per year by building typologies derived from 2000-Watt vision (Hoxha et al., 2016)

Building type	kg CO_2 eq/m²
Residential	11
Schools	11.5
Hotels	11.7
Offices	14
Specialized shops	17.7
Commercial	19.7
Food stores	19.8
Restaurants	20.7

Assuming 60 years of use phase, Hoxha et al. (2016) translate the 2000-Watt target to a 2050 target for different building types, as shown below in Table 2. Depending on the building typology, targets can be presented in a functionally relevant way, such as per full-time equivalent per year for offices, or per occupant per year for residences. The customary use of floor area as the common denominator enables comparisons with existing studies that usually use gross or net floor area, or heated or conditioned area as a basis. Zimmermann et al. (2005) have translated the societal target of 1 ton CO_2-eq per capita to a target of 370 kg CO_2-eq per capita per year for the housing sector.

Pathways to More Meaningful Impact Targets

In informing the design and management of buildings, the focus should be on the most important factors that affect their performance at the pre-construction, construction, use, and end of life stages. To support this process, impact target-setting should be grounded in a solid foundation and should support the identification of high impact efforts of achieving the targets.

One problem with selecting limit and target benchmarks based on reference buildings is that what is considered as a better performance certainly falls short of making the necessary deep cuts in emissions and impacts. The impact of the materials and the majority of the operational consequences of buildings last for decades, even after they are demolished. Any aspiration for significant changes in the built environment by improving the performance of buildings and associated infrastructure should be based on targets that are solid enough to make the quest for sustainable buildings impact-wise highly meaningful. The targets should be closely tied to the survival of humanity in the long run and its collective socio-economic wellbeing in the immediate, short, and medium terms. Anchoring these performance targets on globally relevant science-based concepts is the best path to take.

Two examples of such concepts and frameworks based on global constraints are The Natural Step developed by Robèrt (2000 and 2002) and Broman and Robèrt (2017), based on first-order principles; and the Planetary Boundaries (PB), first developed by Rockström and his team (2009a and 2009b) and later updated by Steffen et al. (2015), based on the notion of safe operating space. While the Natural Step goes beyond the environmental and natural resource aspects to capture the social dimension, the PB framework focuses on the global and regional environmental and ecosystem challenges. As the focus of this chapter is on the environmental performance of buildings, the rest of this section is on the PBs. The idea of using PBs for developing benchmarking values is equally valid for other science-based environmental limits at the appropriate scale. At the time of writing this chapter, the PBs consist of the following nine boundaries with defined thresholds that, if transgressed, risk humanity with cascading catastrophes: climate change, biosphere integrity change, stratospheric ozone depletion, ocean acidification, biogeochemical flows of phosphorus and nitrogen, land-system change, freshwater use, atmospheric aerosol loading, and chemical pollution (later updated as the release of novel entities). Control variables for setting the threshold values and monitoring current states are established for most of the PBs, described briefly below, while research on defining the values for the remaining PBs and exploring the development of additional PBs continues.

Climate change's threshold is set at 350 ppm (parts per million) carbon dioxide concentration in the atmosphere, with an upper bound value of 450 ppm. The concentration of CO_2 in the atmosphere has been increasing every year at an escalating rate over the past decades. According to the US National Oceanic and Atmospheric Administration (NOAA), which monitors different parts of the natural environment including the atmosphere, the annual rate of increase of CO_2 concentration during the early years of measurement averaged around 0.7 ppm, and started to increase to about 1.6 ppm in the 1980s and 1.5 ppm in the 1990s, reaching a 2.2 ppm per year increase during the last decade (NOAA, 2019). For 2018, the average global CO_2 concentration at 410 ppm has by far passed the 350 ppm safe threshold and is heading to the upper bound value of 450 ppm. As buildings are one of the major contributors of CO_2 and other greenhouse gases, they provide opportunities for making deep emission reductions.

The planetary boundary on change in biosphere integrity concerns the loss of biodiversity and ecosystem changes, damages, and extinctions. The control variable used to measure this boundary is the rate of extinction loss per million species-years for which the boundary is set below ten (upper bound value of 100). The value assessed as current at the time of publication of the PBs as 100-1000 extinctions per million species-years breached the lower boundary value by at least ten times (Steffen et al., 2015). Understanding and controlling any building-related impact on biodiversity and ecosystems that potentially contributes to the transgression of this vital boundary is necessary.

The PB on stratospheric ozone depletion is measured using ozone concentration in Dobson Units (DU), which in the pre-industrial era was 290. The boundary is set at a maximum of 5% depletion relative to the pre-industrial level, i.e., 275 DU with a lower bound value of 261 DU. According to The Ozone Hole Inc. (2019), the ozone concentration during the 2018 ozone hole season (July to December) ranged between 102 DU to 264 DU, despite the global action of phasing out ozone-depleting substances such as those used in the air conditioning of buildings, and is on track in helping abate the problem.

The PB on ocean acidification is monitored using the change in the average global surface ocean saturation state of one type of calcium carbonate formed by marine organisms aiming at no or minimal reduction of the carbonate. A maximum of 20% reduction (with an upper bound of 30%) relative to the pre-industrial level is set as a boundary. The level assessed as current at the time of publication of the PBs showed around a 16% reduction. Since ocean acidification is related to climate change, building-related greenhouse gas emissions contribute to the transgression of this PB as they do the climate change PB.

The boundary related to biogeochemical flows of nitrogen and phosphorus is mainly related to fertilizer application. For nitrogen, global industrial and intentional biological fixation measured in terra grams (Tg) of nitrogen per year is used as a control variable, and the boundary is set at a value of 62 Tg with an upper bound value of 82 Tg. The values assessed as current at the time of publication of the PBs are 2.4 times and 1.8 times higher than the lower value and the upper bound value, respectively. Phosphorus is monitored at the global level using its flow from freshwater systems into the ocean, and at the regional level using its flow from fertilizers to erodible soils. The global and regional boundaries are set at 11 Tg (upper bound of 100) and 6.2 Tg per year (upper bound of 11.2), respectively, while the observed values at the time of publication of the PBs were at 22 Tg and 14 Tg. Identifying building materials and life cycle stages of buildings that contribute directly or indirectly to such an undesirable impact will be welcome.

Land system change is measured globally and at a biome level in terms of the area of remaining forested land as a percentage of original forest cover. The boundary at the global level is set at 75% with a lower bound of 54%, and the observed value at the time of publication of the PBs is 62%. The construction sector needs to closely examine its direct and indirect contributions to the depletion of forest lands.

The freshwater use PB is measured globally as the maximum amount of consumptive blue water use in km^3 per year, at the basin level, as the withdrawal of blue water as a percentage of mean monthly river flow. The global boundary is set at 4000 km^3 per year (upper bound value of 6000). The value reported as current at the time of publication of the PBs is at around 2600 km^3 per year. The scale that is more relevant for building-related assessment is the basin level boundaries that are set as 25% (upper bound of 55%), 30% (upper bound of 60%) and 55% (upper bound of 85%) for low-flow months, intermediate-flow months, and high-flow months, respectively. A significant amount of water is consumed and wasted during the operation phase of buildings. The embodied water associated with building products and materials is also an area worth researching.

As shown above, the current values of the control variables for biodiversity and biogeochemical flows are already overpassed (Steffen et al., 2015). At the time of this writing, the boundaries for the atmospheric aerosol loading and chemical pollution (release of novel entities) are not established. One key aspect of the PBs worth considering is the fact that they are developed based on the recognition of their interdependence as part of a one-earth system.

The application of PBs has been explored in different areas of inquiry by different researchers. Whiteman et al. (2013) focused on identifying research priorities for corporate sustainability based on the first work by Rockström et al. (2009a and 2009b). O'Neill et al. (2018) analyzed a selected number of social thresholds that encompass a safe and just space in relation to downscaled planetary boundaries. Earlier, Häyhä et al. (2016) noted the need to translate the global PBs to a decision-making scale using biophysical, socio-economic, and ethical aspects. The use of the PBs in LCA has been explored with a specific focus on life cycle impact assessment methodology (Ryberg et al., 2016). In a recent review paper by Downing et al. (2019), which provides a comprehensive picture of the areas of research and applications using PBs during the first decade since the first PB publication by Rockström et al. in 2009, LCA is identified as a vehicle for translating PBs into smaller scales.

Beyond Communicating Environmental Performance

Once benchmark values are based on absolute performance targets, they can best be utilized by inclusion in existing environmental performance communication vehicles. LCA-based communication instruments such as product category rules (PCRs) and the resulting environmental product declarations (EPDs) are ideal tools for the inclusion of such targets in benchmark performances reported in EPDs. This integration will make the values more useful in informing building design practices based on material and energy specifications in tandem with existing and new tools and processes used in design offices.

EPDs, in general, are developed to communicate transparent and comparable information about the life cycle environmental impacts of products. They allow building material and product manufacturers to communicate the environmental footprints of their products in compliance with international standards such as ISO 14025(ISO, 2006a), ISO 14040 (ISO, 2006b), and ISO 14044(ISO, 2006c). The big advantage of using EPDs is the requirement that they should be independently verified and registered in communicating environmental impacts in a more objective manner. By communicating verifiable environmental information of products, EPDs support science-based choices and stimulate potential continuous improvements. The issuance of an EPD for a building product is a transparent declaration of the life cycle environmental impact of that product. It does not per se imply that the product is environmentally different compared to alternatives, however.

Developing EPDs requires ensuring the consistency, reproducibility, and comparability of the underlying LCA data. The guidelines, requirements, and rules for an even-handed way of dealing with methodological aspects of the relevant LCA, scoping and identifying what should be considered in developing EPDs, are included in PCRs. The PCRs provide the rules and requirements for conducting the LCA used for developing the EPDs for the product categories of interest. As LCA results, in general, are sensitive to assumptions made and quality of data used, EPD-oriented LCAs need to meet the requirements around these aspects, as established in the relevant PCRs, as a way of ensuring the credibility and improving the comparability of EPDs of similar products.

The challenge with LCA is the data availability and quality associated with the inventory of the product system under assessment, and more so in complex systems such as buildings. It is important to be transparent about the magnitude of the uncertainty involved. A good PCR needs to provide quantitative or qualitative data regarding uncertainty estimation in datasets, parameters, and impact factors used in the form of, for example, confidence intervals. One of the merits of both PCRs and EPDs is the fact that their validity is time-bound. In most cases, their re-verification and re-registration should be done within three to five years of publication. New versions are expected to account for new knowledge, data, technology, and performance.

There is a proliferation of PCRs in Europe and in North America that will gradually contribute to the improvement of design offices' access to EPDs of interest. This author took part in the development of the PCR for North American Structural and Architectural Wood Products (covering 15 products) and its update, as well as the development of the first PCR for North American Gypsum Boards (covering 13 gypsum board products). The PCRs were developed in accordance with ISO 14025 (ISO, 2006a) and ISO 14044(ISO, 2006c) with the participation of multiple stakeholders from industry and government. The second version of the structural and architectural wood products PCR was developed by a technical committee of 18 members coming from, among others, the Canadian Wood Council, American Wood Council, Canadian Forest Services, American Forest Services, and Consortium for

Research on Renewable Industrial Materials. During the recent iteration of the PCR, around 20 members from the Canadian Wood Council, American Wood Council, Canadian Forest Services, American Forest Services, Consortium for Research on Renewable Industrial Materials, Building Materials Reuse Association, four universities, and some companies, took part.

Early Design Intervention

The early design stage offers an opportunity of using environmental limit-based performance benchmarks in making the right decisions in good time to implement in later life cycle stages, thereby setting the performance trajectory of the building for decades and beyond. The use of Building Information Modeling (BIM) tools by design offices is increasing, with more expansion expected to occur in the years to come. BIM provides new opportunities for moving from symbolism to virtualization, linking geometric and non-geometric information, and "allowing various forms of automated analysis and simulation, ranging from environmental and structural analysis to cost estimating and scheduling" (Jupp and Singh, 2014, p36). The use of LCA to inform early design stages of buildings is also garnering a significant space in research (Means and Guggemos, 2015; Meex et al., 2018; Cavalliere et al., 2018; Hollberg et al., 2019; Schlegl et al., 2019; Cavalliere et al., 2019; Rezaei et al., 2019; Zemero et al., 2019). The initial independent use of LCA and BIM in building design is paving the way for the emergence of LCA tools connected to BIM tools for an automated pulling of material quantity take-offs for calculating the embodied and life cycle impacts of buildings. This automation significantly reduces the life cycle inventory data collection effort and improves the data quality compared to the conventional way of retrieving data from different sources that suffer from varying data quality aspects. This automation can be done with minimal challenges when design modeling is completed at an advanced level of development (LOD). An LOD 300 or above fit for design development provides a good basis for an LCA of the building design. Conducting LCA calculations at the very early design stage (LOD 100) is challenging as little is known about materials at this stage. In mitigating this problem, Rezaei et al. (2019) recommended and tested the use of probability distributions to cover how different material options would affect the contribution of the different life cycle stages to the overall life cycle impact of their case study building. In comparing the results of LOD 300 and LOD 100, their LOD 300 results fall within the uncertainty range of the LOD 100 results. Such early application of LCA in the design workflow also requires the estimation of the quantity of operating energy, which would normally be based on energy simulation tools that use material data as part of the input files. Estimated values for operating water use, construction stage energy and fuel use, and the potential end of life management are also inputs to the LOD 100-level LCA calculations. Rezaei et al. (2019) handled all these using probability distributions that represent a range of potential values.

The level of detail of life cycle stages, building elements and sub-elements, materials and processes used in PCRs, EPDs, database structures, software tools, design tools, and associated workflows should be compatible. Converting from one level of resolution to another should at least be less cumbersome than it is now. The number of life cycle stages included, and what is covered under each stage at the level of building materials, building elements, and the building, should be consistent to avoid the issue of non-comparable performance data and misalignment (Spirinckx et al., 2018). Such harmonization saves on the time and cost of efforts of conducting new LCAs of building products and materials that have existing standard-compliant EPDs produced at another level of detail.

Design workflows have a central place in providing the right entry point for a positive influence on the design early on. Depending on the level of granularity, a design workflow starts with pre-design activity of programming or strategic definition, preparation, and brief. It then flows from conceptual/initial design through developed design to technical/construction design or final design. Hollberg et al. (2019) proposed a design workflow that takes an initial design into a final design through building-level and building element-level impact calculations, followed by a comparison of impact results with their corresponding building-level and building element-level benchmark values. The building element-level calculation is done only when the building-level test fails or when there is an interest in analyzing improvement potential even after a building-level passes. Based on the proposed workflow, a pass in the building element test leads to a change in the initial design, while a failure can lead to a change in material.

Industry Standards, Guidance, and Software Tools

A reinvigorated design practice based on performance benchmarking driven from environmental limits such as planetary boundaries has the potential to contribute to the achievement of deep cuts in emissions and impacts through better design of new constructions and retrofitting existing building stocks. In supporting this design practice, relevant industry standards, guidance documents, and software tools applicable at different levels and stages should facilitate the development and implementation of the new performance benchmarking systems.

Industry Standards

Industry standards play a critical role in harmonizing the different elements of conducting an LCA of building products and whole buildings. The environmental limits-based benchmarking systems can be used included while updating existing standards and developing new ones. Relevant standards and supporting technical reports developed by the International Organization for Standardization (ISO) and the European Committee for Standardization (CEN) include ISO 21930 (ISO, 2017), EN 15804 (CEN, 2012), EN 15942 (CEN, 2011a), CEN/TR 15941 (CEN, 2010), CEN/TR 16970 (CEN, 2017a), ISO 21931-1 (ISO, 2010), EN 15978 (CEN, 2011b), and CEN/TR 17005 (CEN, 2017b). These can be classified as building product-level standards, building-level standards, and technical report-supporting standards.

Building Product-Level Standards

The standards that focus on building products are ISO 21930 (ISO, 2017), EN 15804 (CEN, 2012), and EN 15942 (CEN, 2011a). ISO 21930 (ISO, 2017) provides the principles, specifications, and requirements for developing EPDs for building products and services, building elements, and integrated technical systems. This standard complements ISO 14025 (ISO, 2006a) by providing specific requirements for EPDs of building products and services. It sets core requirements that make up a core PCR to develop building product EPDs. It provides mandatory requirements for any PCR to be developed based on the standard. ISO 21930 lays out predefined indicators to consider in EPDs and the rules for inventory level and impact assessment level calculations. It defines the processes to include under the life cycle stages that should be considered and how the life cycle stages are organized in terms of information modules. The core elements of EPDs, rules for the development of scenarios, and requirements for reporting relevant non-LCA environmental and technical information, are included in the standard. The standard also specifies the conditions for comparable building product-level EPD information. Though EPDs for business-to-consumer communication under certain conditions are not disallowed by ISO 21930, the standard's primary target is the use of building product EPDs in business-to-business communication.

EN 15804 provides a core PCR for all building products and services to ensure a harmonized way of developing, verifying, and presenting EPDs of building products, construction services, and construction processes. It defines parameters to be declared, the way in which the parameters are organized and reported, and describes the life cycle stages and processes under each life cycle stage included in EPDs. It sets the rules for the development of scenarios and for calculations during the inventory and impact assessment of EPD-oriented LCAs. EN 15804 covers the specification of data quality to apply, and the requirements for reporting additional environmental and health information where applicable. It also sets the conditions to be met for the comparison of building products and construction services-level EPD information. The specification and description of the communication format of the information defined in EN 15804 are covered under EN 15942, which provides a common and consistent format for business-to-business communication.

Building-Level Standards

Currently, existing standards that focus on building-level assessment are ISO 21931-1 and EN 15978. ISO 21931-1 establishes the general framework for improving the quality and comparability of methods for assessing the environmental performance of buildings. It defines and describes issues to consider in the use and development of methods of assessment of the environmental performance of new or existing buildings in their design, construction, operation, maintenance and refurbishment, and deconstruction stages.

EN 15978 specifies the LCA-based calculation method and other quantified environmental information for assessing the environmental performance of new and existing buildings and refurbishment projects. It also provides what to consider when reporting and communicating assessment results. EN 15978 contains, among other things, requirements for the description of the building or refurbishment project under assessment, the relevant system boundary, the procedure for inventory analysis; a list of indicators to calculate; the calculation procedures; the data requirements necessary for the calculation; and the presentation, reporting and communicating of results. This building-level standard requires the assessment to be based on all life cycle stages with data obtained from, among others, EPDs. The assessment covers all building-related products, processes, and services used over the life cycle of the building. EN 15978 excludes the interpretation and value judgments of the assessment results, however.

Technical Report-Supporting Standards

The European standards functioning at the level of buildings and building products are supported by three technical reports that provide information on the technical content of the standards. Technical Report CEN/TR 15941 supports the consistent way of developing EPDs complying with EN 15804 through the use of generic data according to the core PCR. EN 15804 contains the requirements for the use of generic data. The technical report supports EN 15978-based environmental performance assessment of buildings with the help of generic data. Technical Report CEN/TR 16970 provides general guidance for applying EN 15804 and for ensuring consistency in the preparation of PCRs that complement the core PCR of EN 15804. Technical Report CEN/TR 17005 was developed to provide information on relevance, robustness, and applicability of additional predetermined impact categories and indicators for building products and materials. It sets criteria such as relevance, scientific robustness and certainty, and applicability of impact assessment methods in determining the suitability of additional indicators and calculation methods for inclusion in both EN 15978 and EN 15804-compliant assessment. The seven impact categories include global warming, ozone depletion, photochemical ozone formation, acidification, eutrophication, mineral and fossil resource depletion, and non-fossil resource depletion. The additional impact categories included in the technical report are human toxicity and ecotoxicity, particulate matter, land use, biodiversity, water scarcity, and ionizing radiation. Without providing specific figures on levels of uncertainty to be considered, the technical report provides a general assessment of uncertainties related to impact assessment models.

In addition to the aforementioned standards and technical guides, there are currently two relevant upcoming ISO standards that will pave the way for more and consistent use of LCA in general, and EPDs in particular, in design tools. At the time of this writing, ISO/DIS 21678 (ISO, 2019a) and ISO/WD 22057 (ISO, 2019b) are at different levels of development. ISO/DIS 21678 deals with indicators and benchmarks and will focus on outlining principles for the development and use of benchmarks, while ISO/WD 22057 aspires to enable the use of EPDs at a construction works-level using BIM.

Though LCA started with a focus on the environmental dimension recognizing the interdependence of the three pillars of sustainability, the prospect of life cycle sustainability assessment that goes beyond environmental to add economic and social aspects of a building's life cycle is currently increasing. To this end, the relevant ISO standards on building sustainability assessments that started with the environmental dimension are expanding to include the social and economic dimensions.

Guidance Documents

Building product manufacturers and users such as design professionals and specifiers benefit from guidance documents that assist in the realization of uniform and consistent use of industry standards, the use of data and calculation methods, and the interpretation of impact results. Guidance documents provide the right condition for including and ensuring the consistent use of science-based performance benchmarking systems that are informed by first-order environmental limits such as planetary boundaries. The European Commission (EC), as part of its Single Market for Green Products initiative, supports such standardization of the type of data to be used and the methods for collecting, processing, and using life cycle data and making the calculations of impacts. EC's efforts include establishing specific procedures and recommendations included in its product environmental footprint category rules' guidance document (European Commission, 2017). The guide includes a four-step detailed approach to identifying the most relevant impact categories first,

Table 3. Normalization values based on inventory and population for 2010 for the 27 European Union countries (adapted from Benini et al., 2014)

	Impact category	Unit	Normalization factors per person
1	Climate change	kg CO_2 eq.	9.22E+03
2	Ozone depletion	kg CFC-11 eq.	2.16E-02
3	Human toxicity - cancer effect	CTUh	3.69E-05
4	Human toxicity - non-cancer effect	CTUh	5.33E-04
5	Acidification	mol H+ eq.	4.73E+01
6	Particulate matter/Respiratory inorganics	kg PM2.5 eq.	3.80E+00
7	Ecotoxicity for aquatic freshwater	CTUe	8.74E+03
8	Ionizing radiations - human health effects	kBq U235 eq. (to air)	1.13E+03
9	Photochemical ozone formation	kg NMVOC eq.	3.17E+01
10	Eutrophication - terrestrial	mol N eq.	1.76E+02
11	Eutrophication - freshwater	kg P eq.	1.48E+00
12	Eutrophication - marine	kg N eq.	1.69E+01
13	Land use	kg C deficit	7.48E+04
14	Resource depletion - water	m3 water eq.	8.14E+01
15	Resource depletion - mineral, fossil & Renewable	kg Sb eq.	1.01E-01

then the most relevant life cycle stages for the identified impact categories, then the most relevant processes for the short-listed life cycle stages,, and finally the most relevant direct elementary flows for the identified processes. An 80% cumulative contribution is used as a cut for including the relevant impact categories, life cycle stages, processes, and elementary flows. In the identification of the life cycle stages, if an initial screening finds the use stage contributing to 50% or more of the impacts of interest, the 80% test for identified relevant life cycle stages will be limited to the rest of the life cycle. The identification of the 80% impact categories necessitated the application of normalization and weighting methods developed for use with pilot PCRs that comply with the methods and procedures of the EC guidance. The normalization values and the weighting factors used by the EC pilot are shown in Table 3 and Table 4, respectively. Guidance documents can be used to record and communicate new performance benchmark values to practitioners the same way the normalization values and weighting factors are communicated.

The wider adoption and application of guidance on procedures of data collection and use, impact calculations, and result interpretations have the potential to increase the even-handed use of LCA and associated tools and instruments with more robust

Table 4 Weighting values (adapted from Sala et al., 2018)

	Impact category	Weighting factors
1	Climate change	21.06
2	Ozone depletion	6.31
3	Human toxicity - cancer effects	2.13
4	Human toxicity - non-cancer effects	1.84
5	Acidification	6.20
6	Particulate matter	8.96
7	Ecotoxicity freshwater	1.92
8	Ionizing radiation - human health	5.01
9	Photochemical ozone formation, human health	4.78
10	Eutrophication - terrestrial	3.71
11	Eutrophication - freshwater	2.80
12	Eutrophication - marine	2.96
13	Land use	7.94
14	Water use	8.51
15	Resource use, minerals and metals	7.55
16	Resource use, fossils	8.32

benchmarking systems that generate science-based performance values against which designs and proposals can be compared. Once the relevant body of knowledge matures and the broader consensus is reached, the absolute performance benchmark values can be included in mandatory rulebooks such as building codes.

Software Tools

Design offices can use a number of software tools to support their design modeling, production, simulation, and analysis. Tools for 2D CAD such as AutoCAD (Autodesk Inc., 2019a); tools for 3D CAD with no metadata such as AutoCAD® (Autodesk Inc., 2019a), Rhino (Robert McNeel & Associates, 2019) and Sketchup (Trimble Inc, 2019); tools for 3D geometry with metadata such as Allplan (Allplan Inc, 2019), ArchiCAD (Graphisoft, 2019) and Revit® (Autodesk Inc., 2019b); and tools for parametric design such as Dynamo (Autodesk Inc., 2019c) and Grasshopper included in Rhino (Davidson, 2019), are available for use in modeling and production. Design professionals also use different tools for the simulation of energy, daylight, comfort, airflow, and for calculating cost.

LCA software tools are the latest tools entering the universe of design toolboxes. There are generic LCA software tools such as SimaPro (Pré Consultants B.V., 2019) and GaBi (Thinkstep, 2019) for calculating the life cycle impacts of any system, including buildings. There are also a few dedicated tools focusing on the LCAs of buildings such as Athena Impact Estimator (ASMI, 2019), BEES (NIST, 2016), Tally (KT Innovation, 2016) and oneClickLCA (Bionova, 2018). Even fewer are the LCA tools that can easily be interfaced with design tools such as BIM for automated calculation during the design phase. One such tool, Tally, can be used as an add-in tool in Autodesk Revit®. Tally has locked backend material and energy impact data based on US conditions that are inaccessible to the frontend user. The higher level of resolution required for making material choices and informing other design decisions in the early design stage relies on the ability to customize the impact intensity data of inputs such as materials, electricity grid, and transportation to the local conditions. In this regard, the whole building LCA tool Athena Impact Estimator gives a better resolution with regional background data for 15 cities and US average options.

At the time of writing, there is an ongoing effort to integrate the Athena Impact Estimator tool with Autodesk Revit®. The author's research team is testing the system in an ongoing repurposing project of an old library tower at the University of Calgary as a follow up of an early work by Assefa and Ambler (2017). When the LCA for the first study was done, little was known about the fate of the old building largely because of uncertainties about external funding. Hence, the modeling was done using a simplified tool, taking into account the two options on the table—demolishing and building a new building in its place, or repurposing the building for a new use by maintaining the old structure and wrapping it with a new envelope. The study showed that a 20 to 40% reduction could be achieved by repurposing compared to the demolish-build option (Assefa and Ambler, 2017). The repurposing project of converting the old library building into a modern administration building is now underway. Working with the ongoing project provides the opportunity of testing the automated BIM-based LCA in one of the few large-scale projects in Canada with an active double-skin façade coupled with automatic shading and natural ventilation. The double-skin façade creates an insulating blanket of air during extreme outside temperatures and supplies fresh air and cooling during mild outside temperatures. The project is one of the few selected to participate in the Canada Green Building Council's Zero Carbon Building Pilot Program and is expected to contribute to the goal of informing the development of zero-carbon building standards and tools.

When the Athena Impact Estimator is used within the BIM environment, the integration based on demonstrations is expected to significantly reduce the data collection and inputting effort as it capitalizes on BIM files retrieved from project consultants and contractors. The small but gradually increasing body of literature on BIM-LCA alludes to the same merits in simplified data inputting (e.g., Shadram et al., 2016; Soust-Verdaguer et al., 2017; Cavalliere et al., 2019; Jalaei et al., 2019).

Today, design firms are far from seeing the incentive of mainstreaming BIM-LCA in their practice, given that there is not enough demand for LCA-based outcomes from their clients. Design professionals who see the benefit of using BIM-LCA beyond the immediate future will be leading the way in overcoming the challenge of getting the right signals that will make the exercise payoff. This challenge is even bigger when it comes to small firms that would tend to rely on data and tools that are easily accessible at no or low cost. Existing software tools should be developed further to allow for relating building life cycle performance to benchmark values derived from science-based environmental limits and planetary boundaries informed by first-order principles.

CONCLUSION

While there is a lot to learn from the best buildings and best building practices around the world, the right direction is the development and adoption of science-based, absolute—not relative—benchmarking that makes the building sector contribute to keeping humanity well below tipping points. Establishing performance benchmarking systems for buildings based on absolute values and including them in performance communication instruments, standards, guides, and software tools will enable measuring real progress towards a low-impact (e.g., low-carbon) or impact positive (e.g., climate-positive) built environment. The pathways to 80-90% reduction by 2050 needed to meet the 2100 ambition of limiting global temperature increase to 1.5°C are best informed by anchoring performance measurement on the science behind such targets. The successful development and use of such environmental performance benchmarking systems for buildings and building materials necessitate realizing a significant level of stakeholders' engagement, developing datasets and assessment approaches, and charting relevant research directions.

For an impactful and well-coordinated stakeholder engagement, the following roles can be identified as important: mandating, regulating and legislating performance; assessing performance; documenting and communicating performance; verifying performance compliance; rating and certifying performance; and designing for performance. The stakeholders playing roles need to be equipped with the right skillsets associated with the environmental limits and planetary boundaries that inform the development of new performance benchmarking systems. Design professionals stand to play an important role in doing-the-knowing while depending on and influencing the roles of the stakeholders behind the other five roles. Their influence ranges from selecting low impact building materials to determining how the designed building performs during its use stage and end of life stage. This critical role of affecting meaningful changes early in the design process necessitates accounting for all aspects and stages with the right level of detail. Building-level PCRs should be used to guide the development of building product level PCRs in ensuring, for example, consistent definition of the assessment boundary of buildings and deciding the life cycle stages and modules under each life cycle stage to include, as well as the requirement of including benchmarking systems grounded in absolute environmental limits and planetary boundaries. The design practice needs to strive for the realization of meaningful changes in the overall environmental performance of the building sector if significant contributions to the reduction of global, national, and sub-national economy-level environmental impacts are to be made.

Datasets and assessment approaches to be developed in line with environmental limits and planetary boundaries should focus on enabling the identification of transformational improvement opportunities and the implementation of improvement measures. Factors influencing a building's environmental performance during the different stages of its life cycle should be supported by a mix of specific, average, and generic data, depending on the goal and scope of the performance assessment. In addition to conventional sources of data and information, relatively new sources such as BIM and EPDs will be an increasingly significant source of specific data on the embodied segment of building impacts. Established energy performance calculation methods will continue to occupy the center stage of accounting for the operational impact of buildings during the early steps of the design workflow. In terms of impact assessment methods, those that are based on recent knowledge, backed by broader consensus, and have an appropriate level of resolution should be prioritized. For example, IMPACT World+, a globally regionalized method, provides characterization factors at global, continent, country, and native levels (Bulle et al., 2019). Identification of the most influencing material or building element in design offices can be supported by approaches of shortlisting the most important impact categories using relevant normalization values without the need to encroach into the territory of using more subjective weighting factors.

Stakeholders from the public and private sectors playing one or more of the roles identified above need to come together to establish good quality datasets that can be used by design professionals in their design decisions. In the interim, databases for widely and frequently used building products, materials, and systems can be developed with the help of common approaches for data collection and modeling. Developers of design tools need to raise the level of integration between parametric design tools and LCA calculation tools without losing transparency and the ability of users to customize backend data for varying typologies and local contexts. As the integration provides for more seamless interoperability between tools and data, design professionals will be increasingly willing to experiment with new tools and data.

Future research that will inform better design practices in light of the utility of LCA-based knowledge for designing new high-performance buildings and retrofitting existing building stocks needs to focus on creating solid benchmarking systems derived from environmental limits and planetary boundaries. Design guidelines, building performance standards, and building codes should be improved by moving away from focusing on rigid and narrow prescriptive solutions. More solid and grounded performance systems that push for a wide range of solutions that, in aggregate, have a much higher chance of keeping us within safe operating spaces of planetary boundaries are needed. The effort of establishing benchmark values based on planetary boundaries and equivalent first-order principles, and the opening up of building standards and codes to incorporate such values, requires the development of the right expertise and the cultivation of commensurate institutional capacities supported by relevant research.

One pathway of research is to translate the planetary boundary threshold values or their equivalents to a building-level indicator system presented in per floor area or per person, per year values. These values should consider long-term population and built environment expansion projections. With improved data and knowledge, the values can be reviewed on a regular basis at a frequency that matches the pace of major scientific findings on global ecological challenges and socio-economic projections. Regional variations should be considered within a development framework that is anchored in the same planetary boundaries or first-order principles. The adoption of the new benchmarking systems in the early building design stages will require striking a balance between the intricacies of design practice, complying with inflexible building codes, and playing within the limit of foundational thermodynamic constraints.

REFERENCES

Allplan Inc. (2019). *The Allplan solution portfolio*. Retrieved October 25, 2019, from https://www.allplan.com/us_en/products/solution-portfolio/

ASMI. (2019). *Athena Impact Estimator*. Athena Sustainable Materials Institute (ASMI). Retrieved August 4, 2019, from https://calculatelca.com/software/impactestimator/

Assefa, G. (2019). Life Cycle Insights for Creating Sustainable Cities. In Sustainable Cities-Authenticity, Ambition and Dream. IntechOpen. doi:10.5772/intechopen.81633

Assefa, G., & Ambler, C. (2017). To demolish or not to demolish: Life cycle consideration of repurposing buildings. *Sustainable Cities and Society*, *28*, 146–153. doi:10.1016/j.scs.2016.09.011

Autodesk Inc. (2019a). *AutoCAD*. Retrieved October 25, 2019, from https://www.autodesk.com/products/autocad/overview

Autodesk Inc. (2019b). *Revit®*. Retrieved October 25, 2019, from https://www.autodesk.com/products/revit/overview

Autodesk Inc. (2019c). *Dynamo Studio*. Retrieved October 25, 2019, from https://www.autodesk.com/products/dynamo-studio/overview

Benini, L., Mancini, L., Sala, S., Manfredi, S., Schau, E. M., & Pant, R. (2014). Normalisation method and data for Environmental Footprints. Luxembourg: European Union. EUR, 26842.

Bionova. (2018). *OneClickLCA*. Retrieved August 4, 2019, from https://www.oneclicklca.com/construction/

Bionova. (2018). *The Embodied Carbon Review*. Retrieved January 20, 2020, from www.embodiedcarbonreview.com

Broman, G. I., & Robèrt, K. H. (2017). A framework for strategic sustainable development. *Journal of Cleaner Production*, *140*, 17–31. doi:10.1016/j.jclepro.2015.10.121

Cavalliere, C., Dell'Osso, G. R., Pierucci, A., & Iannone, F. (2018). Life cycle assessment data structure for building information modeling. *Journal of Cleaner Production*, *199*, 193–204. doi:10.1016/j.jclepro.2018.07.149

Cavalliere, C., Habert, G., Dell'Osso, G. R., & Hollberg, A. (2019). Continuous BIM-based assessment of embodied environmental impacts throughout the design process. *Journal of Cleaner Production*, *211*, 941–952. doi:10.1016/j.jclepro.2018.11.247

CEN. (2010). *CEN/TR 15941:2010 Sustainability of construction works - Environmental product declarations - Methodology for selection and use of generic data*. Brussels: European Committee for Standardization.

CEN. (2011a). *EN 15942:2011 Sustainability of construction works - Environmental product declarations - Communication format business-to-business*. Brussels: European Committee for Standardization.

CEN. (2011b). *EN 15978: 2011 Sustainability of construction works—assessment of environmental performance of buildings—calculation method*. Brussels: European Committee for Standardization.

CEN. (2012). *EN 15804:2012 Sustainability of construction works—environmental product declarations—core rules for the product category of building products*. Brussels: European Committee for Standardization.

CEN. (2017a). *CEN/TR 16970:2017 Sustainability of construction works - Guidance for the implementation of EN 15804*. Brussels: European Committee for Standardization.

CEN. (2017b). *CEN/TR 17005:2017 Sustainability of construction works - Additional environmental impact categories and indicators - Background information and possibilities*. Brussels: European Committee for Standardization.

Davidson, S. (2019). *Grasshopper®*. Retrieved October 25, 2019, from https://www.grasshopper3d.com/

Downing, A. S., Bhowmik, A., Collste, D., Cornell, S., Donges, J. F., Fetzer, I., ... Mooij, W. (2019). Matching scope, purpose and uses of planetary boundaries science. *Environmental Research Letters*, *14*(7), 073005. doi:10.1088/1748-9326/ab22c9

Eberhardt, L., Birgisdottir, H., & Birkved, M. (2019a). Comparing life cycle assessment modeling of linear vs. circular building components. In Iop Conference Series: Earth and Environmental Science (Vol. 225, No. 1, p. 012039). IOP Publishing.

Eberhardt, L. C. M., Birgisdóttir, H., & Birkved, M. (2019b). Life cycle assessment of a Danish office building designed for disassembly. *Building Research and Information*, *47*(6), 666–680. doi:10.1080/09613218.2018.1517458

Emami, N., Heinonen, J., Marteinsson, B., Säynäjoki, A., Junnonen, J. M., Laine, J., & Junnila, S. (2019). A Life Cycle Assessment of Two Residential Buildings Using Two Different LCA Database-Software Combinations: Recognizing Uniformities and Inconsistencies. *Buildings*, *9*(1), 20. doi:10.3390/buildings9010020

European Commission. (2017). *PEFCR Guidance document, - Guidance for the 14 development of Product Environmental Footprint Category Rules (PEFCRs), version 6.3, December 15 2017*. Retrieved July 25, 2019, from https://ec.europa.eu/environment/eussd/smgp/pdf/PEFCR_guidance_v6.3.pdf

Fouguet, M., Jusselme, T., & Vareilles, J. (2017). Establishing building environmental targets to implement a low carbon objective at the district level: methodology and case study. In *Proceedings of 33rd PLEA International Conference* (*Vol. 3*, No. CONF, pp. 49-56). Academic Press.

Ganassali, S., Lavagna, M., & Campioli, A. (2016). LCA benchmarks in building's environmental certification systems. In *41st IAHS World Congress on Housing Sustainability and Innovation for the Future* (pp. 1-10). PRT.

Graphisoft. (2019). *ArchiCAD*. Retrieved October 25, 2019, from https://www.graphisoft.com/archicad/

Häkkinen, T., Antuña, C., Mäkeläinen, T., Lützkendorf, T., Balouktsi, M., Immendörfer, A., . . . Van Dessel, J. (2012). *Sustainability and Performance Assessment and Benchmarking of a Building – Final Report*. Retrieved August 4, 2019, from http://cic.vtt.fi/superbuildings/. Finland: VTT.

Häyhä, T., Lucas, P. L., van Vuuren, D. P., Cornell, S. E., & Hoff, H. (2016). From Planetary Boundaries to national fair shares of the global safe operating space - How can the scales be bridged? *Global Environmental Change*, *40*, 60–72. doi:10.1016/j.gloenvcha.2016.06.008

Hollberg, A., Lützkendorf, T., & Habert, G. (2019). Top-down or bottom-up? – How environmental benchmarks can support the design process. *Building and Environment*, *153*, 148–157. doi:10.1016/j.buildenv.2019.02.026

Hossain, M. U., & Ng, S. T. (2018). Critical consideration of buildings' environmental impact assessment towards adoption of circular economy: An analytical review. *Journal of Cleaner Production*, *205*, 763–780. doi:10.1016/j.jclepro.2018.09.120

Hoxha, E., Jusselme, T., Brambilla, A., Cozza, S., Andersen, M., & Rey, E. (2016). *Impact targets as guidelines towards low carbon buildings: preliminary concept*. No. CONF.

Innovations, K. T. (2016). *Tally*. Retrieved August 4, 2019, from https://www.choosetally.com/

IPCC. (2018) Summary for Policymakers. In Global Warming of 1.5°C. An IPCC Special Report on the impacts of global warming of 1.5°C above pre-industrial levels and related global greenhouse gas emission pathways, in the context of strengthening the global response to the threat of climate change, sustainable development, and efforts to eradicate poverty. In Press.

ISO. (2006a). *14025: 2006, Environmental labels and declarations - Type III environmental declarations -Principles and procedures*. Geneva: International Organization for Standardization.

ISO. (2006b). *14040: 2006, Environmental management - life cycle assessment - principles and framework*. Geneva: International Organization for Standardization.

ISO. (2006c). *14044: 2010, Environmental management - life cycle assessment - requirements and guidelines*. Geneva: International Organization for Standardization.

ISO. (2010). *21931-1: 2010, Sustainability in building construction. Framework for methods of assessment of the environmental performance of construction works. Part 1: Buildings*. Geneva: International Organization for Standardization.

ISO. (2017). *21930: 2017, Sustainability in buildings and civil engineering works - Core rules for environmental product declarations of building products and services*. Geneva: International Organization for Standardization.

ISO. (2019a). *ISO/DIS 21678 Sustainability in buildings and civil engineering works - Indicators and benchmarks - Principles for the development and use of benchmarks*. Geneva: International Organization for Standardization.

ISO. (2019b). *ISO/WD 22057 Enabling use of Environmental Product Declarations (EPD) at construction works level using building information modeling (BIM)*. Geneva: International Organization for Standardization.

Jalaei, F., Zoghi, M., & Khoshand, A. (2019). Life cycle environmental impact assessment to manage and optimize construction waste using Building Information Modeling (BIM). *International Journal of Construction Management*, 1-18.

Jupp, J. R., & Singh, V. (2014). Similar concepts, distinct solutions, common problems: learning from PLM and BIM deployment. In *IFIP International Conference on Product Lifecycle Management* (pp. 31-40). Springer. 10.1007/978-3-662-45937-9_4

Kennedy, C. A., Stewart, I., Facchini, A., Cersosimo, I., Mele, R., Chen, B., ... Dubeux, C. (2015). Energy and material flows of megacities. *Proceedings of the National Academy of Sciences of the United States of America*, *112*(19), 5985–5990. doi:10.1073/pnas.1504315112 PMID:25918371

Kiss, B., & Szalay, Z. (2019). A Visual Method for Detailed Analysis of Building Life Cycle Assessment Results. *Applied Mechanics and Materials*, *887*, 319–326. doi:10.4028/www.scientific.net/AMM.887.319

Kuramochi, T., Höhne, N., Schaeffer, M., Cantzler, J., Hare, B., Deng, Y., ... Mir, G. U. R. (2018). Ten key short-term sectoral benchmarks to limit warming to 1.5 C. *Climate Policy*, *18*(3), 287–305. doi:10.1080/14693062.2017.1397495

Leoto, R., & Lizarraldea, G. (2019). Challenges in evaluating strategies for reducing a building's environmental impact through Integrated Design. *Building and Environment*, *155*, 34–46. doi:10.1016/j.buildenv.2019.03.041

Lützkendorf, T., Kohler, N., & König, H. (2012). Integrated Life Cycle Assessment: Benchmarks and Uncertainty. In *Proceedings of the International Symposium on Life Cycle Assessment and Construction–Civil engineering and buildings* (pp. 10-12). Academic Press.

Means, P., & Guggemos, A. (2015). Framework for life cycle assessment (LCA) based environmental decision making during the conceptual design phase for commercial buildings. *Procedia Engineering*, *118*, 802–812. doi:10.1016/j.proeng.2015.08.517

Meex, E., Hollberg, A., Knapen, E., Hildebrand, L., & Verbeeck, G. (2018). Requirements for applying LCA-based environmental impact assessment tools in the early stages of building design. *Building and Environment*, *133*, 228–236. doi:10.1016/j.buildenv.2018.02.016

Nault, É., Jusselme, T., & Andersen, M. (2018). Setting contextual life-cycle objectives in urban design: requirements for a decision-support method. In *International Conference for Sustainable Design of the Built Environment SDBE 2018*. Academic Press.

NIST. (2016). *BEES*. Retrieved August 4, 2019, from https://www.nist.gov/services-resources/software/bees

NOAA. (2019). *Global carbon dioxide growth in 2018 reached 4th highest on record*. Accessed on August 2, 2019 https://www.noaa.gov/news/global-carbon-dioxide-growth-in-2018-reached-4th-highest-on-record

O'Neil, D. W., Fanning, A. L., Lamb, W. F., & Steinberger, J. K. (2018). A good life for all within planetary boundaries. *Nature Sustainability*, *1*(2), 88–95. doi:10.103841893-018-0021-4

Pré Consultants, B. V. (2019). *SimaPro*. Retrieved August 3, 2019, from https://simapro.com/

Rezaei, F., Bulle, C., & Lesage, P. (2019). Integrating building information modeling and life cycle assessment in the early and detailed building design stages. *Building and Environment*, *153*, 158–167. doi:10.1016/j.buildenv.2019.01.034

Riera Perez, M. G., Rey, E., Liman, U., Roulet, Y., & Favris-Donzel, A. (2014). *SméO, a sustainability assessment tool targeting the 2000 Watts society*. No. CONF.

Robèrt, K. H. (2000). Tools and concepts for sustainable development, how do they relate to a general framework for sustainable development, and to each other? *Journal of Cleaner Production*, *8*(3), 243–254. doi:10.1016/S0959-6526(00)00011-1

Robèrt, K. H. (2002). *The Natural Step Story, Seeding a Quiet Revolution*. Gabriola Island, BC, Canada: New Catalyst Books.

Robert McNeel & Associates. (2019). Retrieved October 25, 2019, from https://www.rhino3d.com

Rockström, J., Steffen, W., Noone, K., Persson, Å., Chapin III, F. S., Lambin, E. F., ... Nykvist, B. (2009b). A safe operating space for humanity. *Nature, 461*(7263), 472.

Rockström, J., Steffen, W. L., Noone, K., Persson, Å., Chapin, F. S. III, Lambin, E., ... Nykvist, B. (2009a). Planetary boundaries: Exploring the safe operating space for humanity. *Ecology and Society*, *14*(2), art32. doi:10.5751/ES-03180-140232

Roh, S., Tae, S., Kim, R., & Park, S. (2019). Probabilistic Analysis of Major Construction Materials in the Life Cycle Embodied Environmental Cost of Korean Apartment Buildings. *Sustainability*, *11*(3), 846. doi:10.3390u11030846

Ryberg, M. W., Owsianiak, M., Richardson, K., & Hauschild, M. Z. (2016). Challenges in implementing a planetary boundaries based life-cycle impact assessment methodology. *Journal of Cleaner Production*, *139*, 450–459. doi:10.1016/j.jclepro.2016.08.074

Sala, S., Cerutti, A. K., & Pant, R. (2018). *Development of a weighting approach for the environmental footprint*. Luxembourg: Publications Office of the European Union.

Schlegl, F., Gantner, J., Traunspurger, R., Albrecht, S., & Leistner, P. (2019). LCA of buildings in Germany: Proposal for a future benchmark based on existing databases. *Energy and Building*, *194*, 342–350. doi:10.1016/j.enbuild.2019.04.038

Seto, K. C., Dhakal, S., Bigio, A., Blanco, H., Delgado, G. C., Dewar, D., ... Ramaswami, A. (2014). Human Settlements, Infrastructure and Spatial Planning. In O. Edenhofer, R. Pichs-Madruga, Y. Sokona, E. Farahani, S. Kadner, K. Seyboth, ... J. C. Minx (Eds.), *Climate Change 2014: Mitigation of Climate Change. Contribution of Working Group III to the Fifth Assessment Report of the Intergovernmental Panel on Climate Change*. Cambridge, UK: Cambridge University Press.

Shadram, F., Johansson, T. D., Lu, W., Schade, J., & Olofsson, T. (2016). An integrated BIM-based framework for minimizing embodied energy during building design. *Energy and Building*, *128*, 592–604. doi:10.1016/j.enbuild.2016.07.007

Solecki, W., Rosenzweig, C., Dhakal, S., Roberts, D., Barau, A. S., Schultz, S., & Ürge-Vorsatz, D. (2018). City transformations in a 1.5 C warmer world. *Nature Climate Change*, *8*(3), 177–181. doi:10.103841558-018-0101-5

Soust-Verdaguer, B., Llatas, C., & García-Martínez, A. (2017). Critical review of BIM-based LCA method to buildings. *Energy and Building*, *136*, 110–120. doi:10.1016/j.enbuild.2016.12.009

Spirinckx, C., Thuring, M., Damen, L., Allacker, K., Mirabella, N., Ramon, D., ... Röck, M. (2018). Study on the Application of the PEF Method and related guidance documents to a newly office building. *European Union.*

Steffen, W., Richardson, K., Rockström, J., Cornell, S. E., Fetzer, I., Bennett, E. M., ... Folke, C. (2015). Planetary boundaries: Guiding human development on a changing planet. *Science*, *347*(6223), 1259855. doi:10.1126cience.1259855 PMID:25592418

Swilling, M., Hajer, M., Baynes, T., Bergesen, J., Labbé, F., Musango, J. K., & Suh, S. (2018). *The Weight of Cities Resource Requirements of Future Urbanization.* International Resource Panel. United Nations Environment Programme.

Szalay, Z., & Kiss, B. G. (2019). Modular methodology for building life cycle assessment for a building stock model. In R. Caspeele, L. Taerwe, & D. M. Frangopol (Eds.), *Life Cycle Analysis and Assessment in Civil Engineering: Towards an Integrated Vision: Proceedings of the Sixth International Symposium on Life-Cycle Civil Engineering (IALCCE 2018)* (*Vol. 5*, pp. 855-861). London: CRC Press.

Tecchio, P., Gregory, J., Ghattas, R., & Kirchain, R. (2019a). Structured under-specification of life cycle impact assessment data for building assemblies. *Journal of Industrial Ecology*, *23*(2), 319–334. doi:10.1111/jiec.12746

Tecchio, P., Gregory, J., Olivetti, E., Ghattas, R., & Kirchain, R. (2019b). Streamlining the Life Cycle Assessment of Buildings by Structured Under-Specification and Probabilistic Triage. *Journal of Industrial Ecology*, *23*(1), 268–279. doi:10.1111/jiec.12731

The Ozone Hole Inc. (2019). *The Ozone Hole 2018*. Retrieved August 4, 2019, from http://www.theozonehole.com/2018.htm

Thibodeau, C., Bataille, A., & Sié, M. (2019). Building rehabilitation life cycle assessment methodology–state of the art. *Renewable & Sustainable Energy Reviews*, *103*, 408–422. doi:10.1016/j.rser.2018.12.037

Thinkstep. (2019). *GaBi Software - to drive product sustainability!* Retrieved August 4, 2019, from http://www.gabi-software.com/software/gabi-software/

Trimble Inc. (2019). *SketchUp*. Retrieved October 25, 2019, from https://www.sketchup.com/

UNEP. (2013). *Energy efficiency for buildings*. Paris: United Nations Environment Programme.

Vilches, A., Garcia-Martinez, A., & Sanchez-Montanes, B. (2017). Life cycle assessment (LCA) of building refurbishment: A literature review. *Energy and Building*, *135*, 286–301. doi:10.1016/j.enbuild.2016.11.042

Whiteman, G., Walker, B., & Perego, P. (2013). Planetary boundaries: Ecological foundations for corporate sustainability. *Journal of Management Studies*, *50*(2), 307–336. doi:10.1111/j.1467-6486.2012.01073.x

Zemero, B. R., de Lima Tostes, M. E., Bezerra, U. H., dos Santos Batista, V., & Carvalho, C. C. M. (2019). Methodology for Preliminary Design of Buildings Using Multi-Objective Optimization Based on Performance Simulation. *Journal of Solar Energy Engineering*, *141*(4), 040801. doi:10.1115/1.4042244 PMID:30837779

Zimmermann, M., Althaus, H. J., & Haas, A. (2005). Benchmarks for sustainable construction: A contribution to develop a standard. *Energy and Building*, *37*(11), 1147–1157. doi:10.1016/j.enbuild.2005.06.017

Chapter 8
Social Life-Cycle Assessment for Building Materials

Liz Kutschke
University of Minnesota, USA

ABSTRACT

The goal of sustainable design and development is threefold, including economic, environmental, and social sustainability. While there are well-established methods for assessing the economic and environmental performance of products and buildings, the determination of social performance is less clear. This chapter explores the emerging field of social life cycle assessment (S-LCA), particularly as it relates to building materials and construction. This chapter includes 1) an introduction to and overview of S-LCA, summarized case studies of S-LCA; 2) a discussion of the relevance of S-LCA in sustainable design practice and education; 3) an examination of the role of environmental life cycle assessment (E-LCA) in building performance standards and certifications as a model for the incorporation of S-LCA; and 4) a reflection on areas for future research, including the addition of social science theory and practice for methodology, criteria, and metric development.

INTRODUCTION

In the pursuit of sustainable development, project teams are required to manage sustainable materials, buildings, and systems – it is true that one cannot manage what one does not measure, and one cannot improve what one does not manage. While it is impossible to measure all impacts (e.g., economic, environmental, or social), data and analysis of building materials allow for meaningful comparison between alternatives based on specific criteria. Environmental and financial impacts are fairly

DOI: 10.4018/978-1-7998-2426-8.ch008

well understood for material production, use, and disposal; building design, use, and demolition; and other human development activities. The third component of sustainability, social sustainability, is less well understood, measured, and managed. The emerging field of social life cycle assessment (S-LCA) has begun to fill this gap. SLCA is a social impact (both real and potential impact) assessment method that '…aims to assess the social and socio-economic aspects of products and their positive and negative impacts along their product life cycle." (Petti, 2018)

This chapter will include: 1) an introduction to and overview of S-LCA, summarized case studies of S-LCA; 2) a discussion of the relevance of S-LCA in sustainable design practice and education; 3) an examination of the role of environmental life cycle assessment (E-LCA) in building performance standards and certifications as a model for the incorporation of S-LCA; and 4) a reflection on areas for future research, including the addition of social science theory and practice for methodology, criteria, and metric development.

Guidelines for social life cycle assessment of products is the formative document of this growing field that was developed by the United Nations Environmental Program (UNEP) and Society of Environmental Toxicology and Chemistry (SETAC) Life Cycle Initiative in 2009 (Benoit and Mazijn, 2009). This document provides a framework for conducting S-LCA, including guidance on setting the goal and scope, conducting inventory, and identifying and interpreting indicators. The framework is a starting point for developing databases and software to make S-LCA a widely and easily accessible tool for decision-makers across industries. S-LCA is based on the established practice of E-LCA, which has been standardized, broadly acknowledged, and accepted by sustainability researchers and practitioners. S-LCA, on the other hand, is evolving quickly and is thus methodologically inconsistent.

Differences Between E-LCA and S-LCA

Environmental and social life cycle assessments have several characteristics in common – both are based on an ISO framework that includes goal and scope definition, life cycle inventory analysis, life cycle impact assessment, and interpretation. Both types of analysis require huge amounts of data, work in iterative processes, are strengthened by peer reviews, and when completed can be useful tools in decision-making. There are also some key differences between an E-LCA and an S-LCA. The most obvious difference is the topic of study – E-LCA focuses only on environmental impacts and therefore primarily on the physical qualities of the product, its production, and its disposal. On the other hand, S-LCA focuses on the organization-related aspects of the product chain – the actions of the humans producing and managing the product. By focusing only on E-LCA and life-cycle costing, the social component of sustainability is overlooked and unmeasured, and inherently undervalued in decision-making.

The S-LCA Process

The *Guidelines for Social Life Cycle Assessment of Products* (referred to as *Guidelines* throughout the text) identifies possible stakeholders and categories of impact specific to that stakeholder. These basic stakeholders and subcategories lay the foundation for the selection of specific study topics when developing an S-LCA. The established stakeholders and subcategories are as follows:

Table 1. Stakeholder groups and subcategories

Stakeholder	Subcategory
Worker	Freedom of association and collective bargaining
	Child labor
	Fair salary
	Working hours
	Forced labor
	Equal opportunities / discrimination
	Health and safety
	Social benefits / social security
Consumer	Health and safety
	Feedback mechanism
	Consumer privacy
	Transparency
	End of life responsibility
Local Community	Access to material resources
	Access to immaterial resources
	Delocalization and migration
	Cultural heritage
	Safe and healthy living conditions
	Respect for indigenous rights
	Community engagement
	Local employment
	Secure living conditions
Stakeholder	Public commitments to sustainability issues
	Contribution to economic development
	Prevention and mitigation of armed conflicts
	Technology development
	Corruption
Value Chain Actors	Fair competition
	Promoting social responsibility
	Supplier relationships
	Respect for intellectual property rights

Source: (Benoit, 2009)

Goal and Scope

The first, and perhaps most crucial, step in conducting a social life cycle assessment is establishing the goal of the assessment. This includes answering the following – why is the S-LCA being conducted? What is the intended use? Who will use the results? What will be assessed? An S-LCA may be done as a comparative study between products, buildings, scenarios, etc., or the S-LCA may be intended to identify and learn about social 'hotspots'[1] that provide opportunities to reduce negative impacts or promote positive impacts within the life cycle.

The scope establishes the extents of the life cycle and occasionally the geographic boundaries. The scope of the study is directly related to the process chain of the product, which can be established based on economic and/or physical relationships. Generally, this process chain includes the extraction of raw materials, processing and production, transportation, use, and disposal. The identified scope may include the entire process chain or focus on specific portions, as identified by the goal of the S-LCA.

Functional Unit

The functional unit in a social life cycle assessment provides an understandable metric on which to base analysis and comparisons. Additionally, the functional unit is not defined as a specific or static number, but a calculable quantity based in the system being tested (e.g., the water used and reused by a town in a year, the amount of building material required for $1 m^2$ of floor area, or the materials and labor to construct an entire building). Defining the functional unit based on a function, rather than a specific unit, allows for comparison of different means to the same end (e.g., greywater recycling at the home or the neighborhood scale, using concrete or steel for a building's primary structure, or design and construction practices). The definition of the function and functional unit are closely tied to the goal and scope identified for the assessment.

System Boundaries

The system boundaries determine which processes are included in the assessment. As with the functional unit and indicators, the system boundary is closely linked to the goal of the study. The authors of the *Guidelines* suggest using a system boundary borrowed from E-LCA and adapting the system boundary for S-LCA, acknowledging the potential that areas of most impact may be different between an E-LCA and an S-LCA. Questions to consider include: Where are the processes associated with this product located in the world? What or who are the enterprises involved in each of the processes? Who are the other stakeholders (workers/employees, local communities, consumers) associated with the processes? Information for these enterprises, stakeholders, and even processes may not be available, so the system boundary may need to evolve continuously as the study and assessment progress. The system boundary will also aid in the determination of what and where data needs to be collected. Some data may be site-specific, related to a specific enterprise or location, while other data may be generic and based on the larger industry or national reporting.

Life Cycle Inventory Analysis

The inventory phase of an S-LCA is the work of collecting the data to be used in the study. The data needs are determined by the goal and scope previously defined. The data collected in this phase will be used for prioritization, hotspot assessment, site-specific evaluation, and impact assessment or characterization. The *Guidelines* suggest the following operational steps:

1. **Data Collection**: Prioritization, screening, high-level overview data for the location, and hotspot assessment.
2. **Preparing for Main Collection**: Literature review, identifying indicators, and preparing surveys.
3. **Main Data Collection:** Desktop screening for studied organizations, social audits involving organizations and stakeholders, and verification and triangulation of data collected.
4. **Data Needed for Impact Assessment**: Background/baseline data for comparison to and assessment of collected data.
5. **Validation of Data**: Confirm data quality and integrity for both quantitative and qualitative measures, including the validity, relevance, measurement methods, completeness, accessibility, documentation, and uncertainty.
6. **Relating Data to Functional Unit and Unit Process**: Translation of collected data to be expressed per unit of output or functional unit.
7. **Refining System Boundary**: Sensitivity analysis to determine if a change to the study would change the result in a significant way, revised system boundaries if needed.
8. **Data Aggregation**: Collection and maintenance of data in a way that is useful for the assessment phase.

Life Cycle Impact Assessment

The life cycle assessment phase consists of classifying, aggregating, and characterizing the data collected in the inventory analysis. The *Guidelines* outline a procedure that is largely in line with ISO 14044 (2006), with some adaptations made to adapt the assessment to social impacts. The three mandatory steps are:

1. Selecting impact categories and characterization methods.
    a. Impact Categories are those that relate specifically to a theme of interest to the stakeholder (human rights, health and safety, etc.).
    b. Subcategories of impact categories provide a socially relevant characteristic to assess the performance of the impact category and produce indicators that can be measured and assessed.
    c. Categories and subcategories to be assessed may be determined with a top-down approach with broad social and socio-economic issues, and/or a bottom-up approach that provides stakeholders with a summary of inventory information and asks for input on relevant indicators and aggregation methods based on their perspective.
2. Linking inventory data to S-LCA subcategories.
    a. Data collected during inventory analysis is assigned to the stakeholder impact category and subcategories determined in step 1.
3. Determining or calculating subcategory indicator results.
    a. ISO 14044 describes this characterization phase as such: "The calculation of indicator results (characterization) involves the conversion of Life Cycle Inventory results to common units and the aggregation of the converted results within the same impact category. This conversion uses characterization factors. The outcome of the calculation is a numerical indicator result".
    b. There are a variety of ways to define, score, and weigh characterization factors, which should be defined and transparent in the assessment report.

Life Cycle Interpretation

The final phase of the Social Life Cycle Assessment is the interpretation phase. The *Guidelines* recommend the following steps, augmented from the steps outlined in ISO 14044:

1. Identification of significant issues.

   a. Documentation of important social findings and critical methodological choices, including limitations and assumptions made and key concerns identified.
2. Evaluation of the study, including consideration of completeness and consistency.
   a. Performance of a critical review with some range of qualitative and quantitative approaches, documentation of all evaluation processes, and actions taken for transparency and verifiability.
   b. Completeness evaluation: Assessment of relevant crucial issues and data collected, evaluation of indicators and metrics.
   c. Consistency evaluation: Assessment of appropriateness of modeling and methodological choices based on goal and scope.
3. Engagement with stakeholders.
   a. Report on the participation and involvement of stakeholders in the study.
4. Conclusions, recommendations, and reporting.
   a. Report of results in a fully transparent manner, including identification of all assumptions, rationales, and choices made.
   b. Recommendations made to formulate future actions.
   c. Results reported should reflect the intended audience, the stated goal and scope, and the significant issues that emerged from the study and assessment.

Relevance for Practice and Education

The general goals of social life cycle assessments fall into two general categories – decision support and hotspot identification. The results of a social life cycle assessment may reveal the restorative potential that goes beyond the direct goal of the S-LCA.

Decision Support

The process and results of an S-LCA can be used in a decision-making process based on the social impacts related to the various product life cycles. There are two primary areas in which decision-making by a specifier using S-LCA can effect change or create pressure for companies. The first is the production levels of companies: increasing production levels of one company by specifying their product, and presumably decreasing production levels of another. The second is the conduct of companies: specifying products based on the favorable conduct towards stakeholders and pressuring other companies to do the same.

These two mechanisms for change create three possible types of decision-making social life cycle assessment: (1) a consequential S-LCA, which aims to influence production levels in companies; (2) an educative S-LCA, which aims to influence both the production levels and company conduct; and (3) a lead firm S-LCA, intended to influence the conduct of the company (Jorgenson, 2012). Each of these types of S-LCA has unique goals, scopes, analyses, and assessment practices. As S-LCAs are conducted, the overall goal of supporting decision making to effect change should be held in mind, particularly when selecting stakeholders and impact categories.

Hotspot Identification

Hotspot analyses are, by nature, more generic than an in-depth social life cycle assessment. Moreover, hotspot analyses are often performed as a predecessor to or as part of a full S-LCA. Hotspot assessments provide information about where in a product life cycle, human rights, or workers' rights may be violated or harmed and reveal opportunities for improvement. The process and results of a hotspot analysis can generate high-level identification of potential issues using generic data, which can provide new insights and areas of study to decision-makers and to stakeholders.

Restorative Potential

When considering the outcomes of a social life cycle assessment, the focus tends to be on the negative impacts and qualities of the assessed topic. There is potential to expand the scope of an S-LCA to include positive impacts and restorative potential. Assessing socially positive impacts can help communities (and other stakeholders) identify objectives that maximize positive results, which may be more important than minimizing the damage (Di Cesare, 2016). Greg Norris of The International Living Future Institute (ILFI) introduced the concept of a 'Handprint' as what a product gives to the world, in contrast to a footprint of what the product has 'taken' from the world. Handprints are measured in the same units as footprints and can be created in two ways: by preventing, avoiding, or reducing the impact of footprints that would have occurred without intervention; or by creating positive benefits that do not increase the footprint and that would not have otherwise occurred. These changes are relative to the business-as-usual baseline.

Manufacturers can create Handprints in their own processes and influence positive impacts in their supply chains. The concept has been primarily applied to environmental impacts such as energy and water use throughout the product life cycle, and could be expanded to apply to social impacts. The Handprinting processes for environmental impacts is based on a life cycle assessment and hotspot analysis, which generates a set of results. These results are examined and interpreted with a goal of finding ways to reduce the footprints and increase Handprints. The practice of Handprinting is integral to ILFI's Living Product Challenge, which will be discussed in more detail later in this chapter.

The *Guidelines* provide a general framework that has been used and adapted in many ways. The next section describes five case studies of various scales that utilize the *Guidelines* framework in ways that are tailored to their area of study.

Case Study Summaries

The case studies reported in this article were selected because of their relative relationship to the architecture, engineering, and construction fields. They represent a variety of scales (system, building, material), scopes, and approaches to the general steps of an S-LCA. For each case study, the stated goal, scale, type of evaluation, scope, functional unit, stakeholders, stakeholder engagement method, indicators, data collection method, and data analysis method are identified. Because this study is focused on process rather than outcome, the results of the S-LCA case studies are omitted.

The first two case studies utilize an analytical hierarchy process (AHP) to characterize and weigh indicator data. The analytical hierarchy process is briefly described as follows: AHP: A multi-criteria decision analysis technique capable of dealing with both qualitative and quantitative data. The process is based on organizing the problem in a hierarchical structure, with the overall goal at level 1 and various criteria arranged in subsequent levels. By using a preferential scale comparison of pairs of the hierarchy elements, a ranking of the relative importance of each indicator and category can be achieved. The preferential scale comparisons are made to convert verbal judgments into numerical values on an established scale, then the resulting values are mathematically considered, and the priority level of each indicator is established.

Case Study 1

A Comparative Social Life Cycle Assessment of Urban Domestic Water Reuse Alternatives *Tamar Opher, Aviad Shapira, Eran Friedler 2017*

Goal: "The goal of this S-LCA is to compare the societal impacts of four alternative water management approaches, hypothetically implemented in an Israeli urban environment." (Opher et al., 2017)

Scale: Regional system.

Type of Evaluation: Comparative, four possible future scenarios including one business-as-usual scenario.

Scope or System Boundary of Evaluation: Hypothetical medium-sized city, systems of potable water production and supply, wastewater and greywater conveyance and treatment, reclaimed water supply, and water reuse.

Functional Unit: The supply, reclamation, and reuse of water consumed by the hypothetical city during one year.

Stakeholders: Society, local community, and consumers.

Stakeholder Engagement Method: Twenty Expert interviews, in-person or via phone/internet. Experts served as a proxy for named stakeholders, and areas of expertise included: environmental engineering, environmental economy, social science, and water sector regulation.

Indicators: Indicators were established based on the study of water management systems and the geographic region, and are outlined in Table 2.

Data Collection Method: Expert interviews consisting of a fixed set of questions derived from the AHP process. Six sets were compared, each with a two- or three-step question: (1) Which of the criteria is more important in regard to (the parent node in the hierarchy)? (2) To what extent? This resulted in 14 total questions, and interviews took approximately one hour.

Data Analysis Method: Analytical hierarchy process.

Discussion: This case study focuses on highly specific and unique indicators that are relevant to the specific geographic area and the system of study (water management). While these indicators are based on local data, they could be adapted to study water management systems in other geographic regions, and the general process could be adapted to study other systems. The stakeholder engagement method of expert interviews as a proxy for the named stakeholders (public, community, and consumers) may lack the nuance that could have been achieved with a combination of top-down and bottom-up approaches to data collection. This stakeholder approach could be improved by engaging stakeholders directly after the initial data inventory and analysis to verify or modify the impact categories and subcategories analyzed. The analytical hierarchy process used to characterize the collected data is an effective method for handling the quantitative and qualitative indicators studied, particularly in the case of a comparative analysis of future situations.

Case Study 2

Social Life Cycle Assessment for Material Selection: A Case Study of Building Materials *Seyed Abbas Hosseinijou, Saeed Mansour, Mohsen Akbarpour Shirazi 2013*

Goal: "The goal of this case study is to perform a comparative assessment of the social and socio-economic impacts in the life cycle of concrete and steel as building materials in Iran in order to identify the best socially sustainable options." (Hosseinijou et al., 2013)

Scale: Building materials

Table 2. Urban domestic water reuse stakeholders, categories, sub-categories and indicators

Stakeholder (Level 2)	Category (Level 3)	Sub-Category (level 4)	Weight (%)	Indicator	Type of Indicator
Public	Water-saving		29.6	Water saved	N
	Equity / fairness		9.3	Water supply equivalence	Q
Community	Community engagement		12.0	Reclamation system scale	Q
	Local employment		3.4	Maintenance work hours	N
	Urban landscape		12.6	Reclaimed water availability	Q
Consumer	Health concerns	Level of contact with reclaimed water	6.5	Types of reuse	Q
		Source of reclaimed water	4.6	Type and origin of reclaimed water	Q
		Trust in supplier and technology	4.9	Type of operating company	Q
	Finance		9.2	Household water expenses	N
	Convenience	Supply reliability	5.2	Pipeline length	N
		Consumption habits	2.7	Required precautions	Q

(N = quantitative indicator, Q = qualitative indicator)

Type of Evaluation: Comparative, two building materials: concrete and steel.

Scope of Evaluation: Geographic scope includes the entire Islamic Republic of Iran, with site-specific analysis in the northern regions of Iran in the provinces of Golestan, Mazanderan, and Gilan. The material analysis scope is cradle-to-grave.

Functional Unit: Amount of material (concrete or steel) needed for 1m^2 of floor area. This floor area is located in a hypothetical building requiring 0.8t concrete and 0.1t steel per 1m^2 floor area, with an assumed life span of 50 years.

Stakeholders: Workers/employees, local community, society (national), end consumer, and value chain actors.

Stakeholder Engagement Method: Expert and stakeholder interviews. Experts included participants from Provincial Offices of Industry, Mine, and Trade, Offices of Economic and Finance Affairs, Regional Offices or Departments of the Environment, Regional Offices of Natural Resources, Regional Waste

Management Organizations, Golestan University, University of Guilan, a local cement production company, a local steel production company, and a trade association of construction workers in the region.

Indicators: Indicators were identified by hotspot analysis and expert interviews, and they are listed in Table 3.

Data Collection Method(s): Direct observation, company documents, interviews with managers, staff, and workers.

Data Analysis Method: Analytical hierarchy process

Discussion: This case study demonstrates an iterative process for determining the indicators to study. The indicators listed in Table 3 were deemed to be most relevant after a hotspot analysis of 28 indicators during eight life cycle categories. Of those, twenty indicators across four life cycle categories were selected for further study. The initial list of indicators was generated by a top-down approach and based on the identified subcategories in the *Guidelines.* Hotspot analysis was conducted based on market data provided by the individual companies participating in the study. Next, the final indicators were selected based on hotspot analysis and stakeholder and expert interviews. This approach incorporates both top-down and bottom-up approaches and may result in a more robust assessment and report. Similar to the previous study, an analytical hierarchy process was utilized to characterize and compare the indicators, including quantitative and qualitative results.

Case Study 3

Development of Social Sustainability Assessment Method and a Comparative Case Study on Assessing Recycled Construction Materials *Md. Uzzal Hossain, Chi Sun Poon, Ya Hong Dong, Irene M. C. Lo, Jack C. P. Cheng 2017*

Goal: "The aim of this case study is to assess the social sustainability of commonly used construction materials (such as aggregates)."(Hossain et al., 2017)

Scale: Building material component.

Type of Evaluation: Comparative, between recycled and natural aggregates including crushed stone, river sand, construction and demolition waste, and waste glass.

Scope of Evaluation: Cradle-to-grave.

Functional Unit: Undisclosed.

Stakeholders: Material producers, workers/employees, traders, the general public, government, and society.

Stakeholder Engagement Method: Online survey via Google Forms. Respondents were asked to identify and select subcategories for stakeholder groups. Those surveyed include academics, producers, recyclers, users, government officials,

Table 3. Building Material Production Stakeholders, Subcategories, Indicators, and Product Life Stages.

Stakeholder Category	Subcategory	Inventory Indicators	Indicator Type	Raw Material Acquisition	Production of Construction Material	Construction	Disposal / Recycling
Workers	Freedom of association and collective bargaining	Respect to freedom of association and collective bargaining	Q	7	7	7	3
	Health and safety	Occupational accidents	N	7	7	7	3
	Fair salary	Living / non-poverty wages	N	7	7	7	3
Local Community	Access to material resources	Changes in land ownership / land use	Q	5	7	n/a	5
		Levels of industrial water use	N	3	7	7	3
		Extraction of Material Resources	Q	7	n/a	n/a	n/a
	Safe and healthy living conditions	Burden of disease	Q	1	5	n/a	n/a
		Pollution levels	N	7	9	5	5
		Waste generation	N	5	9	7	5
	Local Employment	Job creation	N	1	1	1	1
		Use of local labor	N	1	1	1	1
		Use of technology that generates employment	Q	1	1	1	1
		Presence of local supply network	Q	1	1	1	1
	Local community acceptance	Presence of complaints by local community	Q	5	7	n/a	5
	Cultural Heritage	Protection of cultural heritage	Q	5	7	n/a	n/a
Society	Technology development	Technology development	Q	1	1	1	1
		Technology transfer	Q	1	1	1	1
		Research and development	Q	1	1	1	1
	Contribution to economic development	Contribution of product to economic process	N	1	1	n/a	n/a
	Suppliers development	Use and support of national suppliers	Q	1	1	n/a	1

(9 = very negative effect, 7 = negative effect, 5 = lightly negative effect, 3 = indifferent effect, 1 = positive effect.)

(N = quantitative indicator, Q = qualitative indicator)

Source: (Hosseinijou, Mansour, & Shirazi, 2013)

and members of the general public. Detailed interviews were conducted to collect data for various indicators during the operational research phase. These interviews were with human resource departments, employees/workers, industry engineers, relevant researchers, community members, suppliers, and users of the materials.

Indicators: Indicators are based on guidelines from UNEP/SETAC, Global Reporting Initiative, Hong Kong Business Environmental Council Limited, and they are outlined in Table 4.

Data Collection Method: Qualitative research based on expert interviews, to prioritize subcategories and inventory indicators. Operational research based on field survey to collect case-specific data.

Data Analysis Method: 'Social Sustainability Grading Model' (Hossain et al., 2017) consisting of six stakeholder categories, 30 impact sub-categories, six endpoint categories, and a resulting sustainability index (single score). Sub-categories are weighted based on hotspot identification by the experts interviewed. This weighting creates a prioritization in the scoring system based on the level of importance. The final score or index number will fall between 0.00 and 1.00, a range between 'highly unsustainable' and 'highly sustainable.'

Discussion: Of the case studies presented, this study utilizes the greatest number of subcategories (28) and indicators (109), while focusing on the smallest scaled topic (building material component). The process of reducing these indicators and subcategories to a single score between zero and one may be an oversimplification that harms the potential of this type of study as a comparative assessment. It may be more useful for decision-makers to see scores for various impacts or stakeholders, as decisions about building materials and construction are rarely made based on a single factor. The stakeholder engagement process is robust; however, the hotspot determination by experts alone may slant the results in ways that do not mirror the reality of the industry. Including stakeholders directly in the hotspot identification for weighing of impacts may result in a more comprehensive and rigorous study. Furthermore, the format of this study could be applicable for the study of other construction materials and components of a similar scale, provided the study is tailored to the specific industry and geographic region being assessed.

Case Study 4

A Social Life Cycle Assessment Model for Building Construction in Hong Kong
Ya Hong Dong, S. Thomas Ng 2015

Goal: "The goal of this case study is to test Social-Impact Model of Construction and examine the social impacts of a standard public rental housing project (Dong & Ng, 2015)

Scale: Single building

Type of Evaluation: Single score

Scope of Evaluation: Cradle-to-end of construction activities

Table 4. Recycled construction material use stakeholders, subcategories, and indicators

Stakeholder	Sub Category	Indicator	Type
Producer	Health and Safety	Level of the potential safety concern with product transportation and handling	Q
		Is the product harmful to health through its life cycle	Q
		Safety indicators on product	Q
		Consumer complaints about health and safety	Q
		Total annual number of user complaints regarding health and safety	N
		Labels or requirements for health and safety	Q
		Potential emissions or leakage/discharge during life cycle	N
		Potential to prevent emissions or leakage/discharge during life cycle	N
	Use Stage	Any relevant policy for use stage responsibility for materials producer	N
		Any future risk of accident/health complications for final user	N
	Product and Service Labeling	Product component parts or ingredients suitably labeled	N
		Compliance with international accounting practices and regulatory requirements	N
		Transparency of certification standards, labels, or special indices	N
		Sustainability report regarding social and environmental life cycle impact assessment	N
		Certification or company sustainability rating transparency	N
	End of Life	Management effort to address end-of-life options	N
		Level of attention to end-of-life impacts	Q
		Buy-back or recycling programs	N
	User Satisfaction	Materials / products maintain international / national quality standards	N
		Level of standard achieved	N
		Practice related to user satisfaction (e.g., survey)	N
		Level of user satisfaction	N
Worker / Employee	Fair Salary and Employee Characteristics	Demographic information of the workforce	N
		Average hourly salary	N
	Hours of Work	Normal weekly working hours	N
		Overtime availability	N
		Overtime pay	N
		Mutually (company and employee) agreed upon overtime availability and pay	N
	Health and Safety	Level of safe and healthy workplace	Q
		Number/percentage of injuries or fatal accidents in the past year	N
		Regard for health and safety or employees accommodations and food	Q
		Accident insurance/medical insurance and reimbursement	N
		Health and safety committee in factory/industry	N
		Recorded cases of health and safety issues within last year	N
	Social Security and Social Benefits	Profit-sharing for employees	N
		Bonus structure for employees	N
		Social benefits – retirement, disability, dependents coverage, paid leave	N
		Difference of social benefits between part-time and full-time employees	Q
	Forced Labor	Existence of forced (paid or unpaid) labor	N
		Abuse of employees – mental or physical	N
	Training and Education	Training and education for employee skill improvement	N
		Average hours/year of training by employee and by employee category	N
		Percentage of employees receiving regular performance and career development reviews	N
Traders of Material	Fair Competition	Presence of unfair business practices	N
		Compliance with legislation regarding unfair business practice	N

continued on following page

Table 4 continued

Stakeholder	Sub Category	Indicator	Type
	Intellectual Property Rights	Intellectual property right to the materials or product	N
		Local IP utilization	N
	Supplier Relationships	Relationship with suppliers	Q
		Supplier relationships comply with agreements and regulations regarding exchanges and trade among organizations	Q
		Absence of coercive communication	N
		Suppliers are paid on time	N
	Promoting Social Responsibility	Monitoring, auditing, training events to promote social responsibility internally and to suppliers	Q
		Membership in organizations that support social responsibility throughout the supply chain	N
		Company's social responsibility efforts	Q
General Public	Community Engagement	Direct involvement with or financial support of community projects	Q
		Meetings with community stakeholders regarding products, social awareness, sharing knowledge, etc.	Q
		Level of community engagement	Q
		Community engagement in recycling and economic benefits	Q
	Local Employment	Local community development by training local employees in technical and transferable skills	N
		Policy for local employment	N
		New job creation	N
	Safe and Healthy Living Conditions	Product safety and health impacts for the community (use phase)	N
		Generation of pollution or other hazards to health in the local community	N
		Operational accidents and structural failures with potential to impact the local community	Q
		Management to minimize health and safety issues	Q
		Reducing health hazards by reducing waste	Q
		Reducing nuisance conditions	Q
		Increase nearby property values	N
	Public Opinion of Materials/ Products	Reports of complaints regarding material/product	N
		Public awareness of materials/products	Q
		Level of public satisfaction	N
		Level of public acceptance	N
Government and Society	Commitment to Sustainability	Existence of company commitment to sustainability	N
		Collaboration within sector regarding sustainability	Q
		Legal obligation for sustainability reporting	N
	Contribution to Economic Development	Health of company	Q
		Stability of companies size and operations	Q
		Contribution to economic development of country	N
		Revenue earnings by sector	N
		Imported materials	Q
		Taxation on employee and manufacturer	N
		Reduction of waste volume and associated costs	N
		Reduction of disposed land and associated costs	N
	Technology Development	Investment in technology development or transfer, research for efficient technologies	Q
		Level of technological innovation	Q

continued on following page

Table 4 continued

Stakeholder	Sub Category	Indicator	Type
	Support from the Government	Types of support available/received from government	Q
		Scale of support for industry	Q
		Future government support of sustainability goals	Q
Relevant Stakeholders (Socio-Environmental Performance)	Materials (natural and recycled)	Raw materials – recycled, renewable, virgin	N
		Raw associated process materials	N
		Semi-manufactured goods	N
		Packaging materials	N
	Energy Use and Water Consumption	Energy consumption within organization	N
		Source of electricity	N
		Energy consumption outside of organization	N
		Reduction of energy consumption	N
		Total water withdrawal with source	N
		Percent of water recycled and reused	N
	Emissions	Industry emissions	N
		Emissions from outside industry (i.e., transport)	N
		Emissions reduction target	N
	Solid Wastes and Effluents	Effluent discharge and treatment	N
		Solid waste management	N
		Hazardous waste management and significant spills	Q
		Reports of complaints by local residents	N
		Potential effects on local community from solid waste and effluents	Q
	Biodiversity	Surrounding area of industry	Q
		Significant impacts of activities, products, and services impacts on biodiversity	Q
		Habitat protection or restoration	N
		Conserve natural resources	Q

(N = quantitative indicator, Q = qualitative indicator)

Functional Unit: Building - Public rental housing project with a standard layout and precast elements.

Stakeholders: Workers, local community, and society.

Stakeholder Engagement Method: Questionnaire surveys were sent via email and mail. The questionnaire included a section to rate the importance of subcategories, which was used to weight categories in the analysis process. The questionnaire also included a section studying the social impacts of nine environmentally friendly practices, asking respondents to indicate the subcategories that would be positively or negatively impacted by the environmentally-friendly practice. Respondents include members of the following sectors: government, private developers, consultants, contractors, supplier/manufacturers, carbon auditors, academia, and others.

Table 5. Building construction stakeholders, subcategories, indicators, and normalized value ranges

Stakeholder	Subcategory	Indicator	Normalized Value Range	Ranked Importance
Worker	Freedom of association, collective bargaining	FACB rights violations	-1~1	12
	Child labor	Percentage of child labor	-1~1	13
	Fair salary	Comply with minimum regulation	1 if compliant, -1 if non-compliant	3
	Working hours	Hours worked weekly	1 if <60h, -1 if >60h	5
	Forced labor	Percentage of forced labor	-1~1	10
	Equal opportunities / discrimination	Social institutions and gender index	-1~1	7
	Health and safety	Fatality rate	-1~1	1
Local Community	Access to material resources	Improved sanitation facilities - % of population with access	-1~1	9
	Cultural heritage	Status of cultural resource	1 if protected, 0 if no change, -1 if damage	11
	Safe/healthy living conditions	Reliability of police services	-1~1	2
	Community engagement	Index of transparency of policymaking	-1~1	8
	Local employment	Unemployment rate	-1~1	6
Society	Public commitments to sustainability issues	Obligation on public sustainability reporting	-1,0,1	4

Source: Dong & Ng, 2015

Indicators: Indicators were determined based on UNEP / SETAC *Guidelines*, and they are presented in Table 5.

Data Collection Method: Questionnaire survey, interviews, database research, and national statistics.

Data Analysis Method: In line with ISO 14040 (Life Cycle Assessment Principles and Framework) structure of characterization, normalization, and weighing. Characterization refers to the process of converting subcategories into interpretable indicators, normalization is the rescaling of indicators to a comparable range, and weighing is revising the normalization based on the importance of each subcategory and indicator. By calculating each weighted score and stakeholder, a single score can be determined for the social life cycle assessment.

Discussion: Like the previous case study, this example also produces a single score to indicate the social-impact model of construction. This score ranges from -5 to 5 and is determined by on-site construction practices and by materials used in the building project. The score is determined based on the normalized indicator value and the relative weight determined by stakeholders. Unlike the previous case study, just thirteen subcategories with one indicator each were studied. The scope of evaluation for this case study is also smaller than other examples in that it excludes the use and disposal phases of this building. This omission is concerning given the potential social implications of residents inhabiting the building for a number of years, and the eventual removal and disposal of the structure. Additional potential stakeholders for this study would be future residents of the building, and members of the demolition and disposal industries.

Case Study 5

Development of a Social Impact Assessment Methodology for Recycling Systems in Low-Income Countries Sandra Aparcana, Stefan Salhofer 2013

Goal: "The goal of this study is the assessment of recycling systems based on formalization in terms of social impacts, in comparison to informal recycling systems in low-income countries." (Aparcana & Salhofer, 2013)

Scale: System, recycling activities of recyclable waste collection, and manual preprocessing.

Type of Evaluation: Comparative, between potential formalized recycling implementation and the informal recycling business-as-usual.

Scope of Evaluation: Recycling system in Peru.

Functional Unit: The amount of household recyclable waste collected by one house for one year – 60kg/inhabitant/year.

Stakeholders: Informal recyclers, municipalities, recyclers, NGOs.

Stakeholder Engagement Method: Questionnaire – 56 questions, combination of closed and open-ended.

Indicators: Indicators were selected from a literature review of documented social problems of informal recyclers in low-income countries, and they are presented in Table 6.

Data Collection Method: Questionnaire – 56 closed and open-ended questions. The questions aim to obtain precise answers able to be assigned a 0 or 1 to indicate compliance or non-compliance with the social criterion.

Data Analysis Method: Stakeholder scores from the questionnaire (0 or 1) are averaged to a decimal between 0 and 1. An average score of 0.5 or less indicates the social criterion is not met, while an average score of 0.5 or higher indicates the social criterion is met. These results are then rounded to 0 or 1 to indicate the status of the criterion. The score for each subcategory is calculated such

Table 6. Impact categories, subcategories and indicators

Impact Category	Impact Subcategory	Indicator	Type
Human Rights	Child labor	No child labor	Q
	Discrimination	Formal policy against discrimination	Q
		No income differences between women and men	Q
	Freedom of association and collective bargaining	Presence of collective bargaining	Q
Working Conditions	Working hours	Fulfillment of overtime agreed in working contracts	Q
	Minimum income, fair income	Average income according to legal framework	Q
		Absence of non-agreed income deductions	Q
		Regular payment for workers	Q
		Minimum payment according to legal framework	Q
	Recognized employment relationships and fulfillment of legal social benefits	Existence of legal working contracts for all workers	Q
		Access to legal social benefits	Q
		Access to further social support programs for workers	Q
	Physical working conditions	Absence of work accidents	Q
		Formal policy about occupational health and safety	Q
		Vaccination for workers	Q
		Training programs for workers on occupational health and safety	Q
		Access to preventative health care program for workers	Q
		Presence of medical equipment in workplace for worker use	Q
		Absence of disease related to waste handling	Q
		Appropriate working equipment	Q
	Psychological Working Conditions	Willingness to continue working in the same company/sector	Q
		Work satisfaction	Q
		Willingness to be trained for work activities	Q
Socio-Economic Repercussions	Education	Educational level of children from families of recyclers	Q
		No school absence of children from families of recyclers	Q
		Existence of educational programs for self-development	Q

(N = quantitative, Q = qualitative)
Source: Aparcana, & Salhofer, 2013

that if all indicators receive a score of 1, the subcategory retains the score of 1. If one or more indicators in a subcategory receive a score of 0, the subcategory receives a score of 0 to indicate that the social criteria of that subcategory have not been met. This results in an indication of which subcategories, and therefore which aspects of the studied problem, are favorable and which are not favorable.

Discussion: Researchers selected the subcategories and indicators selected in this study in a unique way – by a review of literature documenting specific social problems of informal recyclers in low-income countries. This results in semi-quantitative questions with answers that can be translated to a number for characterization and assessment. The questionnaire developed for stakeholders was made of 56 closed- and open-ended questions designed to collect relevant information. The questionnaire was further specialized in that the informal recyclers were only ones asked about the physiological working conditions, as they are the only stakeholder with firsthand experience in those categories. This study placed heavy emphasis on the workers involved in the informal recycling system and the potential formal systems. The structure of the study results in scores for several subcategories in several scenarios, presenting a nuanced view of the potential outcomes of formalized recycling.

These case studies illustrate a variety of approaches to the practice of social life cycle assessment, all grounded in the basic process outlined in the *Guidelines*. Some stakeholders, impact categories, subcategories, and indicators are consistently applied to these studies, while others are highly specific to the industry and geographic region of interest. All of the presented studies offer a methodology that can be tested and re-tested as the field of social life cycle assessment gains use and recognition. In the field of environmental life cycle assessment, the adoption, production, and use of LCA has been supported by the inclusion of LCA and associated study in building performance standards and certifications. The next section examines how E-LCA has been applied in Leadership in Energy and Environmental Design (LEED), Minnesota's Sustainable Building Guidelines, the UK's Building Research Establishment Environmental Assessment Method (BREEAM), and the Living Product Challenge.

POTENTIAL TO INCORPORATE S-LCA INTO SUSTAINABLE BUILDING PERFORMANCE STANDARDS AND CERTIFICATIONS

E-LCA in Standards and Certifications

One driver of environmental life cycle assessment uptake and popularity is the use of E-LCA in building performance certifications and standards, both voluntary and mandatory. A brief review of the E-LCA requirements in three standards: LEED v.4, Minnesota Sustainable Building Guidelines, and BREEAM UK, follows. These standards were selected based on their widespread use and reputations as leaders in sustainable building program development and implementation.

LEED v.4 Materials and Resources (MR) – Building Life Cycle Impact Reduction

Intent: "To encourage adaptive reuse and optimize the environmental performance of products and materials." (LEED v.4)

Requirements

Option 4: Whole Building Life Cycle Assessment

Project team should conduct a life cycle assessment of the project's structure and enclosure that demonstrates a minimum 10% reduction, compared with a baseline building, in at least three of the six following categories, one of which must be global warming potential:

- Global warming potential (greenhouse gases) in $C0_2e$
- Depletion of the stratospheric ozone layer, in kg $CFC\text{-}_{11}$
- Acidification of land and water sources, in moles H+ or kg SO_2
- Eutrophication, in kg nitrogen or kg phosphate
- Formation of tropospheric ozone, in kg NO_x, kg O_3 eq, or kg ethane
- Depletion of nonrenewable energy resources, in MJ.

MR – Building Product Disclosure and Optimization – Environmental Product Declarations

Intent: "To encourage the use of products and materials for which life cycle information is available and that have environmentally, economically, and socially preferable life cycle impacts. To reward project teams for selecting products from manufacturers who have verified improved environmental life cycle impacts." (LEED v.4)

Requirements

- **Option 1:** Project team should use at least 20 different permanently installed products from at least five manufacturers that meet one of these disclosure criteria:
    - Life cycle assessment that is product specific and conforming to ISO 14044.
    - Environmental product declaration conforming to ISO 14025, 14044, and EN 1504 or ISO 21930. EPD may be industry-wide or product-specific.
    - Other US Green Building Council (USGBC) approved program.
- **Option 2:** Project team should use products that comply with criteria below for 50%, by cost, of the total value of permanently installed products.
    - Third-party certified products that demonstrate an impact reduction below industry standard in at least three of six climate change impact categories.
    - Product certified in USGBC program.

LEED utilizes life cycle assessment at the building scale, asking design teams to assess the structure and enclosure and demonstrate a reduction in three of six categories. A similar approach could be taken to a social life cycle assessment, with established social impact categories taking the place of environmental impact categories. At the building material scale, disclosure of environmental impacts is required. There is an option to specify products that disclose environmental impacts or to specify and install products that perform better than the industry average in environmental impact categories. Similarly, this requirement could be adapted to social life cycle assessment by either requiring disclosure of social impacts or specifying and installing products that demonstrate a reduction in negative social impacts.

Minnesota Sustainable Building Guidelines

Materials and Waste-M.1 – Life Cycle Assessment

Intent: "To use life cycle analysis to quantify and minimize the environmental impact of building materials which have significant effects on global warming, air pollution, water pollution, energy consumption, and waste." (Minnesota Sustainable Building Guidelines)

REQUIREMENTS

Whole Building Life Cycle Assessment

The project team should demonstrate a reduction in life cycle global warming potential of the building's construction materials through design and construction decisions. Next, the project team should document this reduction with a whole building LCA model. The reduction approach may be taken at the scale of the building, an assembly, or a material.

Product Life Cycle Assessments

- Project teams should specify at least five different permanently installed products sourced from at least five different manufacturers that meet one of the disclosure criteria below:
    - Product-specific life cycle assessment conforming to ISO 14044
    - Environmental product declaration (EPD) conforming to ISO 14025, 14040, 14044, and EN 15804 or ISO 21930. EPDs may be industry-wide or product-specific.

Minnesota's Sustainable Building Guidelines require an E-LCA at the building scale and allows reductions to take place at the scale of the building, the assembly, or the material. There is also a requirement for installing products with life cycle assessments or environmental product declarations. These products are not required to meet any specific level of performance; they are simply required to disclose environmental impacts. This requirement could be adapted to social life cycle assessment by requiring projects to reduce social impacts at the building scale and to install products that have social life cycle assessments completed. Similarly, no specific level of performance for individual product disclosures would be required, simply a reporting practice with social and socio-economic impacts.

BREEAM UK Mat 01 – Environmental Impacts from Construction Products – Building Life Cycle Assessment

Intent: "To reduce the burden on the environment from construction products by recognizing measures to optimize construction product efficiency and the selection of products with low environmental impact over the life cycle of the building." (BREEAM UK)

Requirements

- The project team should complete a building LCA during concept design and compare it to the BREEAM LCA benchmark.
- During concept design, the project team should identify opportunities for reducing environmental impacts:
  - Analysis of two to four significantly different design options
- The project team should complete a building LCA during technical design and compare it to BREEAM LCA benchmark.
- During technical design, project team should identify opportunities for reducing environmental impacts:
  - Analysis of two to three significantly different design options

Mat 02 – Environmental Impacts from Construction Products: Environmental Product Declarations

Intent: "To encourage availability of robust and comparable data on the impacts of construction products through the provision of EPD."(BREEAM UK)

Requirements

- The project team should specify construction products with Environmental Product Declarations that meet a set of criteria for validity and timeliness.

BREEAM requires iterative analysis of environmental life cycle assessments, including in concept design and technical (schematic) design. This process allows for the fine-tuning of the building itself and the materials specified as the design progresses. Additionally, the requirement of at least two design options and the baseline model ensures a rigorous exploration that creates real reductions. An S-LCA requirement could be implemented in a similar way, with an initial assessment of the social impact of the building in the concept design stage and a more refined analysis during schematic design. This approach would require the development of a baseline for the specific region, possibly in the form of a hotspot analysis.

LEED, BREEAM, and Minnesota B3 are just a few of the building standards and certifications that utilize environmental life cycle assessment. Green Globes, German Sustainable Building Certification, Haute Qualité Environnmentale (France), Milijöbyggnad (Sweden), Minergie (Switzerland), and Qatar Sustainability Assessment System all utilize E-LCA data in some form. According to one survey, 50% of green building professionals who responded claimed to integrate life cycle assessment into their studies as a result of BREEAM requirements. Moreover, 80% of respondents viewed the integration of LCA into their practice as a positive (Bruce-Hyrkäs, 2018). As with many building performance assessments and tools, the E-LCA is integrated into sustainable building standards and certifications as a decision-making tool intended to deliver an efficient, cost-effective, and environmentally conscious building, yet the social aspects that complete the goal of sustainability are underemphasized.

Some building performance certifications and standards have begun to explore the incorporation of social life cycle assessment. Namely, LEED has introduced a pilot credit (a credit outside the standard requirements of the program that fulfills a requirement for 'innovation') that encourages integrated analysis, and the International Living Future Institute's Living Product Challenge, which combines environmental and social performance requirements. These two approaches are described in more detail below:

LEED v.4 Pilot Credit – Integrative Analysis of Building Materials

Intent: "To encourage the use of products and materials for which life cycle information is available and that have environmentally, economically, and socially preferable life cycle impacts. To inform decision making by project teams by rewarding building material manufacturers that share life cycle health, safety, and environmental information about their products." (LEED v.4 Pilot Credit)

Requirements

Project teams should specify at least three different, permanently installed products that have a documented qualitative analysis of the potential health, safety, and environmental impacts of the product in five stages of the product's life cycle (product assembly/manufacturing, building product installation, product use, product maintenance, and end of product life/reuse). The team should analyze and consider the following:

- Intended and reasonably anticipated uses of the product.
- Potential hazardous exposures.
- Product service life.
- Waste generation and reuse potential.
- Contributions to health, safety, and the environment, including improvements to occupant safety, air and water quality, material reuse, energy efficiency, and carbon mitigation..

Catalog impacts in applicable areas from the following:

Human health impacts:

- Carcinogenicity.
- Mutagenicity/genotoxicity.
- Reproductive and developmental toxicity.
- Acute toxicity.
- Eye and skin irritation.
- Aspiration hazard.
- Chronic toxicity – skin and respiratory effects.
- Systemic toxicity – organ effects.
- Air purification/filtration or positive impacts to indoor air quality.

Occupant safety impacts:

- Passive survivability.
- Functionality, including access/egress.

Environmental Impacts:

- Air pollution abatement.
- Bioaccumulation.
- Persistence.
- Acute and chronic aquatic toxicity.
- Water use.
- Energy use.
- Greenhouse gas emissions
- Solid waste generation.
- Biodiversity, habitat, and ecosystems.

This pilot credit is an attempt to select building materials that have holistic sustainability and consider the social impacts of products alongside environmental and economic qualities. There is an increased focus on human health throughout the product life cycles, and there are some specific considerations for the use phase. Moreover, there is a lack of focus on human rights and working conditions throughout the product life cycles, in contrast to the focus of the S-LCA case studies examined in the previous section. A potential expansion of these criteria would be to add a category of consideration for human rights, with subcategories based on those identified in the *Guidelines* and tested in case studies.

Living Product Challenge – 2.0

01 – Responsible Place

- Products are made in factories responsibly located with respect to ecology. Onsite landscaping promotes healthy ecosystems and habitat without the use of petrochemical fertilizers or pesticides.

03 – Living Economy Sourcing

- Products are made from materials sourced locally to support the local economy and reduce the environmental and social impacts of transportation.

04 – Water Footprint

- On-Site: Manufacturers must determine how much water is used on-site and identify opportunities to decrease use or improve use and disposal.
- Life cycle: Manufacturers must conduct a cradle-to-gate Life Cycle Assessment to assess and document the five processes that make the largest contribution to water use, and demonstrate that their water footprint is lower than the industry average.

05 – Net Positive Water

- One hundred percent of water needs at the final manufacturing facility met by captured rainwater or other closed-loop water system, including recycling industrial water.
- The project team should create a water Handprint that is greater than the water footprint using one or more of the following strategies:
  - Conserving or recapturing water across the life cycle of the product that exceeds the base case.
  - Engaging customers and other product users to achieve water conservation/restoration.
  - Working outside the supply chain to reduce potable water consumption or harvest potable water.

06 – Energy Footprint

- On-site: Manufacturers must describe sources and uses of energy at the production facility, and identify opportunities for decreasing demand and increasing on-site production of renewable energy.
- Life cycle: Manufacturers must conduct a cradle-to-gate Life Cycle Assessment to assess and document the five processes that make the largest contribution to fossil energy use, and demonstrate that their fossil energy footprint is lower than the industry average. The fossil energy footprint reflects the use of fossil fuel energy across the supply chain.

07 – Net Positive Energy

- One hundred and five percent of the energy used to produce the product at the final manufacturing site generated on-site by renewable energy.
- The project team should create a Handprint that exceeds the footprint through on-site or supply chain innovations to reduce combustion-based energy use through one or more of the following strategies:
    - Conserving or generating renewable energy across the life cycle of the product.
    - Engaging with users to achieve on-going conservation through improved use of the product.
    - Working outside the supply chain to reduce combustion-based energy consumption or generate renewable energy.

10 – Human Thriving

- Facilities should contribute to an active and healthy lifestyle for employees, and foster the innate human-nature connection by completing the following:
    - Providing sufficient and frequent human-nature interactions and encourage a healthy and active lifestyle.
    - Demonstrating no reported deaths or serious injuries related to the final manufacturing of the product in the last year.
    - Demonstrating that there are programs in place to support the health and wellbeing of employees.
    - Providing a mechanism for employee feedback regarding facility conditions.

15 – Ethical Supply Chain

- Manufacturers should commit to responsible processes throughout supply chains and business operations. To demonstrate that the promotion of human rights exists across the supply chains, manufacturers must:
    - Performing human rights due diligence for top ten priority suppliers (by spending).
    - Identifying the most critical social risks associated with each supplier and the leading standards/certifications that address those risks.
    - Giving preference to suppliers that obtain relevant certification or conduct an audit and address identified social risks.

The Living Product Challenge is a stand-alone certification for building projects, intended to integrate into building scale programs. This standard includes consideration of environmental and social impacts across the product life cycle, including a focus on health, safety, and human rights throughout the supply chain.

Some imperatives include a Handprinting initiative, as previously discussed. These Handprint actions are intended to outweigh the footprint and create a net-positive effect on the category. This type of certification would be supported by the continued growth of the S-LCA field, and it may provide a template for developing a building-product-specific assessment system for broad use.

Incorporating S-LCA into Building Standards

The previous section suggests some possibilities for the integration of social life cycle assessment into building performance certifications and standards and points to examples of attempts. While this area of study has enormous potential to develop in terms of methodology, rigor, and databases, there are some challenges for making S-LCA a common practice. The primary issues identified with the practice of S-LCA – the real-world applications like the case studies presented previously – are disparate impact assessment methods and scattered product classifications that make benchmarking a challenge (Subramanian, 2018). One approach to a more unified impact assessment method would be agreement upon analytical research tools (analytical hierarchy process, multi-criteria decision analysis, etc.) that complement the S-LCA impact assessment process and deliver more robust results and decision-making support.

Additionally, including stakeholders' input in the characterization and weighing process (as was done in some of the previous case studies) could lead to specific and relevant results that a generic or equally weighted impact assessment may not achieve. Complex products, those with many sub-products and those that change frequently, pose another challenge to S-LCA. Company information, intellectual property, and privacy are also all concerns that make data acquisition difficult. There is also a lack of benchmarking within the S-LCA field, as case studies have had scattered topics, methods, and results. Despite all this, and while it would be difficult to make S-LCA a mandatory requirement for industries, it can be hoped that in the coming years, S-LCA will follow the path of E-LCA. This path would allow S-LCA to develop into a more mature decision-making tool, with more case studies conducted, and the capacity for synthesis of results and data for sectors, industries, even individual actors (Subramanian, 2018).

FUTURE RESEARCH

The field of S-LCA is growing at a rapid rate and in a variety of directions. This review has focused primarily on the architecture, engineering, and construction fields, and it has examined literature from practitioners of LCA. As the field and practice of S-LCA expand, it will become important to study the efforts of other fields of product manufacturing, including automotive, food, and electronics. Equally important is connecting with other disciplines and experts in the assessments and reviews. For example, one obvious field to engage is that of social science, particularly for consideration of social and socio-economic impacts, and different ways and means for determining and quantifying those impacts.

Incorporation of Social Science Research

One potential lens to examine social life cycle assessment in combination with environmental life cycle assessment is the doughnut model, a visual framework for sustainable development, developed by Kate Raworth in 2017. The doughnut describes the "safe and just space for humanity" between the ecological ceiling of environmental impacts and the social foundation of human rights. Environmental impacts include ozone layer depletion, climate change, ocean acidification, chemical pollution, nitrogen and phosphorus loading, freshwater withdrawals, land conversion, biodiversity loss, and air pollution. Additionally, the social foundation of human rights includes energy, water, food, health, education, income and work, peace and justice, political voice, social equity, gender equality, housing, and networks.

The parallels between the ecological ceiling and E-LCA are clear, as are the similarities between the social foundation and the areas of study for social life cycle assessment. This framework is based on the idea that sustainable economic development takes place within the environmentally safe and socially just space for humanity and strikes a balance between the needs of the planet and people. This doughnut framework could be used to evaluate the case studies presented above by mapping indicators in the assessment to the social foundations identified, and combined with a similar study of E-LCA and the listed environmental impacts. This type of approach could form the basis of an LCA methodology that combines social and environmental impacts that results in useful comparative analysis for building products or could form a new way of developing and assessing products based on the potential to provide for social needs while limiting ecological impact. Either approach would necessitate the use of a combination of qualitative and quantitative indicators and units that attempt to capture the full spectrum of sustainability.

Tool Development

Several software tools exist for environmental life cycle assessment of buildings and building materials, and while they are far from perfect, the tools do provide useful information for decision making. As early as the initial publication of the *Guidelines* in 2009, there has been an acknowledgment of the potential for software that supports the practice of social life cycle assessment. The *Guidelines* are intended to establish a skeleton of the S-LCA process and have been interpreted in many ways. The development of a tool for carrying out S-LCA will require additional research and some consensus on methodology relating to data collection and input, impact assessment, and results reporting, but software that functions similarly to E-LCA tools is conceivable. Furthermore, the data collected and used for any type of software analysis would need to be routinely and robustly reviewed and may require new or novel approaches to data collection.

CONCLUSION

This chapter acts as a high-level introduction and overview of the growing field of S-LCA, including a brief description of the process, case studies testing the process, and some potential ways forward to increase the quantity and quality of S-LCA. The fields of architecture, engineering, construction, research, and policymaking have the potential to drive the use and effectiveness of S-LCA through building standards and certifications, codes and requirements, and continuing education throughout industries and supply chains. This topic is complicated and messy, but necessary in the pursuit of buildings, developments, and systems that meet the goal of true economic, environmental, and social sustainability.

REFERENCES

Aparcana, S., & Salhofer, S. (2013). Development of a social impact assessment methodology for recycling systems in low-income countries. *The International Journal of Life Cycle Assessment*, *18*(5), 1106–1115. doi:10.100711367-013-0546-8

Benoit, C., Andrews, E., Barthel, L., & Beck, T. (2009). *Guidelines for Social Life Cycle Assessment of Products*. United Nations Environment Programme.

Bruce-Hyrkäs, T., Pasanen, P., & Castro, R. (2018). Overview of Whole Building Life Cycle Assessment for Green Building Certification and Ecodesign through Industry Surveys and Interviews. *Procedia CIRP*, *69*, 178–183. doi:10.1016/j.procir.2017.11.127

Building Research Establishment. (2018). *BREEAM UK New Construction – Technical Manual.* Author.

Di Cesare, S., Silveri, F., Sala, S., & Petti, L. (2018). Positive impacts in social life cycle assessment: State of the art and the way forward. *The International Journal of Life Cycle Assessment*, *23*(3), 406–421. doi:10.100711367-016-1169-7

Dong, Y., & Ng, S. (2015). A social life cycle assessment model for building construction in Hong Kong. *The International Journal of Life Cycle Assessment*, *20*(8), 1166–1180. doi:10.100711367-015-0908-5

Hossain, M., Poon, U., Dong, C., Lo, S., & Cheng, Y. (2018). Development of social sustainability assessment method and a comparative case study on assessing recycled construction materials. *The International Journal of Life Cycle Assessment*, *23*(8), 1654–1674. doi:10.100711367-017-1373-0

Hosseinijou, S., Mansour, A., & Shirazi, S. (2014). Social life cycle assessment for material selection: A case study of building materials. *The International Journal of Life Cycle Assessment*, *19*(3), 620–645. doi:10.100711367-013-0658-1

International Living Future Institute. (2018). *Living Product Challenge 2.0.* Author.

Jørgensen, A., Dreyer, L., & Wangel, C. (2012). Addressing the effect of social life cycle assessments. *The International Journal of Life Cycle Assessment*, *17*(6), 828–839. doi:10.100711367-012-0408-9

Minnesota B3. (2018). *Buildings*. Benchmarks, & Beyond.

Opher, T., Shapira, A., & Friedler, E. (2018). A comparative social life cycle assessment of urban domestic water reuse alternatives. *The International Journal of Life Cycle Assessment*, *23*(6), 1315–1330. doi:10.100711367-017-1356-1

Petti, L., Serreli, M., & Di Cesare, S. (2018). Systematic literature review in social life cycle assessment. *The International Journal of Life Cycle Assessment*, *23*(3), 422–431. doi:10.100711367-016-1135-4

Raworth, K. (2014). Defining a safe and just space for humanity. In State of the World 2013: Is Sustainability Still Possible? (pp. 28-38). Island Press-Center for Resource Economics.

Subramanian, C., Chau, C. K., & Yung, W. K. C. (2018). Relevance and feasibility of the existing social LCA methods and case studies from a decision-making perspective. *Journal of Cleaner Production*, *171*, 690–703. doi:10.1016/j.jclepro.2017.10.006

USGBC. (2014). *LEED* (Vol. 4). Reference for Building Design and Construction.

USGBC. (2019). *LEED* (Vol. 4). Pilot Credit Library.

ADDITIONAL READING

Anand, & Amor. (2017). Recent developments, future challenges and new research directions in LCA of buildings: A critical review. *Renewable and Sustainable Energy Reviews, 67*, 408-416.

Moltesen, A. P., Bonou, A., Wangel, A., & Bozhilova-Kisheva, K. (2017). Social life cycle assessment: An introduction. In Life Cycle Assessment: Theory and Practice (pp. 401-422). Springer International Publishing.

Petti, L., Serreli, M., & Di Cesare, S. (2018). Systematic literature review in social life cycle assessment. *The International Journal of Life Cycle Assessment, 23*(3), 422–431. doi:10.100711367-016-1135-4

Suckling, J., & Lee, J. (2017). Integrating Environmental and Social Life Cycle Assessment: Asking the Right Question. *Journal of Industrial Ecology, 21*(6), 1454–1463. doi:10.1111/jiec.12565

Sureau, S., Mazijn, B., Garrido, S., & Achten, R. (2018). Social life cycle assessment frameworks: A review of criteria and indicators proposed to assess social and socioeconomic impacts. *The International Journal of Life Cycle Assessment, 23*(4), 904–920. doi:10.100711367-017-1336-5

Wu, R., Yang, D., & Chen, J. (2014). Social Life Cycle Assessment Revisited. *Sustainability, 6*(7), 4200–4226. doi:10.3390u6074200

ENDNOTE

1 Social hotspot: A unit process located in a region where a particular situation may be considered a problem, risk, or opportunity. Themes of interest may include human rights, working conditions, cultural heritage, poverty, disease, political conflicts, indigenous rights, or other (Benoit & Mazijn, 2009).

Chapter 9
Improving the Weather:
On Architectural Comforts and Climates

Andrew Cruse
The Ohio State University, USA

ABSTRACT

This chapter proposes an approach to thermal comfort that increases occupant pleasure and reduces energy use by connecting architecture's material and environmental dimensions. Today's dominant thermal comfort model, the predicted mean vote (PMV), calls for steady-state temperatures that are largely unrelated to building design decisions. A more recent alternative approach, the adaptive thermal comfort (ATC) model, ties comfort to outdoor conditions and individual experience. Yet reliance on HVAC technology to provide building comfort hampers how such ideas are integrated into building design. This chapter outlines the historical background of the PMV and ACT models to understand the current status of thermal comfort research and practice. It then uses four recent buildings to outline how the insights of adaptive comfort research can be translated to bespoke comforts through spatial, material, formal, and other design strategies.

INTRODUCTION

Architecture improves the weather. That is, the indoor climate created by buildings is intended to increase human health, productivity, and happiness while decreasing disease and discomfort. We are more comfortable inside than out. Yet, given comfort's fundamental role to architecture, architects typically outsource its creation to mechanical engineers, who in turn base their designs on the Predicted Mean

DOI: 10.4018/978-1-7998-2426-8.ch009

Vote (PMV) comfort standard. This standard restricts comfort to a narrow band of temperature and humidity that requires significant amounts of energy to maintain.

Global climate change is heightening our collective consciousness of comfort. It is becoming increasingly difficult to separate stable interior climates from changing exterior ones. This growing awareness of comfort has led some building researchers to question the dominant PMV approach. The Adaptive Thermal Comfort (ATC) model is a promising step in this direction. ATC holds that, by allowing indoor climates to change in ways that reflect outdoor ones, the design team can improve building comfort and reduce energy use. ATC currently exists largely as a building standard. Its observations have yet to be embraced by architects as design opportunities to shape building environments. Similarly, ATC researchers have not fully considered how the formal, spatial, and material intelligence architects bring to the design process can further advance the goal of a more comfortable, less energy-intensive architecture. This chapter connects ideas from adaptive comfort research to examples of building design. It demonstrates how architects can consciously improve the weather in ways that relate architecture's material dimensions to its environmental ones.

The 20th-century engineering definition of comfort focused the attention of building professionals on a narrow yet politically effective definition of *comfort*. This limited definition became a discursive vehicle for naturalizing comfort into a matter of technical and legal fact captured in building standards and embodied in the buildings they governed. Yet, when seen historically, comfort appears as a sociotechnical concept that links cultural norms, expectations, and differences on one hand with technical and natural systems on the other. This history shows that the concept of comfort is a rich locus of ideas that is both fundamental to architecture and part of larger interdisciplinary debates. Comfort reflects how the built environment affects health and creates pleasure. It manifests cultural conventions, from temperature preferences and seasonal traditions to notions of privacy and gender identity. It reflects important technical developments that have accelerated since the 19th century and architectural traditions that have faded during this same period. Perhaps most importantly for the challenges facing architecture today, building comfort reflects the social construction of energy use (Shove, 2003). As such, understanding how comfort came to be what it is today provides insights into how it can continue to evolve to address today's unique challenges.

This chapter briefly traces the development of architectural comfort from the 19th century through the one-size-fits-all PMV comfort standard to the bespoke comforts provided by an adaptive approach. It has a hybrid structure. The first part, which outlines the development of a comfort standard in America, is a historical account based on scientific and industry literature regarding comfort. What emerges is a narrative of changing comfort ideals rather than the fixed, "scientifically" determined ideal comfort. The second part is structured around four contemporary buildings that embody ideas of adaptive comfort without explicit reference to current comfort research. These examples flip the narrative of thermal comfort in the 20th and early 21st centuries, defining it through architectural design rather than scientific research. They demonstrate how comfort can be operative in design in ways that affect poetics and performance. The chapter's hybrid approach is intended to overcome the seeming difficulty of translating comfort research into practice. As prominent researcher Richard de Dear recently noted, "A somewhat depressing deduction from the comfort research community is that instead of innovation in indoor environmental design assimilating our incrementally expanding knowledge of human thermal comfort, the technology itself is driving our concepts and models of thermal comfort" (2011, p. 10). Indeed, the dominance of a technological narrative around comfort skews much architectural writing about the interior environment. To cite a well-known example, the term *well-tempered* figures prominently in the title of Reyner Banham's *The Architecture of the Well-Tempered Environment* (1969). Yet the book's focus was an optimistic assessment of the technology of comfort rather than the evolving environmental ideas of comfort itself. To that end, this chapter demonstrates that, for architects to help improve the weather outside of buildings, we must pay more attention to the weather we help to create inside of them.

AN OUTLINE OF ARCHITECTURAL COMFORT STANDARDS

The development of thermal comfort standards follows a similar trajectory to other engineering standards. The construction historian Bill Addis outlines this process as progressing from empirical rules that capture collective experience to design rules grounded in scientific understanding, and then to scientific rules that specify performance in quantitative terms (2007, p. 610). For comfort, the empirical rules began with building ventilation. In the second half of the 18th century, public health researchers established an experimental link between disease outbreaks and air quality. Prevailing medical theories of the day held that foul or "vitiated" air resulted from people breathing inside enclosed spaces. As one ventilation engineer put it, "Our breath is our own worst enemy" (Leeds, 1871, p. 13). In order to improve the environmental quality inside buildings, early building researchers, including doctors, chemists, and engineers, focused on quantifying ventilation rates, that is, the amount of fresh air that needed to be supplied to indoor spaces (Addington, 2001).

The English engineer Thomas Tredgold was one of the first to propose ventilation and temperature guidelines based on a scientific understanding of human physiology and building physics. In his *Principles of Warming and Ventilation* (1824), Tredgold combined conventional wisdom with his own experimental results, providing practical "rules" for architects and engineers to calculate, among other things, the amount of fresh air needed to ventilate interior spaces properly. His recommendation of four cubic feet of fresh air per minute per person was one of the earliest quantitative building ventilation requirements (Tredgold, 1824, pp. 69–73). Tredgold's discussion of ideal building temperature was less specific. He based his recommendations on building type and related them to exterior temperature. Tredgold advised that during the winter, dwellings be kept at 60°F, hospitals at 50°F, and prisons left unheated, and during the summer, interior temperatures be limited to 5°F above exterior conditions by using fans to increase air movement.

In the latter half of the 19th century, Hermann Rietschel applied the newly understood principals of thermodynamics to ventilation and temperature standards. As chair of heating and ventilation at the Royal Technical School in Berlin, Rietschel held the first academic position in the field and established one of the first academic research laboratories to study these issues. He published his findings in German in 1894, and they appeared in English in 1927 as *Heating and Ventilating: A Handbook for Architects and Engineers*. There Rietschel gives ventilation rates based both on rules of thumb and on his own research in an auditorium affiliated with his laboratory. These rates were generally higher in the summer, presumably to increase cooling by increasing airflow (Rietschel, 1927, p. 129). Despite Rietschel's greater understanding of the mechanisms of heat transfer, he based his approach to temperature standards on "customary" practice. In the *Handbook*'s opening chapter, Rietschel provided a list of desirable indoor temperatures that, like Tredgold's, was organized by building type. Churches and jails (for "overnight use only") were to be kept at 50°F; lecture rooms, prisons, and business offices at 65°F; and bathrooms at 72°F. In a brief commentary on these temperature guidelines, Rietschel recognized the impact of humidity, air movement, and radiant heat on comfort conditions without quantifying how they would alter target temperatures. He ended the chapter in an otherwise scientifically based book by stating ideal comfort conditions could be described by the popular expression "warm feet and cool head" (Rietschel, 1927, p. 4).

Tredgold's and Rietschel's approaches to the indoor environment combined scientific rules and research with common practice to propose building ventilation rates and temperature guidelines. While standard ventilation rates were quantitatively specific, temperature standards were based on common practice and varied depending on use and the weather outside. It would take both technological developments and social forces at the start of the 20th century for these factors to coalesce into the idea of a comfort standard expressed in quantitative terms and based on scientific rules.

ASHVE Comfort Research

The American Society of Heating and Ventilating Engineers (ASHVE) was formed in 1895, in part to establish a scientific basis for heating and ventilation practice in the United States. This effort included incorporating the new technology of air conditioning into engineering practice. Initially developed for humidity-sensitive industrial processes such as printing and food processing, air conditioning was increasingly being used in spaces designed for human occupancy such as department stores and cinemas. Unlike industrial processes, where interior climate standards were set based on the efficient production of a product, comfort cooling for people had no objective metric on which it could be established. ASHVE members saw the development of such a comfort standard as an important goal for their organization. As a commentator in an industry journal, *The Weather Vein,* noted, "One of the Engineer's most effective safeguards against loss of public confidence is the development of fixed, rational standards and determination of precise finite values that he may express himself with an exactness which ensures understanding" (Cooper, 1998, p. 70). In building public trust through such an engineering standard, ASHVE took comfort standards to the next stage outlined by Addis, with scientific rules used to specify performance in quantitative terms.

The research used to establish such a comfort standard was conducted in a climate chamber at ASHVE's Research Bureau, which opened in 1919 at the Bureau of Mines in Pittsburgh, Pennsylvania. *Climate chambers*, also called *psychrometric*, *environmental*, or *air-conditioned chambers*, were well-insulated, windowless, and air-conditioned rooms in which temperature and humidity could be precisely set and independently controlled. They developed in the early 20th century as an essential piece of laboratory equipment for the new academic field of industrial hygiene. Industrial hygiene researchers studied how the environmental conditions found in factories and on job sites affected workers' health, from extremes of temperature and humidity to chemical exposures (Sellers, 1977, pp. 141–186). Climate chambers allowed industrial hygiene researchers to recreate industrial environments in a laboratory setting, where the response of subjects could be closely studied. Unlike the extreme industrial environments of blue-collar work, ASHVE researchers used their climate chamber to identify ideal conditions for white-collar office environments.

The premise for ASHVE's research was that comfort resulted from the heat balance between a body's interior environment and the external environment surrounding it. This idea built on French physiologist Claude Bernard's concept of the "*milieu intérieur,*" the body's internal environment, which was separate from the environment in which it lived (1879). The American Walter Cannon later elaborated on this idea of self-regulation with the concept of *homeostasis* in his poetically named book, *The Wisdom of the Body* (1932). Homeostasis described how a living organism maintained a steady-state internal condition despite being part of a dynamic, open system. Although Bernard and Cannon related their idea to a variety of biological functions, ASHVE researchers were interested in how it applied to temperature regulation in the body. They reasoned that when the heat

Figure 1. Comfort zone for humans at rest, 1924
(Credit: ASHVE [1924]. Celebrate thirty years of progress. Journal of the American Society of Heating and Ventilating Engineers, 30[3], n.p.)

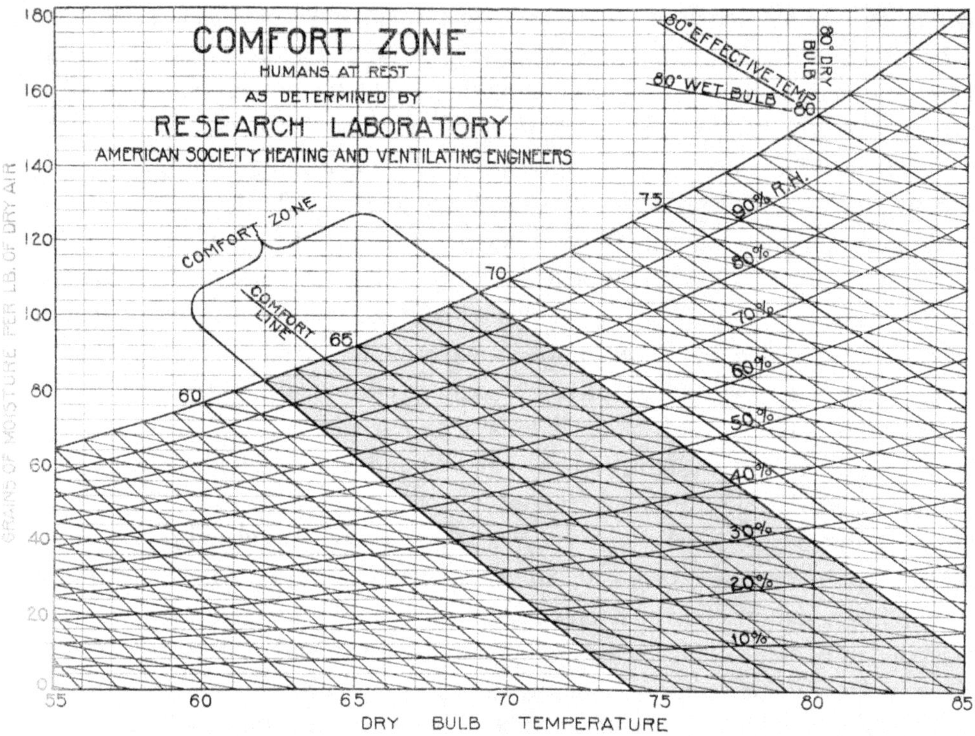

produced by metabolism was balanced with a room's environmental conditions, people felt comfortable. Based on this connection between a body's physiological functioning with the building's mechanical system, lab researchers hypothesized that there must be different combinations of temperature and humidity that produced equally comfortable conditions. To test this hypothesis, researchers had test subjects stay in their climate chambers under a variety of environmental conditions. Initial experiments tested different combinations of dry bulb temperatures and relative humidity, which they synthesized into the new scale of "effective temperature." (Later experiments tested other environmental parameters, including air speed and radiant heat.) Upon exiting, subjects were asked to rate how comfortable they found those conditions. By 1924, researchers published a Comfort Zone Chart that organized their results graphically onto a psychrometric chart. This visualization showed a limited band of effective temperature in which half of the subject felt comfortable, as well as a comfort line at which 97% of them considered themselves comfortable (ASHVE, 1924).

Buildings soon began to resemble climate chambers where interior environments were thermally separated from exterior ones. This trend played out in greater attention to building wall insulation (Moe, 2014), airtightness, and most visibly, windows. As one ASHVE commentator memorably put it, when a building has operable windows and no air conditioning, employees tend to blame nature for their discomforts. But when a building has air conditioning, they blame the building's management (Leopold, 1947, p. 304). Tenants in the Milam Building in San Antonio, Texas (1928), the first air-conditioned office building, were prohibited by lease agreements from opening their windows because it made managing the air conditioning system too difficult.

Individual comfort control was instead surrendered to a building engineer who was responsible for managing the comfort of all building occupants. In more extreme examples, windows themselves were eliminated from buildings. The first windowless office building was built in 1935 for the Hershey Chocolate Corporation in Hershey, Pennsylvania. Touted as "the latest contribution of science to business" and for its energy efficiency, the building was thought to create a healthier work environment at lower costs. Weather information was communicated to building occupants through colored lights beneath wall clocks.

By 1938, ASHVE fulfilled one of the original goals for the Research Bureau when, together with the American Society of Refrigerating Engineers (ASRE), it issued the "Code for the Minimum Requirements for Comfort Air Conditioning." The document set out minimum design requirements for air-conditioned spaces, including design temperatures and relative humidity ranges. In doing so, it translated the scientific research conducted by ASHVE into engineering practice. By putting the comfort conditions in the language of a building standard, the groups aimed for their large-scale adoption by the building industry. (ASHVE merged with ASRE in 1959 to form ASHRAE, the American Society of Heating, Refrigerating and Air-Conditioning Engineers.) By 1966, ASHRAE revised and reissued this material as *ASHRAE Standard 55-66: Thermal Comfort Conditions*. This ASHRAE standard later served as a model for the International Organization for Standardization's comfort standard, *ISO 7730: Ergonomics of the Thermal Environment.*

Predicted Mean Vote Comfort Research

Despite ASHVE's and later ASHRAE's extensive comfort research, conclusions about ideal comfort conditions became less certain. By 1961, their comfort chart included the qualification that "there is no precise physiologic observation by which comfort can be evaluated" (Nevins, 1961, p. 68). This uncertainty led to growing frustration among the organization's members, who wanted clear direction upon which to base their mechanical designs. By the early 1960s, ASHVE recognized

Figure 2. Milam Building, San Antonio, TX, 1925. George Willis was the architect and Willis Carrier oversaw the design and installation of the air conditioning system. (Credit: Author's collection.)

Figure 3. Modern office building, Hershey, PA, 1935. Designed by D. Paul Witmer.
(Credit: Author's collection.)

that the comfort zone had lost the authority it once held and began planning a comprehensive reevaluation.

This goal was realized by the end of the decade with Danish researcher P. O. Fanger's Predicted Mean Vote (PMV) comfort model, a more sophisticated and decidedly more useful model than had been previously proposed. Working at Kansas State University's Institute for Environmental Research and later at the Technical University of Denmark, Fanger established a comfort model based on climate chamber research and useful to the building industry. Like ASHVE before him, Fanger based his work on the premise that comfort resulted from the heat balance between a body and its environment. Unlike the graphic method of depicting comfort on a psychrometric chart that included two variables, dry bulb temperature and relative humidity, Fanger's "comfort equation" addressed six: dry bulb temperature, relative humidity, mean radiant temperature, and air velocity, as well as two personal variables—activity level and the insulation value of clothing. Using early computer technology, researchers could solve this equation to determine which combinations of variables created heat balance between the subject and environment. Fanger also went a step further, reorganizing this data to predict how many people would find specific combinations of conditions thermally acceptable. Referred to as Predicted Percentage Dissatisfied (PPD), Fanger considered conditions comfortable when

80% or more of occupants were satisfied with the building's interior climate. Importantly, such satisfaction reflected a state of thermal neutrality, where building occupants did not object to interior environmental conditions, rather than one of thermal pleasure. In his book *Thermal Comfort* (1970), Fanger claimed there were no national, geographic, or seasonal variations in these ideal comfort conditions. In fact, he asserted that any deviation from a quasi-steady-state thermal environment would only result in a greater number of people being dissatisfied (Fanger, 1970, pp. 85–95).

In 1992, Fanger's PMV model was adopted into the *ASHRAE 55* comfort standard. With this addition, it became a scientific rule that specified performance in quantitative terms, the last stage outlined by Addis. Starting in the 1960s, ASHRAE's focus on comfort shifted from laboratory research into ideal comfort conditions to political advocacy, promoting its comfort standard within the building industry. As one ASHRAE member commented, "Adoption and implementation [of standards] are fundamental to achieving the purpose addressed by ASHRAE standards because until a standard is formally codified and adopted it is voluntary and has no major bearing on technology or construction practice" (Comstock, 1995, p. 106). It would take another research method, one that expanded upon the heat-balance basis for comfort, to challenge the dominance of the PMV model.

NEW PERSPECTIVES ON THERMAL COMFORT

As Fanger was advocating for the thermal neutrality of sealed building enclosures, another group of researchers was investigating an adaptive approach to comfort that linked interior conditions to exterior weather. This approach developed from field studies in which researchers asked building occupants to assess how comfortable they found their interior environments. Unlike the laboratory-based climate chamber work, field studies reflected the real-world conditions found in existing buildings with a diversity of populations. Generally, these studies demonstrated a connection between preferred interior climates and the weather outside. A group of ASHVE researchers had conducted similar field studies in the early 1940s and were aware that they showed different preferred comfort conditions than had been found with climate chamber research. A 1942 report on the field studies recognized that comfort preferences could vary based on age and gender and that people accustomed to different exterior climates had different interior comfort expectations. The report concluded, "The impossibility of pleasing everyone is evident in air conditioned buildings" (Chester, 1942, p. 122). It would take several decades for such observations to be formulated into a new comfort model.

Figure 4. The PMV comfort model loses its predictive ability in naturally ventilated buildings. While in a centrally controlled HVAC building (left), predicted comfort conditions match what was observed in field studies; in a naturally ventilated building (right), field studies showed building occupants preferred a wider range of temperatures than predicted by the PMV model
(Credit: de Dear, R. J., & Brager, G. S., [1998]. Developing an adaptive model of thermal comfort and preference. ASHRAE Transactions, 104[1], 145–167.)

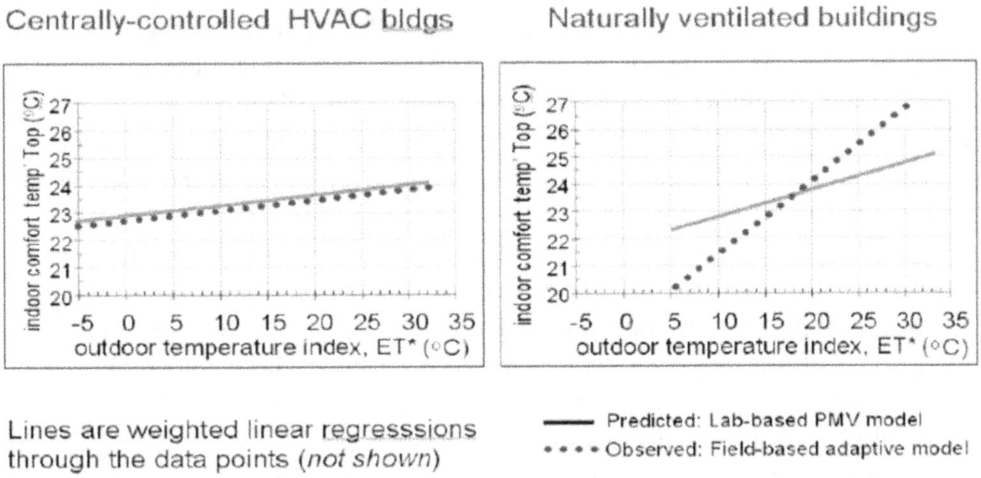

Adaptive Thermal Comfort

Starting in the 1970s, a new generation of university-based comfort researchers began to develop the Adaptive Thermal Comfort (ATC) model. Led primarily by Fergus Nicol and Michael Humphreys in England, and later Gail Brager at the University of California, Berkeley, and Richard de Dear in Australia, researchers found that occupants of naturally ventilated buildings considered themselves comfortable over a wider range of temperatures than were anticipated by the PMV model (de Dear & Brager, 1998; Nicol, Humphreys, & Roaf, 2012).

In other words, the PMV model lost its predictive ability when interior climates were connected to exterior ones. To explain this discrepancy, ATC researchers proposed that there were adaptive factors affecting comfort beyond the body's heat balance with the surrounding environment. These researchers attributed thermal adaptation to three different processes: behavioral adjustment, physiological acclimation, and psychological expectation. Behavioral adjustments included changes a person might make to their immediate environment, such as opening a window, taking off a sweater, or taking a break from work in the heat of the day. Physiological adjustments were changes in how the body reacts to the thermal environment. These ranged from acclimatization that can happen in a matter of weeks to genetic changes that occur on an evolutionary time scale. Psychological changes are the result of an altered perception of, and reaction to, thermal information based on past experiences and expectations. The adaptive approach recast building occupants as active agents in shaping their comfort through these processes rather than as passive recipients of comfort conditions as implied by the PMV model.

ATC researchers also demonstrated that connecting interior comfort conditions to exterior weather saved energy. An adaptive comfort approach reduced building energy use since it allowed temperatures to float over a wider range that more closely followed exterior conditions. Later research qualified that, in commercial buildings, for each 1.8°F shift out of the thermally neutral zone, the energy required to condition interior spaces was reduced by roughly 7% (Arens, Humphreys, de Dear, & Zhang, 2010, p. 5).

Like ASHRAE researchers before them, proponents of ATC looked to building standards as a way to turn scientific research into building practice. But, unlike the PMV model, the ATC standard has not been widely adopted in practice. Due largely to the efforts of Brager and de Dear, the Adaptive Thermal Comfort model was included as an alternative compliance method starting in the 2004 version of the *ASHRAE 55* standard. However, the language adopted in the standard specified that the adaptive option was only applicable to buildings without mechanical cooling systems. It further stipulated that the standard could be used only when the prevailing mean outdoor temperature fell between 50°F and 92°F, and the heating system was not in operation. Such restrictions significantly limited when and where the ATC model could be used. They also suggested a crude bifurcation between air-conditioned buildings designed with the PMV standard and naturally ventilated ones designed with the ATC model. This false dilemma stymied design teams exploring hybrid or complementary uses of these two approaches, such as mixed-mode buildings that can operate with either natural or mechanical ventilation, depending on exterior conditions. Instead of becoming part of a design choice about how to integrate comfort decisions into the form and operation of the building, *ASHRAE 55* suggested that comfort was the choice of a single standard.

Alliesthesia

Despite the limited application of ATC through *ASHRAE 55*, adaptive comfort researchers have continued to expand our understanding of the thermal environment in ways that are suggestive to architects. Recent research has examined the physiological basis of thermal pleasure through the idea of *alliesthesia*. *Alliesthesia* describes how a stimulus can be experienced as either pleasant or painful, depending on a subject's internal state. Initially proposed by the French physiologist Michel Cabanac, the term comes from *allios*, meaning "changed" and *esthesia*, meaning "sensation" (1971). In psychological terms, *alliesthesia* describes the mechanism of perception—the interpretation of a stimulus—rather than that of sensation—the recognition of a stimulus. Take, for example, two subjects, one with an elevated body temperature and the other with a lowered body temperature. If both subjects were to put their hands into water at 110°F, both would sense that the water was warm. However, the subject with the lowered body temperature would find the perception of warmth pleasant due to their lower body temperature, while the subject with the elevated body temperature would find it unpleasant due to their higher body temperature. As Cabanac writes, "Pleasure is actually observable in transient states, when the stimulus helps the subject return to normothermia," that is, normal body temperature (2000, p. 41).

Adaptive comfort researchers became interested in alliesthesia to explain why people could have different reactions to the same environmental conditions. That is why occupants of naturally ventilated spaces found specific conditions acceptable while occupants of air-conditioned buildings found these same conditions unacceptable. Behaviorally, physiologically, and psychologically adaptive processes explained how this discrepancy happened; alliesthesia explained why.

In reconsidering Cabanac's work, comfort researchers have extended his observations to distinguish between spatial and temporal alliesthesia (Parkinson, de Dear, & Candido, 2016). *Spatial alliesthesia* refers to the experience of different thermal stimuli on parts of the body, such as the "warm feet and cool head" that Hermann Rietschel had used to describe ideal comfort. It recognizes the skin as a "sensory bellwether" that anticipates the future thermal state of the body. *Temporal alliesthesia* refers to temporary variations in thermal experiences over relatively short periods. Such experiences form part of the thermal texture of everyday life, such as moving from a patch of shade into the sun on a winter walk, or rounding a street corner and stepping into a cool breeze on a humid summer day. We can also find such transient thermal experiences in building spaces such as vestibules, lobbies, and even parking garages that serve as transition spaces between interior and exterior climates.

An important conclusion drawn from alliesthesia research is that thermal pleasure is unattainable in a thermally neutral state. That is, the static conditions prescribed by the PMV model are incapable of providing thermal pleasure because they are based on maintaining a static heat balance between the body and the environment. In fact, the spatial and temporal variations that alliesthesia describes as sources of pleasure are discouraged by the PMV model. Vertical air temperature differences, asymmetrical radiant fields, and local convective cooling (labeled "drafts") are considered "localized discomfort" and should be eliminated under PMV conditions. Alliesthesia, on the other hand, shows that there needs to be a thermal difference, however small, between subjects and their thermal environment for the subjects to experience pleasure.

In a certain sense, the insights of comfort research over the past 50 years have moved in the opposite direction from the PMV model found in ASHRAE's comfort standard. Adaptive thermal comfort and alliesthesia have linked interior comfort to exterior weather, energy savings, and thermal pleasure, while the PMV model has restricted comfort to the energy-intensive, artificial conditions of thermal neutrality. As Richard de Dear has lamented, our expanding knowledge of thermal comfort has unfortunately not translated into innovative environmental design. Instead, design teams rely largely on building equipment to drive concepts of comfort. Yet this is not entirely true. Some contemporary architectural practices are connecting architecture's material qualities with its environmental ones. Instead of the one-size-fits-all approach to comfort of Fanger's model, these architects are designing bespoke comforts that connect the physical but invisible aspects of comfort to a building's formal, spatial, and material choices in ways that improve the weather both inside and outside of buildings.

BESPOKE COMFORTS

Comfort is typically seen as an unremarkable dimension of architecture. When historians and critics attempt to define what is essential to architecture, what makes it different from other arts, they often focus on form, space, and materials. Comfort is not seen as an essential quality. This lack of attention may be related to the modern habit of connecting comfort largely with mechanical equipment rather than with architectural design. A seldom-noted paradox of architectural modernism is that while new materials, such as steel, reinforced concrete, and plate glass, visually connected indoor and outdoor spaces, contemporary environmental technologies thermally separated indoor and outdoor climates. Regardless of its origin, this limited understanding of comfort is ripe for redefinition. In the second half of this chapter, I shift narrative registers from a historical account of comfort research to a reading of four contemporary projects and their different design-based approaches to comfort. These projects are emblematic of how some architects are creating bespoke comforts that are integral to a building's design. I chose this set of examples in part because of their diversity. They have different forms, sizes, material assemblies, clients, programs, sites, and geographic regions. Such diversity demonstrates that designers can think with comfort in many different contexts. I also chose these projects because each has received the attention of design juries and critics, some of whom acknowledge the importance of comfort to the individual project. In highlighting the role of comfort in each project and connecting the projects to other works by the architects and by others, I argue that comfort can and should be seen as an essential architectural quality. Such a reading recognizes the cocreation of architecture and its environment and follows a vein of recent scholarship in architectural history that incorporates insights from the environmental humanities (Hochhäusl et al., 2018).

Combining these historical accounts and project narratives is an attempt to bridge the gap between comfort research and architectural design. Comfort standards have clearly been an effective technique in promoting the wide-scale adoption of a steady-state thermal comfort model. But they have been considerably less effective in promoting a more nuanced idea of adaptive comfort. Approaching comfort through architectural design offers another avenue to advance this important set of ideas. These examples highlight how comfort is an essential quality of architecture in two important and related ways. First, they position comfort as a cultural value for architecture. This insight recalls Lisa Heschong's argument in her soft manifesto, *Thermal Delight in Architecture*. There she argues that the thermal environment has the potential for "sensual, cultural roles and symbolism that need not, indeed should not, be designed out of existence in the name of a thermally neutral world" (Heschong, 1979, p. 17). Framing comfort as a cultural value that is socially defined moves it beyond the largely scientific and technical understanding of the term that has grown over the past century to encompass a broader, historically nuanced meaning. While Heschong illustrated her appeal primarily with biological and anthropological examples, here I extend her appreciation of thermal delight to contemporary architectural projects where the overlapping concerns of comfort and construction can be clearly seen. Second, by making explicit connections between the interior climate and the exterior one, comfort can be another way to bring architecture into larger environmental discussions about climate change beyond current approaches to energy efficiency. In changing the indoor climates in ways that reduce energy use and improve comfort, architects can simultaneously help to positively improve the weather outside by reducing carbon emissions associated with buildings.

Life Science Building (2024)

The proposal for a new university Life Science Building by the French practice Bruther in collaboration with the Belgian firm Baukunst engages adaptive ideas of comfort through its formal organization and material selection. Won through an invited competition, the project is located on the campus of the University of Lausanne, next to the Swiss Federal Institute of Technology, and close to SANAA's Rolex Learning Center (2010). This strategic location, with views south toward Lake Leman and north toward the Alps, was chosen to unite the two campuses using a shared program dedicated to research and teaching.

Figure 5. Schematic plan for the new Life Science Building (Bruther and Baukunst, 2024 anticipated). The plan is organized as an environmental onion, with the spatially nested layers of thermal control related to programmatic organization
(Credit: Bruther.)

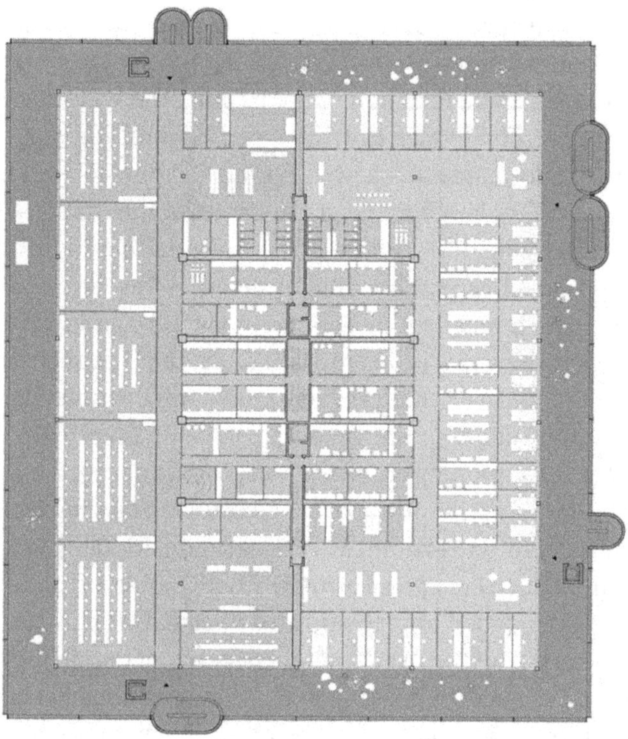

The building is a compact 290,000-square-foot cubic volume over seven floors. Floors 2 through 6 contain a central laboratory zone surrounded by a wide ring of classrooms, meeting rooms, and offices. These are evenly divided in two, with one half dedicated to teaching chemistry and biology, and the other to neuroscience and microbiology research. This intermediate zone is surrounded by a wide perimeter ring for circulation and public spaces that features several sculptural stairs for inter-floor circulation. The six floors of classrooms and offices intersect three double-height perimeter spaces, which bring light into this intermediate ring of the plan through internal windows. The top floor houses an auditorium and restaurant under a series of narrow gable roofs. The building is clad with a largely transparent greenhouse-like skin with large operable panels and integrated sun shading. The building's structure and services are clearly separated to allow for future reconfiguration as teaching and research needs change, a decision for which the architects cite the influence of British high-tech architecture, specifically the work of Richard Rogers.

Figure 6. Proposed view into the perimeter zone of the Life Science Building (Bruther and Baukunst, 2024 anticipated). Part winter garden, part social condenser, this buffer space allows for the mixing of interior and exterior climates
(Credit: Bruther.)

The Life Science Building is organized like an environmental onion, where nested layers of climate control relate to spatial and programmatic organization. Here the laboratories, which have the strictest environmental requirements, are located in the center surrounded by the classrooms, meeting rooms, and offices with looser requirements. These, in turn, are encircled by the perimeter zone, which serves as a permeable layer between the exterior and interior climates. Part winter garden, part social condenser, the perimeter zone mixes the climatic and the communal. This multi-height area thermally buffers indoor spaces from less favorable weather outside, or it can be opened through operable windows on the façade to mix these two climates. Bruther designed similar compact, flexible spaces in its award-winning New Generation Research Center in Caen, France (2015). Beyond Bruther's own buildings, these climatically flexible spaces recall the work of the French firm Lacaton & Vassal. Their use of industrial materials and "extra space," from the small-scale Latapie House (Floirac, France, 1993) to recent public housing transformations in Bordeaux (2016), has influenced a younger generation of architects who consider the interior climate as part of a design brief.

The Life Science Building's layered organization and permeable perimeter provide a variety of adaptive comfort opportunities. The operable façade allows occupants to adjust the immediate climate in this perimeter. Additionally, as the sun moves across the sky during the day, making brighter and warmer areas and putting other areas into a cooler shade, people can experience a range of climates by merely walking around the perimeter. Moving radially through the different layers of the environmental onion also creates different comfort conditions. Researchers, who spend the majority of their day in the central windowless laboratory's controlled climate, could alleviate their thermal boredom by leaving the lab, going through the intermediate ring and into the perimeter zone, with each successive layer bringing greater thermal texture. Moving in the opposite direction, students coming to class or lab would experience a moment of acclimation in the perimeter zone before moving into the more protected central areas. While comfort researchers could describe such experiences with the terms *temporal alliesthesia*, *behavioral*, and *physiological adaptation*, here the architects have foregrounded comfort for a diverse population of faculty, researchers, and students as an essential feature of their project's spatial, formal, and material organization.

International House Sydney (2017)

A university building with diverse programs and environmental requirements offers many opportunities for bespoke comforts, but commercial office space is much more homogeneous, and the industry is resistant to non-PMV comfort conditions. ASHRAE and ISO standards contain three categories of environmental quality for mechanically cooled office buildings. The higher the class, the smaller the deviation allowed from "ideal" environmental conditions and the greater the amount of energy required (Arens, Humphreys, de Dear, & Zhang, 2010). Given the more challenging parameters of this project type, the International House Sydney (2017) by Tzannes takes an innovative approach to indoor comfort through its material choice of timber.

The International House Sydney was the first commercial engineered-timber building of its size and type in Australia. It was built and is managed by Lendlease, an international construction, property, and infrastructure group headquartered in Sydney. The site, located on Darling Harbor, was formerly a dockland where timber wharves and warehouse buildings stood. It now sits on the western edge of the city's central business district. The 75,000 square-foot building has five floors of

Figure 7. Exterior view of the International House Sydney (Tzannes, 2017) in the new Barangaroo area of the city's central business district
(Credit: Ben Guthrie.)

office space supported by spruce and beech glulam columns and beams, with CLT (cross-laminated timber) walls and floors. This wooden structure sits on top of a two-story commercial base with a conventional concrete frame and a colonnade made from raked solid ironbark columns. The timber frame is clad with a low-iron glass façade that showcases the wood within. Interior stairs are located on the building perimeter and are designed to be an inviting alternative for vertical circulation. On the interior, the wood structure dominates, while exposed ducts, conduits, and pipes hung from the ceiling are painted a matte black. Chilled radiant beams provide cooling. Together with a decoupled outside air supply system, these systems create a more varied interior climate than the typical air-conditioned office.

The exposed timber frame forms the building's exterior image and helps to shape its interior environment. Viewed in the context of downtown Sydney, the wood contrasts with the surrounding palette of aluminum and steel, creating the building's unique identity. In its marketing literature, Lendlease claims that the building's wood interior is beneficial to the health and well-being of its tenants. There are some indications that wood has a positive psychological impact, but its effect on the comfort of building occupants is not well studied (Nyrud & Bringslimark, 2010). Recent research has looked at how people experience building materials through a combination of tactile and visual senses. It has shown that the impression of warmth corresponds largely to visual perception, which is directly affected by color and indirectly affected by a material's thermal effusivity, that is, how quickly heat transfers from one material to another. (Wastiels, Schifferstein, Heylighen, & Wouters, 2012). Architect and theorist Kiel Moe has argued that a material approach to comfort that links tactile and visual perception on the one hand, with physical material properties like thermal effusivity on the other, points toward a larger agenda for energy in architecture through "thermodynamic depth" (2014, pp. 188–227). Certainly, given the increasing use of wood in large-scale commercial and residential buildings, architects need to ask how the experience of this "sustainable" material affects thermal comfort. Comfort researchers could also contribute to this investigation by considering how psychological adaptation to wood may affect the perception of comfort. While the International House Sydney relies on mechanical systems to connect interior and exterior climates (albeit unconventional ones), Tzannes makes a compelling case for the role of material experience as an underexplored avenue to provide novel comforts in the conservative context of commercial real estate.

House for Trees (2014)

The Vietnamese architect Vo Trong Nghia has built a substantial body of work that finds adaptive opportunities for thermal pleasure in the country's tropical climate, one historically considered "uncomfortable" by comfort researchers and the air conditioning industry (Chang, 2016). Nghia built the House for Trees (2014) in the Tan Binh district, one of Ho Chi Minh City's densest residential neighborhoods. The single-family house consists of five individual volumes, informally arranged around an irregularly shaped interior courtyard. These volumes also act as "pots" for large trees, which give the house its name. Each structure contains a single program on each of its two levels. These include a kitchen, dining room, and library on the first floor and separate bedrooms on the second. The external walls are made from concrete cast in bamboo-lined formwork, giving them a highly textured surface, and a withe of local brick lines the interior. The extended family can use the outdoor spaces for more than circulation, occupying these unprogrammed areas as they choose and as the weather allows.

Nghia links the House for Trees to Ho Chi Minh City's tropical urban climate through what he calls "green architecture" (Harikae, 2016, p. 8). This approach directly integrates plants into a building's appearance and performance. Yet Nghia's ambition for the idea extends beyond individual projects. With rapid urbanization, cities in Vietnam have lost much of their open space. Nghia calculates that Vietnam has the lowest amount of green space per capita of any East Asian city (2016). With his "green architecture," Nghia wants to use buildings to bring greenery back into

Figure 8. Exterior view of the House for Trees (Vo Trong Nghia Architects, 2014). Each of the house's five volumes acts as a planter for large trees that shade the buildings and garden spaces beneath. The unprogrammed courtyard spaces serve a variety of uses as part of the house's matrix plan
(Credit: © Hiroyuki Oki.)

cities in a way that reaffirms their tropical character. Nghia has built many such "green" buildings, from private homes like the recent Breathing House (2019) and the Stepping Park House (2018) to the larger Atlas Hoi An Hotel (2016) and the Nanoco office building (under construction). The House for Trees was one of the first such projects, and one for which he developed a novel plan strategy.

The domestic garden typically acts as an outdoor foil to interior spaces, while at the House for Trees, the garden and interiors work together to form a matrix-like plan. In his essay on domestic space, the critic and historian Robin Evans contrasted Renaissance matrix plans, where enfilade connections between rooms provided for social mixing, with 19th-century corridor plans, where separate rooms fostered privacy and independence (1997). Evans argued that matrix plans encourage "passion, carnality, and sociality" and that corridor plans encourage "segregation" of people and uses. Pavilion-type matrix plans can be found elsewhere, most notably in SANAA partner Ryue Nishizawa's Moriyama House (2005) in Tokyo (Cruse, 2019). (Nghia received his architectural training in Tokyo.)

At the House for Trees, Nghia uses the matrix plan to extend his idea of "green architecture" to integrate architecture and landscape and to create a variety of adaptive comfort opportunities. The rooftop trees shade the volumes and the outdoor spaces below, providing welcome coolth while the buildings' thermal mass further buffers the tropical conditions. Exterior spaces connect to interior ones through doors and large operable panels. Moving between inside and outside, residents acclimate to small but significant yearly changes in the weather. Living space can expand or contract as furniture is moved between inside and out during different times of the year. Even the green roofs are occupiable as part of this domestic landscape. When the exterior climate becomes too extreme, such as during the summer rainy season, the interiors can become thermal refuges by closing the doors and turning on the air conditioning.

Beyond Nghia's impressive portfolio, a younger generation of Vietnamese architects—including Tropical Space, Sanuki Daisuke Architects, a21studio, and NishizawaArchitects—are making buildings marked by the meteorological effects of a tropical climate. Historically, colonial powers have taken a negative view of tropical regions, seeing them as disease-ridden with economically backward, lethargic populations. With a postcolonial perspective, these Vietnamese architects are using their buildings, and the adaptive opportunities for comfort they provide, to create positive images and experiences of this same climate. In doing so, they demonstrate that comfort reflects climatic and cultural conditions.

Figure 9. Atrium view looking toward the main entrance of Harquitectes' Lleialtat Santsenca (2014)
(Credit: Harquitectes.)

Lleialtat Santsenca Civic Center (2014)

Harquitectes' Lleialtat Santsenca civic center (2014) transforms a historical structure in Barcelona's Sants neighborhood by connecting adaptive reuse to adaptive comfort. Built within the shell of a former workers' cooperative, the architects kept the building's civically important façade while removing the roof and interior walls. They then reorganized the interior using a linear atrium to create an open circulation spine under a new translucent roof. The atrium is lined on one side by the existing load-bearing masonry wall. On the other side, Harquitectes inserted a scaffold-like steel and wood structure that encloses fixed programmatic spaces, including meeting rooms, classrooms, and a large performance hall.

The building's new roof connects the atrium's interior climate with the mild Mediterranean climate outdoors. Its gabled shape spans the shell of the existing building with open web steel joists. These joists are covered with translucent polycarbonate panels on the south-facing slope and opaque insulated metal panels on the north. In the winter, sunlight passing through the polycarbonate helps to warm the atrium, which has no active heating system. In the summer, hot air is exhausted through operable panels at the roof's peak, driving the stack effect and drawing in air below to ventilate the atrium. Solar curtains beneath the roof can be used to adjust the interior climate, either to reflect away unwanted solar radiation or to trap warm air below. Relying on the roof to mediate between indoor and outdoor environments works well in a dense urban context where adjacent buildings and other constraints limit what can be done with the wall surfaces. Harquitectes later used a similar lightweight performative roof on top of a largely historic masonry shell at their Cristalleries Planell civic center (2016), also in Barcelona.

The climate of the Lleialtat Santsenca atrium is used to buffer the more tightly controlled program space inside the steel and wood addition. This new volume has an air conditioning system for heating and cooling and sits under its own interior roof. Depending on the weather outside and conditions in the atrium, either tempered exterior air from the atrium is drawn into the rooms, or the rooms exhaust conditioned air into the atrium where it is drawn out of the building through the roof. Formally, this relationship inverts the nested layers of an environmental onion, where less controlled rings surround a central protected zone. Here the central area of the atrium acts like a lung for the building, connecting the exterior climate to the controlled spaces along the building's edge. Neighborhood residents walking into the Lleialtat Santsenca from the street would first experience the thermal pleasure of temporal alliesthesia in the atrium's tempered climate. Depending on their plans, they might remain there, using one of the informal meeting areas along the length of the atrium. Or they might go into one of the climate-controlled rooms for an event or performance.

In their essay "Organizing Matter," the four founders of Harquitectes explain their approach to using the solid material of architecture to shape the fluid flows of air and energy. One can easily understand this material-focused approach as creating a variety of climates and comforts. Referring to spaces like the Lleialtat Santsenca atrium, they write, "We like these intermediate spaces, where users feel a proximity to exterior conditions, but with added improvement" (Harquitectes, 2016, p. 151). Here the building's material and formal organization literally improves the weather. This organization of matter, which combines adaptive reuse and adaptive comfort, recalls critic and architect Luis Fernández-Galiano's metaphorical description of an "architecture of rehabilitation" as an "architecture of the second principle" (2000). This comparison between the energetic process of entropic degradation and the reuse of old buildings focuses attention on how the embodied energy of architecture also embodies the cultural values of memory and history. At the Lleialtat Santsenca, adaption is an energetic and cultural process.

The four examples discussed here create bespoke comforts by integrating adaptive ideas with a diversity of design approaches. These range from the compact formal organization of the Life Science Building to the dispersed one of the House for Trees. The material palette can be limited as in the International House Sydney or expansive like the Lleialtat Santsenca civic center. The concept of comfort even stretches beyond the immediate context of the buildings to say something about larger climatic and social conditions, as in the case of tropical Vietnam and historical Barcelona. All the projects provide a uniform level of comfort through artificial climate control when and where it is needed, but each creates connections between interior and exterior climates in unique ways that are integral to their specific conditions.

Seen collectively, these visually appealing projects demonstrate that adaptive comfort can contribute to architecture's image-based aesthetic culture. One should not underestimate the importance of giving a contemporary visual identity to the experience of comfort through projects such as these. As historians of technology remind us, there is no innovation without representation. In order for adaptive ideas to gain attention within the design community, architectural comfort needs to be recognized and celebrated through such images.

CONCLUSION

Architecture improves the weather through attention to comfort. To accomplish this goal, architects need to understand comfort in new, more expansive ways. During the 20th century, comfort became understood through the narrow but politically effective PMV model. This model prescribed a limited band of environmental conditions that a majority of people were expected to find thermally neutral. Such conditions were provided almost exclusively by mechanical equipment and largely without the design input of architects. Spread through the wide-scale adoption of ASHRAE's comfort standard, this limited understanding of comfort became naturalized as a matter of technical and legal fact. Comfort researchers unhappy with this limited idea of comfort developed an adaptive approach that reduced energy demands while allowing for experiences of thermal delight. The emergence of the ATC model highlights the fact that comfort is historically conditioned by social circumstances as well as by technical developments. Comfort ideals change over time and when considered by different groups.

The historical account and contemporary examples presented here demonstrate how architects think about comfort in ways that are different from but complementary to how comfort researchers think about it. Ideas originating in the research community should be of interest to the design community. Spatial and temporal alliesthesia can engage architects' spatial and temporal intelligence and imagination in the design process. Behavioral adaptations rely on architectural choices such as flexible plan configurations and operable façade elements. And psychological adaptation can be triggered by diverse environments within a building's interior climate. Similarly, essential elements of architecture such as massing, spatial organization, and material choices deserve to be considered by researchers for their effects on comfort. Clearly, there needs to be more dialogue between these two groups, establishing a common ground between their different areas of expertise. For architects, consciously addressing comfort as a fundamental design consideration can expand their professional purview to shape interior environments directly and to improve buildings' energy performance. For comfort researchers, understanding how adaptive comfort opportunities are embedded in building design can help to direct research in ways that are more readily translated into practice. Collectively such work counters the dominant narrative that energy savings result in reduced comfort while sharing a commitment to a more thermally pleasurable and less energy-intensive architecture. In recognizing these common goals, both groups can work to bring this expanded notion of comfort to a wider group that includes manufacturers, clients, and building occupants.

Comfort is good to think with. A design-based approach to comfort can improve the impoverished sensoria commonly created in building interiors. It can activate architects' environmental imagination to propose novel design solutions that address the pressing challenges of climate change. And it can demonstrate not only how architecture improves the weather but how the weather can improve architecture.

REFERENCES

Addington, D. M. (2001). The history and future of ventilation. In J. D. Spengler, J. M. Samet, & J. F. McCarthy (Eds.), *Indoor air quality handbook* (pp. 2.1–2.16). New York, NY: McGraw-Hill.

Addis, B. (2007). *Building: 3000 years of design, engineering and construction.* New York: Phaidon Press.

Arens, E., Humphreys, M. A., de Dear, R., & Zhang, H. (2010). Are 'class A' temperature requirements realistic or desirable? *Building and Environment*, *45*(1), 4–10. doi:10.1016/j.buildenv.2009.03.014

ASHRAE. (1960). *Heating, ventilating, air conditioning guide*. New York, NY: ASHRAE.

ASHVE. (1924). Celebrate thirty years of progress. *Journal of the American Society of Heating and Ventilating Engineers, 30*(3).

Banham, R. (1969). *The architecture of the well-tempered environment*. Cambridge, MA: MIT Press.

Bernard, C. (1879). *Leçons sur les phénomènes de la vie communs aux animaux et aux végétaux*. Paris: J. B. Baillière. doi:10.5962/bhl.title.103133

Cabanac, M. (1971). Physiological role of pleasure. *Science. New Series, 173*(4002), 1103–1107. PMID:5098954

Cannon, W. B. (1932). *The wisdom of the body*. New York, NY: W. W. Norton. doi:10.1097/00000441-193212000-00028

Chang, J.-H. (2016). Thermal comfort and climatic design in the tropics: An historical critique. *Journal of Architecture (London), 21*(8), 1171–1202. doi:10.1080/13602365.2016.1255907

Chester, T. (1942). Comfort with summer air conditioning. *ASHVE Transactions, 48*, 107–122.

Comstock, S. (1995). *Proclaiming the truth: An illustrated history of the American Society of Heating, Refrigerating and Air-Conditioning Engineers, Inc.* Atlanta, GA: ASHRAE.

Cooper, G. (1998). *Air-conditioning America: Engineers and the controlled environment, 1900–1960*. Baltimore, MD: Johns Hopkins University Press.

Cruse, A. (2019). The temperature of a SANAA. In B. Wilke (Ed.), *Ryue Nishizawa / SANAA: Grace Farms* (pp. 62–69). New York, NY: Applied Research and Design.

de Dear, R. (2011). Revisiting an old hypothesis of human thermal perception: Alliesthesia. *Building Research and Information, 39*(2), 108–117. doi:10.1080/09613218.2011.552269

de Dear, R. J., & Brager, G. S. (1998). Developing an adaptive model of thermal comfort and preference. *ASHRAE Transactions, 104*(1), 145–167.

Evans, R. (1997). Figures, doors and passages. In *Translations from drawing to building* (pp. 54–91). Cambridge, MA: MIT University Press.

Fanger, P. O. (1970). *Thermal comfort: Analysis and applications in environmental engineering*. Copenhagen, Denmark: Danish Technical Press.

Fernández-Galiano, L. (2000). *Fire and memory: On architecture and energy* (G. Cariño, Trans.). Cambridge, MA: MIT Press.

Harikae, A. (2016, July). Vo Trong Nghia Architects. *Architecture and Urbanism, 550.*

Harquitectes. (2016). Organizing matter. *2G: Harquitectes, 74*.

Heschong, L. (1979). *Thermal delight in architecture*. Cambridge, MA: MIT Press.

Hochhäusl, S., Lange, T., Adams, R. E., Barber, D. A., Bierig, A., Cupers, K., ... Vronskaya, A. (2018). Architecture and the environment. *Architectural Histories, 6*(1), 1–13. doi:10.5334/ah.259

Houghten, F. C., & Yaglou, C. P. (1923). Psychrometric chart with equal comfort lines or effective temperatures for still air. *ASHVE Journal, 29*, 165–176.

Leeds, L. (1871). *Treatise on ventilation*. New York, NY: John Wiley & Sons.

Leopold, C. S. (1947). Conditions for Comfort. *ASHVE Transactions, 53*, 295–306.

Moe, K. (2014). *Insulating modernism*. Boston, MA: Birkhäuser. doi:10.1515/9783038213215

Nevins, R. (1961). Psychrometrics and modern comfort. *ASHRAE Transactions, 67*, 609–621.

Nghia, V. T. (2016, December). *Lifestyles in Asia in the future—Ideas from history, culture and the environment* [Video file]. Retrieved from https://www.youtube.com/watch?v=PSdD16ox48s

Nicol, F., Humphreys, M., & Roaf, S. (2012). *Adaptive thermal comfort: Principles and practice*. New York, NY: Routledge. doi:10.4324/9780203123010

Nyrud, A. Q., & Bringslimark, T. (2010). Is interior wood use psychologically beneficial? A review of psychological responses toward wood. *Wood and Fiber Science, 42*(2), 202–218.

Parkinson, T., de Dear, R., & Candido, C. (2016). Thermal pleasure in built environments: Alliesthesia in different thermoregulatory zones. *Building Research and Information, 44*(1), 20–33. doi:10.1080/09613218.2015.1059653

Rietschel, H. (1927). *Heating and ventilating: A handbook for architects and engineers* (C. W. Brabbee, Trans.). New York, NY: McGraw-Hill.

Sellers, C. (1977). *Hazards of the job: From industrial disease to environmental health science*. Chapel Hill, NC: University of North Carolina Press.

Shove, E. (2003). *Comfort, cleanliness and convenience: The social organization of normality*. New York, NY: Berg.

Tredgold, T. (1836). *Principles of warming and ventilation*. London, UK: M. Taylor.

Wastiels, L., Schifferstein, H. N. J., Heylighen, A., & Wouters, I. (2012). Relating material experience to technical parameters: A case study on visual and tactile warmth perception of indoor wall materials. *Building and Environment*, *49*(1), 359–367. doi:10.1016/j.buildenv.2011.08.009

Yaglou, C. P. (1932). How to use the effective temperature index and comfort charts. *ASHVE Transactions*, *38*, 411–423.

ADDITIONAL READING

Brager, G. S., & de Dear, R. J. (1998). Thermal adaptation in the built environment: A literature review. *Energy and Building*, *27*(1), 83–96. doi:10.1016/S0378-7788(97)00053-4

Brager, G. S., & de Dear, R. J. (2003). Historical and cultural influences on comfort expectations. In R. J. Cole & R. Lorch (Eds.), *Buildings, culture and environment: Informing local and global practices* (pp. 177–201). Oxford, England: Blackwell. doi:10.1002/9780470759066.ch11

Chang, J.-H., & Winter, T. (2015). Thermal modernity and architecture. *Journal of Architecture (London)*, *20*(1), 92–121. doi:10.1080/13602365.2015.1010095

Chappells, H., & Shove, E. (2004, March). *Comfort: A review of philosophies and paradigms*. Retrieved September 2019 from http://www.lancs.ac.uk/fass/projects/futcom/fc_litfinal1.pdf

Crowley, J. (2001). *The invention of comfort: Sensibilities & design in early modern Britain & early America*. Baltimore, MD: Johns Hopkins University Press.

de Dear, R. J., Akimoto, T., Arens, E. A., Brager, G., Candido, C., Cheong, K. W. D., ... Zhu, Y. (2013). Progress in thermal comfort research over the last twenty years. *Indoor Air*, *23*(6), 442–461. doi:10.1111/ina.12046 PMID:23590514

Murphy, M. (2006). *Sick building syndrome and the problem of uncertainty: Environmental politics, technoscience, and women workers*. Durham, NC: Duke University Press. doi:10.1215/9780822387831

Prieto, E. (2014). The welfare cultures: Poetics of Comfort in the Architecture of the 19th and 20th Centuries. In F. Graf & G. Marino (Eds.), *Les dispositifs du confort dans l'architecture du XXe siècle* (pp. 58–67). Lausanne, Switzerland: Presses Polytechniques et Universitaires Romandes.

Shove, E., Walker, G., & Brown, S. (2014). Material culture, room temperature and the social organization of thermal energy. *Journal of Material Culture*, *19*(2), 113–124. doi:10.1177/1359183514525084

KEY TERMS AND DEFINITIONS

Adaptive Responses: Factors that affect thermal comfort beyond the body's heat balance with the surrounding environment. There are generally considered to be three adaptive responses: behavioral adjustment, physiological acclimation, and psychological expectation. The response recasts building occupants as active agents in shaping their comfort rather than as passive recipients of comfort conditions.

Climate Chamber: A windowless room in which temperature and humidity can be precisely set and independently controlled. Also called *air conditioned*, *environmental*, or *psychrometric* chambers.

Heat Balance: A mechanistic approach to comfort based on the idea that when the heat produced by metabolism is balanced with a room's air temperature and humidity, people feel comfortable.

Predicted Mean Vote (PMV) Comfort Model: Developed by comfort researcher P. O. Fanger, the PMV comfort model equates comfort with the heat balance between a body and its immediate environment. It is based on the four environmental variables of dry bulb temperature, relative humidity, mean radiant temperature, and air velocity, as well as the two personal variables of activity level and the insulation value of clothing. It is typically associated with cool, dry still air provided by air conditioning in which people are largely sedentary and wearing typical business attire.

Predicted Percentage Dissatisfied (PPD): An estimation of how many people will find PMV comfort conditions thermally satisfactory. Danish comfort researcher P. O. Fanger, who developed the PMV comfort model, considered conditions acceptable when 80% or more of occupants were satisfied with the building's interior climate. Importantly, such satisfaction reflected a condition of thermal neutrality, where building occupants did not object to interior environmental conditions, rather than one of thermal pleasure.

Spatial Alliesthesia: The experience of different thermal stimuli on parts of the body as captured by the folk expression describing comfort as resulting from "warm feet and cool head."

Temporal Alliesthesia: The experience of temporary variations in thermal experiences over relatively short periods of time, such as moving from a patch of shade into the sun on a winter walk, or rounding a street corner and stepping into a cool breeze on a humid summer day.

Chapter 10
Post-Normal Material Practice:
"Building" as Inquiry

Jacob Wayne Mans
University of Minnesota, USA

ABSTRACT

This chapter explores building as a form of material inquiry. The process of building generates new ideas and questions that remain latent in unbuilt designs. These ideas and questions are uniquely trans-scalar and boundary spanning as compared to material inquiries that focus on isolated material attributes. Building projects embed material inquiries within the open systems that make up our environment. Thinking about material performance in this way can co-produce political, social, economic, and ecologic relationships that extend design agency beyond the artifact.

INTRODUCTION

Buildings consist of energy, concentrated into matter, organized into materials, processed into assemblies, used as construction units and/or system components to form building systems further configured into spaces that serve social and/or environmental purposes. This trans-scalar material description connects molecular formation to territorial performance (and vice versa) through feedback loops that facilitate energy flows across scales. If the concern surrounding architecture is that architects are not developing solutions that have a large enough positive impact (or a small enough negative impact) toward the current environmental crisis, then perhaps we need to *re-evaluate* the impact of architecture toward a system scale that does.

This essay explores *building* as a form of material inquiry. It is concerned with *scale* in general over any particular metric. It seeks a way to evaluate materials

DOI: 10.4018/978-1-7998-2426-8.ch010

through the extended relationships they create rather than their intrinsic properties. Scale not only refers to a thing's physical size, but to a thing's relational influence; to its network. So, while a building is a relatively small-scaled physical thing, the process of building creates a very large-scaled relational network. A network that is socio-technical in its nature and that *co-produces* political, social, economic, and ecologic relationships through the production of a building. Measuring and designing for the improved performance of these extended relational networks through the co-production of systemic improvements that relate such things as economic opportunity, political representation, social equity, and/or ecological regeneration shifts the agency of the architect toward much larger scales of systems requiring a recharacterization of architectural impacts.

The motivation behind constructing a post-normal material practice is twofold. First, it is to validate and curiously consider the immense amount of pragmatic, theoretical, and performative material knowledge attached to architecture's extended relational networks—knowledge that extends beyond research institutes, academic faculties and professionals. The expert and non-expert alike are participants in this discourse, with the non-expert playing a critical role in establishing new ideas that disciplinary researchers and professionals struggle to articulate. Second, is to consider the opportunities that an architecture optimized to improve the performance of these larger systems has on the physical formation and materials of the this we build.

This process focuses on articulating the right question rather than providing a particular solution for the pre-existing architectural question of "green building." There is a substantial difference between getting the answer to the wrong question right and working on the right question but getting the wrong answer. The latter is particularly useful, the former—deceptively misleading. The difference often lies in whether one is seeking system efficiency or system effectiveness (Ackoff, 2003). This essay is concerned with the systems that architecture engages and claims that "a building" (noun) is actually the wrong question, due in part to its preoccupation with its own operational efficiency. It poses the question of "building" (verb), which is relational, complex, and can be concerned with the operational effectiveness of communities.

This chapter does not define an ideal scale at which architecture becomes "effective", but rather a method for questioning our assumptions about what architecture can and should be optimized to *impact.* The chapter consists of three parts. The first part focuses on the conceptual framework of a post-normal material practice. It focuses on three key concepts central to the argument: *post-normal science*, *abductive reasoning*, and *ignorance as a design method.* The second part presents three *built* projects through a post-normal framework. The rationale for reporting on these projects is to examine the impact-based variables and insights that emerged through the process of building and impacts co-produced through the projects. The essay concludes with a section calling for a more critical discourse on *built* work as a collective research project that spans professional/academic realms as well as expert/non-expert boundaries.

PART 1: A POST-NORMAL MATERIAL PRACTICE

A post-normal material practice becomes necessary when making material decisions that seek to influence a particular kind of problem set—*wicked problems* (Rittel & Webber, 1973). "Wicked problems hold the key to the most consequential breakthroughs of the 21st Century," writes John Kao (2007) in *Innovation Nation,* and that is true for the built environment as it is for every other area of human activity. Although today's "wicked problems crowd us like piranha," as Marty Neumeier (2009) observes in *The Designful Company,* the discipline of architecture needs to embrace them, not only because they comprise most of the problems worth solving, but also because designers know how to address such ill-defined, complex challenges such as pollution, overpopulation, inequality, and climate change. How might we reconsider material performance in this context? What opportunities for co-production emerge when we open our material inquiries to more broadly consider wicked problems?

Architects deal with wicked problems and socio-technical solutions in almost every project. This said the profession still too often accepts as normal the drive-in capitalism and in science to either treat wicked problems as insoluble or as tame problems in disguise, resulting in solutions that either ignore or repress the real complexity of the most pressing challenges we face.

Post-Normal Science

A concept that is particularly useful when considering how to operate on material inquiries that contain high levels of uncertainty is "Post-Normal Science" (PNS). The concept of PNS developed in the early 1990s (Funtowitz & Ravetz, 1993) as an expansion of a scientific critique from the 1960s (Kuhn, 1962) that characterized "normal" science as being overly constrained to studying problems with relatively low uncertainty and relatively narrow systems boundaries. This critique described the progression of normal scientific inquiry from the enlightenment onward as being focused on dividing systems into smaller and smaller increments (isolated from the at large system to minimize uncertainty) that are studied by more and more esoteric experts and specialists. Recognizing that normal science was uninterested or unable to establish inquiries within the levels of uncertainty attached to non-linear, far from equilibrium, systems (Kay et al, 1999) and/or the uncertainties attached to the application of normal science toward human policy decisions (Funtowitz & Ravetz, 1993), the concept of "post-normal science" (PNS) was developed. PNS established an intellectual framework for developing scientific inquires that operated under the assumption of unpredictability, incomplete control, and a plurality of legitimate perspectives.

A "Post-Normal" approach to science and policy, rooted in the wicked-problem-solving capacity of design, offers us a way to grapple with that complexity, and to achieve the consequential breakthroughs, that we need to when operating at larger scales. Some of the conditions that characterize a post-normal situation include irreducible complexity, deep uncertainties, multiple legitimate perspectives, value dissent, high stakes, and urgency of decision-making. PNS calls for extended peer communities encompassing broader notions of knowledge, uncertainty management, and acknowledgment and management of multiple valid perspectives. Unlike normal science, the goal is not to attain certain knowledge. The goal of PNS is quality, a more robust 'science for policy' (Ravetz, 1987). This approach requires long-term relationships with diverse communities and points of view. It adjusts its methods and alters its materials based on local conditions, cultures, and resources. In addition, it combines different disciplines in a diversity of places in order to learn from as many other perspectives as possible. The post-normal approach outlined here focuses on the adoption of a set of practices that increase the number and disciplinary complexity of collaborative relationships.

This complexity expands the scale of system boundary considered during the co-production of architectural interventions and opportunities that extend beyond buildings. When we apply the PNS framework to architecture, we see similar patterns. Contemporary practice has not leveraged advances in information and design technology to operate on problems with higher levels of uncertainty. Instead, it has specialized in narrowing bands of performances in order to optimize isolated aspects of "a building's" performance.

Abductive Reasoning

If PNS frames the type of extended relational network needed to research material inquiries with increased levels of uncertainty, abductive reasoning shapes the methodology used to conduct research on those inquiries. This form of reasoning can structure an alternative form of empirical inquiry that moves away from inductive methods and explicit knowledge toward a design-oriented process that can shape important design questions (possibly framed by other nested empirical inquiries) and identify opportunities for design intervention within them.

For many of the projects outlined below, had they "gone to plan," many of the key insights derived from their making would never have materialized. In fact, they would never have taken place had their research stayed within the bounds of inductive or deductive reasoning. It is not quite right to say that the projects failed, but rather that they progressed from scenario to improved scenario over and over again. Abductive methods require a perpetual state of inquiry; conclusions are drawn only inasmuch as they serve to perpetuate the inquiry. For Charles Sanders Peirce, “in pure abduction, it can never be justifiable to accept the hypothesis otherwise than as an interrogation (Peirce & Buchler, 1955).” As designers, we are comfortable with this practice, but the empiricist in us puts its foot down on the methodological inconsistencies that this scenario-hopping induces. Thus, abductive reasoning pairs well with PNS; the latter accounts for the uncertainties produced by the former. Over the short lives of these built projects described below, they have changed location, size, number, form, material, and occupancy. Each shift was a redirection that the scale of inquiry at the very core of these projects that progressed the research from deductive theses to inductive experiments and finally toward abductive arguments. Pierce used the following syllogisms to help illustrate the differences between deduction, induction, and abduction (Peirce et al., 1992):

Deduction

rule: all beans from this bag are white
case: these beans are from this bag
result: these beans are white

Induction

case: these beans are from this bag
result: these beans are white
rule: all beans from this bag are white

Abduction

result: these beans are white
rule: all beans from this bag are white
case: these beans are from this bag

Shank (1998) expands Pierce’s syllogisms to clarify the points needed to understand the abductive model.

Deduction

rule: [it is true that] all beans from this bag are white
case: [we know that] these beans are from this bag
result: [certainly, it is true that] these beans are white

Induction

case: [we know that] these beans are from this bag
result: [we have observed that] these beans are white
rule: [probably then] all beans from this bag are white

Abduction

result: [we have experienced that] these beans are white [but this experiences lacks any real meaning for us]
rule: [the claim that] all beans from this bag are white [is meaningful in this setting]
case: [therefor it is both plausible and meaningful to hypothesize that] these beans are from this bag

Before moving on, it is worth expanding on these different kinds of inferences so it is clear how and why they differ.

First, *deduction* refers to an inference based on a reference to a general law or principle: these beans are white because I know them to have come from a particular bag that contains white beans. Deductive inferences can only correlate claims to pre-existing truths that are either obvious or self-evident.

Second, *induction* refers to an inference made from a series of individual observations that move toward a broader generalization: all the beans in this bag are white because the particular beans in my hand are white, and I know them to have come from this bag. In order to have confidence in an inductive inference, one has to understand the phenomena that surround it sufficiently. The scientific method is inductive and uses "scientific theories" to describe the phenomena that surround its inferences (Shank, 1998). Induction always carries some level of uncertainty even when all the conditions of the inference are met—there is no way of knowing that there are not blue beans in the bag by there being white beans in my hand.

Third, *abduction* refers to an inference that moves from a unique observation that is explained by a logical rule to determine a common explanation for the observation: I observe that the beans are white and because all beans from the bag are white, I guess that these beans are from the bag. What is important about abduction is that like induction, it doesn't require reference to a pre-existing truth, but unlike induction, it is not concerned with accounting, through some theory, for all the phenomena that surround the inference. It is a kind "hunch." Abductive inference is the cognitive method that we use to explain the unique observations we make at any given moment. In design, it reflects the combinatory efforts we make cognitively to link disassociated variables to one another—abductive inference describes the creative process (Fisher, 2016).

Making a distinction when inquiring about materials to either deduce, induce, or abduce is important. It affects our perception of the material, the perceived success of the research attached to it, and our willingness to shift an inquiry as new, often unexpected and sometimes unwanted, knowledge is gained. The nimbleness that abduction affords is critical for structuring material research that is *searching* for the most effective scale of impact. Prior to being built, the network is narrow and the scale of inquiry small. However, as the research unfolds, through the process of building, the network expands, relationships are structured, and an appropriate scale of impact is defined.

Operationalized Ignorance

Having framed both the kind of inquiry and the methodology used to research it, the question still remains: how to do we operate on the variables of material systems that we do not fully know or that we know contain high levels of uncertainty? How do we structure a design whose impact is defined through its production and prior to its definition? Generally, we approach unknown and uncertain variables by excising them from our system boundary, isolating them within our system boundary, or outsourcing them to an external consultancy. The option not generally considered is to *try* and understand them openly within our system boundary, which we resist because we are constrained to operate within our pre-established silos of expertise and under our own predetermined concepts of performance.

Knowing a material system fully and completely is impossible. When operating on material systems, we hold only partial knowledge (some of which is likely inaccurate or misunderstood). We also know that there are parts of material systems that are unknown to us (known unknowns), and we should acknowledge to the rest of the system that we are ignorant about (unknown unknowns)—likely the vast majority of the system. In every case, we are infinitely more *ignorant* than we are knowledgeable, which in and of itself suggests a particular course of action (Vitek & Jackson, 2008).

The term *ignorance*, as used here, is a pragmatic assessment of the real state of knowledge that we need to acknowledge if we are to, more effectively, impact the world. It is not meant to comment on the misapplication or omission of well-established facts, but rather to critique our diminished ability to characterize the complexity and uncertainty of complex systems in order to identify potentially consequential future states (Roy & Zeckhauser, 2015). States that we find through abduction. The end goal is not to replace ignorance, but to cope with it. To acknowledge that we should expect the unexpected. While this sounds inefficient, it is so only if our intention is to operate at small scales and isolate ideas of architectural performance to "a building." If our goal is effectiveness, then iteration and repositioning are efficient steps in finding the right scale at which to optimize our initial designs. There is such a thing as "usable ignorance" that informs how we operate on complex systems by refocusing our interventions toward feedback loops (Ravetz, 1987)—dynamic change-makers within a dynamic system—rather than submitting to the status quo and static system states.

Rather than turning away from uncertainty and questions that relate to the dynamic processes of building, architects should develop practices to expand the relational networks that allow us to positively impact large-scale systems of planetary significance through the co-production of "a build." The philosopher Anna Peterson, in drawing a connection between ignorance and relational ethics states that:

In an ignorance-based model, we no longer assume that we can know what people will do and what will result from the actions of various parties, including ourselves. In the absence of such knowledge, moral decisions and actions are grounded in relationships and in people's individual capacities to persevere in the face of uncertainty. (Peterson, 2008, p. 133)

The shift toward relational research and pedagogical models is critical for operating within extended relational networks. The goal of collaboration is not simply to augment or buttress existing expert knowledge but to shift the way we think about architectural performance in general. There is a difference between *extracting* information on a system *from* an extended relational network and *shifting* how one understands a system by engaging with an extended relational network (Freire, 1970). This analogy extends to the difference between positioning oneself as an expert and positioning oneself as a non-expert collaborator. The expert extracts knowledge to validate his or her preexisting expertise; the non-expert shifts his or her expertise in response to outside knowledge. Given the complexity and uncertainty of building, it seems prudent for the architect to position him or herself as a non-expert (Mans, 2017) when positing these kinds of "post-normal" questions for architecture.

Together the ideas embedded in *post-normal science, abductive reasoning*, and *operationalized ignorance* provide a structured way to frame material performance, one that is deeply rooted in the design thinking process and is actively seeking to find ways to optimize material performance at scales where it can impact wicked problems. If architecture is to achieve the planetary significance that architects ascribe to it, its performance needs to be optimized at scales that have sufficient energetic power and ecological significance to achieve this goal. In this context, "a building" is not a sufficient scale of inquiry to support such a claim.

PART 2: BUILT PROJECTS

The act of building is a form of research that reveals new material opportunities. This section reports on three *built* projects. As currently constructed, the projects consist of seven small buildings that, when viewed through a post-normal material framework, enable new material opportunities generated through the co-production of architectures optimized toward the performance of political, social, economic, and ecologic relationships.

A few general notes about the projects. The building projects have emerged over the course of roughly three years and span teaching, research, and practice contexts. I note this because building as a research practice is cumulative and transferable. As feedback loops are identified through projects, they fundamentally shift how one frames the system's boundary of future projects. Not all projects identify *new* feedback loops, but knowledge and identified feedback loops do compile and transfer over time. The projects discussed are physically small and prototypical in scale.

The Littleton Trials

Location: Littleton, Massachusetts

Project Partners: David Kennedy, Benjamin Peek, the Northeastern Lumber Manufacturers Association, the Softwood Lumber Board, the New England Forestry Foundation, the Massachusetts Department of Natural Resources, Harvard Energy and Environments Lab

Project Year: 2016

The Littleton Trials (TLT) consist of a set of small wood structures constructed to probe the influence of construction logic and species on the thermal performance of wood buildings. This project emerged out of a broader interest in the thermal performance and was eventually re-focused on wood given the range of densities across different wood species that can be sourced and configured into wood constructions. At a basic level, the project sought to understand how the physical variables of density (Forest Products Laboratory, 2010) and specific heat capacity (Radmanovic et al., 2014) influence the thermo-physical exchanges of heat between wood materials and their environments (Hameury & Lundström, 2004).

This inquiry was modeled after an experiment conducted in Trondheim, Norway (Moe, 2014) in the 1930s in which several small huts were designed and built using different materials and construction logics and then outfitted with sensors to measure their thermal performance. TLTs consisted of two small huts (Figure 1) built of four distinct mass timber construction logics (Nail Laminated Timber, Dowel Laminated Timber, Nail Cross Laminated Timber, and Dowel Cross Laminated Timbers). These panels were made of Spruce, Pine, Fir south and were modified with secondary constructions of various hardwood species layered on top of them. Thermal data from the different panels were collected using heat flux sensors (Figure 2) and infrared camera (Figure 3) to calculate conductivity and effusivity (Mans et al., 2018). The measurements we recorded generally aligned with project values we calculated based on the thermal wood literature.

Figure 1. Huts from the little trials, May 2016

But not exactly;

...the numerical ranges that accompany the results represent material defects and environmental inconsistencies within the test sampling. Our abductive line of thermal reasoning allowed us to speculate about the design potential of the material

Figure 2. TLT heat flux sensor setup, April 2016

variations within these ranges rather than isolating and removing them as material defects. Instead of optimizing a thermal solution based on a theoretically idealized material, our project focused on the sub-optimizations that could positively result from the material variations with wood. The concept of thermally tuned mass timber panels emerged from this line of questioning. The basic concept looks at strategically placing layers of higher density wood (our selected species include ash, birch, beech, and black locust) with higher volumetric heat capacity on the interior face of a panel and lower density wood (our selected species included eastern white pine, SPFs, and eastern hemlock) with lower conductivity on the exterior face of the panel. (Mans et al., 2018, p. 197)

Figure 3. TLT infrared camera setup, April 2016

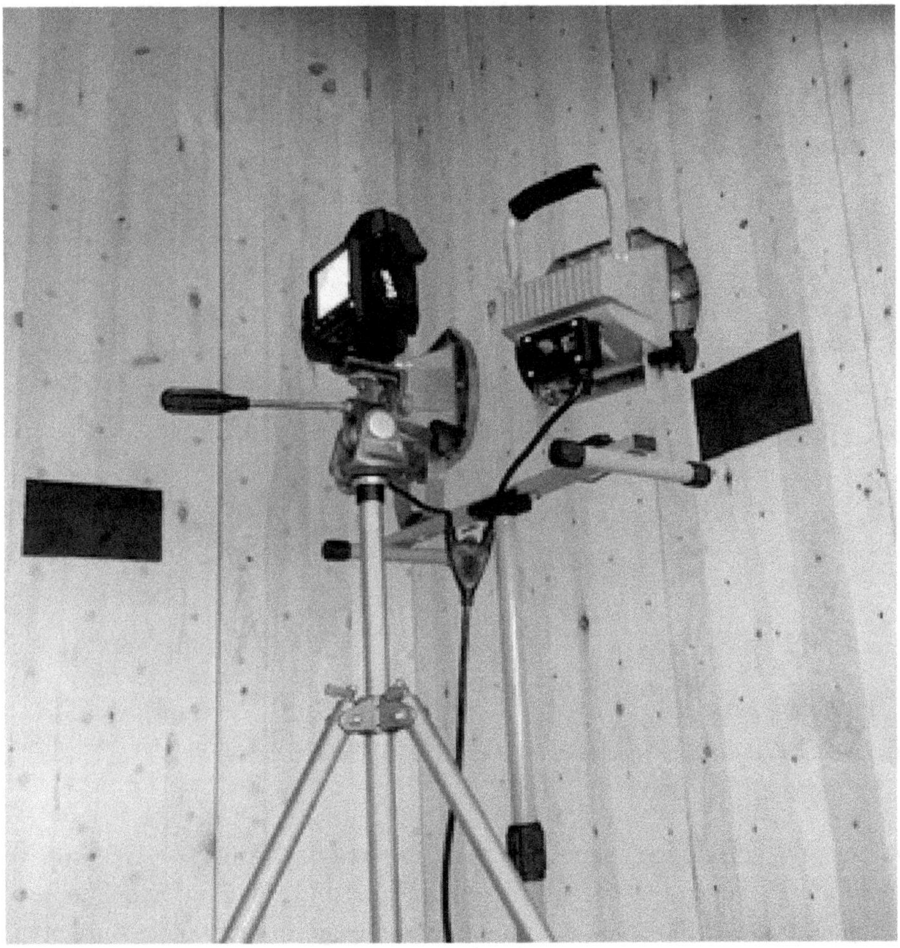

What was initially suspected as setup errors and defects in the test samples began to redirect the research in profound ways. The "defect" became a design opportunity. We initially perceived the defect as a negative because it caused our results to deviate from those suggested by the literature (Figure 4). However, there is nothing inherently negative about an increase in a material's density. Characterizing the material in this way is conditional—based on ideas of wood grading that devalue material based on structural imperfections. Thermally these defects represent an opportunity to shift performance.

Had the project had remained on its initial "thermal experiment" trajectory, the impact would have remained at scale of the small huts we had built. However, as the huts were being built the project, we began to understand their impact differently.

Figure 4. LTL thermal Image of wood sample showing variations in wood density, May 2016

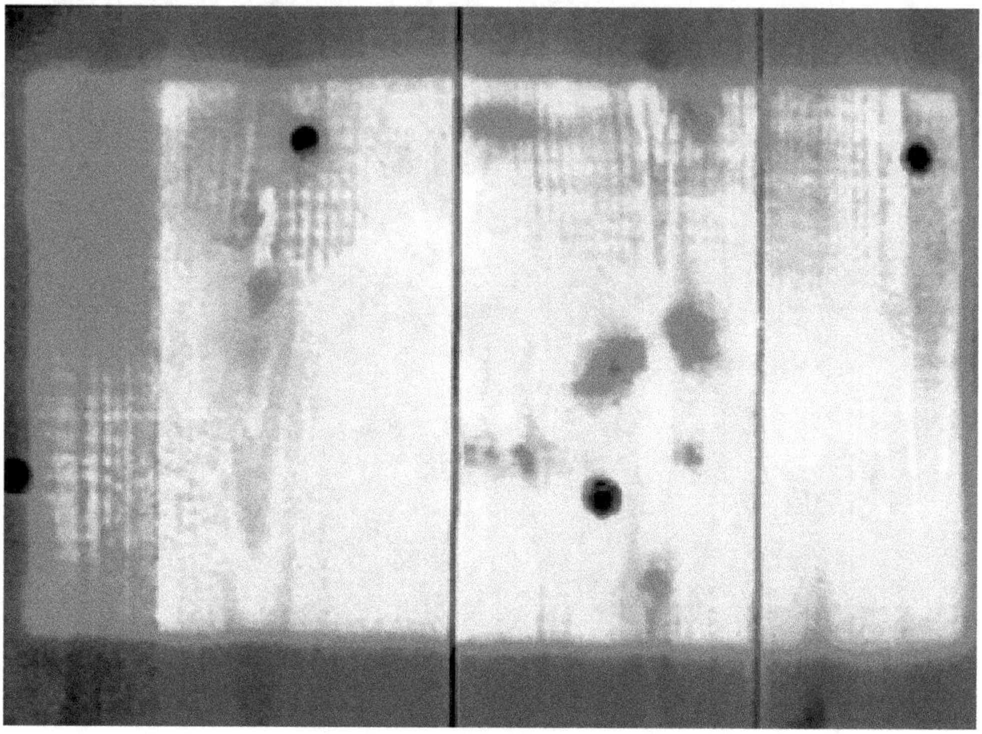

An interesting, and maybe not obvious thing to point out, is that constructing even a small building takes a considerable amount of time and planning. Even more so for this project, because the construction logics we tested in the project were atypical and one-off in both their setup and execution. This meant that for the better part two months, the team effectively lived within the construction site. Located on the New England Forestry Foundation (NEFF) property in Littleton, MA, the construction of the project doubled as a kind-of ad hoc forest management residency amongst the NEFF community who were actively working with landowners in the region on the development of forest management plans. By matter of sheer proximity, the project created a unique relationship between foresters and architects that shifted the scale at which we came to define the potential performance of wood buildings.

The key abductive moment, the critical association made between seemingly dissimilar understandings of "wood" as material, returned to this idea of the "defect." Both forester and architect related the term to diminished value; however, our value systems were quite different. The architectural defect related to structural capacity—a break in grain structure that weakened the material physically. The forestry defect

Figure 5. Red Maple harvest with foresters from the New England Forestry Foundation, April 2016

related to marketability and the economics of whether it was affordable to manage a particular forest—no market = no management. The forest is a vastly more powerful system than a building. The project sought to tap into this larger forest system by reframing the performance of wood architecture through a forest management regime that we came to understand through our relationship with NEFF.

The question became, "how do we optimize a wood building increase the performance of a forest?" When considering the forest at the scale of the individual tree, species selection and adjacent stand density are important variables. We walked the forest and identified and marked low-valued (red maple) trees that were in the immediate proximity to more valued species (eastern white pine) (Figure 5) with the goal of harvesting to re-direct solar and nutrient access to the higher-valued species. The local disturbance of falling one tree becomes a local opportunity for another, which can now grow faster and accumulate more value. We observed the impact of selective tree thinning on sections of the forest and discussed with foresters the variables considered when making a decision to harvest a particular tree now or letting it grow to increase its value, which in turn increased the risk of the tree getting damaged between this and a subsequent thinning and thus losing its value. We saw how selective patch cuttings at various levels of succession operated with one another; cuttings of less than half an acre in size that yielded a mosaic pattern across the forest.

We associated this management scale with our material research. Both were nuanced and required considerable planning, and both were selective at a species level. For this type of management to be feasible, the forest, as an economic resource (for the landowner), needed to, at a minimum, match the cost of management and harvesting. The cost of harvesting is dependent on stumpage value plus the harvester's markup. One way to achieve this is to harvest higher-valued species in order to subsidize the harvest of lower-valued species. Another option is to create new markets that add value to undesirable species so that landowners can afford to harvest and sell them without having to rely on the harvest of higher-valued species. In either case, managing a forest to improve its ecological performance, requires extracting value from it.

Of the two options outlined above, we focused on the development of new markets. The under-valued species that we focused on was red maple. One reason for this low economic value is the species' tendency to have a large number of material defects. In general, red maple is a relatively dense wood that contains a high level of material defects that relates to a relatively low corresponding structural capacity. We translated these properties into panel designs that provide strength and lower conductivity on the exterior of the panel (where it is needed), and as higher thermal heat capacity on the interior of the panel (where it is needed) (Figure 6). The panels were a hybrid of SPFs NLT and hardwood nCLT that were structurally sub-optimized at the building-scale (through the inclusion of lower quality hardwoods) to better optimize the building's performance at the scale of the forest (through the development of new wood markets that would spur increased management and ecological performance of the forest).

The project, as initially framed, asked the wrong question of wood. Its optimization was limited to "a building," which in turn limited its impact. The Post-Normal framework shifted the question toward an architecture that performed as a market driver for extracting lower quality material from the forest in support of increased management. Critical to redefining this question was the creation of the extended relational network developed with the NEFF that provided the necessary outside expertise needed for the architectural project to shift scales. This abductive process did not erase the initial thermal inquiry; it still exists, only nested within a scale of inquiry that is now more powerful and potentially impactful.

Anoka County Learning Kiosks

Location: *Anoka, Minnesota; Cloquet, Minnesota*

Figure 6. Thermally tuned mass timber panel from TLT, May 2016

Project Partners: *Anoka County, University of Minnesota Cloquet Forestry Center, SmartLam, Crosby Hill Plumbing and Heating, Savanna Pallets*
Project Year: 2017 to present

Building on the ideas of The Littleton Trials, but shifting location to Minnesota, the Anoka County Kiosk project's initial inquiry sought to develop a local wood architecture based on increasing the value of domestically harvested timber. Unlike the Littleton Trials, the project was not established as a thermal experiment but as a two-phase design-build studio. The first phase focused on a cohort of students developing individual research projects (Figure 7) that unpacked several trans-scalar variables of wood across Minnesota. The studio explored topics of wood production, wood species, wood decay, wood building vernaculars, and tall and small wood buildings to establish a shared discourse around local wood architecture. The research was interested in exploring the optimized alongside the sub-optimized, new construction methods alongside the old, and long-term solutions alongside short-term. The phase ended with students developing speculative design interventions based on their research.

The second phase of the studio started with a new cohort of students extracting ideas from these projects and applying them to the program for our design-build project.

Figure 7. Student Research project investigating wood processing cycle and wood decay

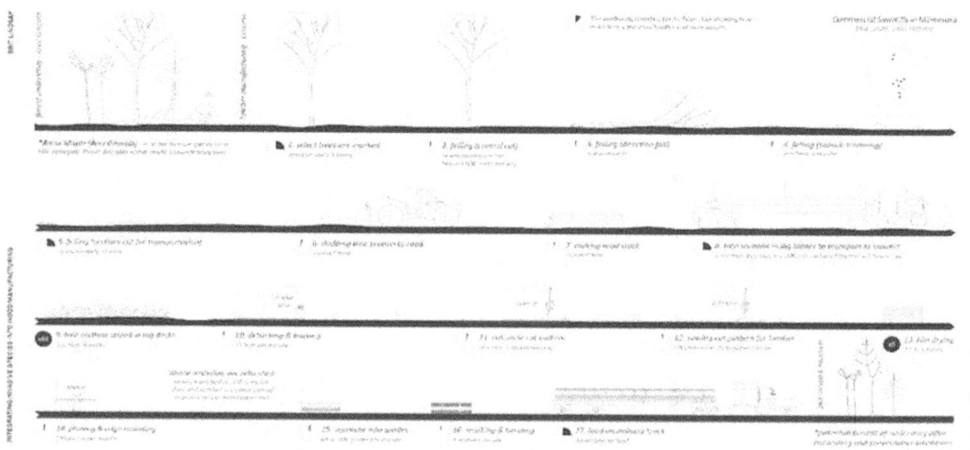

The goal behind this two-phased pedagogy was to help seed the development of an extended relational network within the studio. Students developed, and subsequent students adopted, research that extended far beyond the scale of a building as a way of conceptualizing a design that they would later build. The program of the building was a set of small portable learning kiosks, but the design was optimized to perform effectively at the scales of wood product production and logistical sourcing. To this end, the studio developed a hybrid construction system that combined mass timber, heavy timber, and light-weight wood framing. The design created a co-dependent architecture (Figure 8) that leveraged the qualities of individual construction systems to yield a wood architecture that would utilize a greater range of wood products of decreasing economic value. Mass timber was leveraged for its strength and shear capacity; heavy timber was utilized for the ability to incorporate whole logs, to connect to vernacular patterns, and to increase the building's longevity and durability; and the light-weight wood frame was utilized for its ability to increase formal complexity easily. The final design consisted of a system of three portable wood teaching kiosks that Anoka County could transport across their properties to facilitate educational programming (Figure 9).

From the conceptual phase of this project, the studio had the goal of co-producing something larger than "a building." Like the Littleton Trials, the Anoka County project looked at utilizing low-quality wood and developing new wood markets, but quickly expanded beyond it. The solution of designing a building out of hybrid panels was quickly seen as one component within a larger and materially inclusive construction system that could utilize a wider range of wood "units" (i.e., heavy timber, 2x4, wood panel, etc.).

Figure 8. Construction logic diagram of the Anoka County Learning Kiosk project

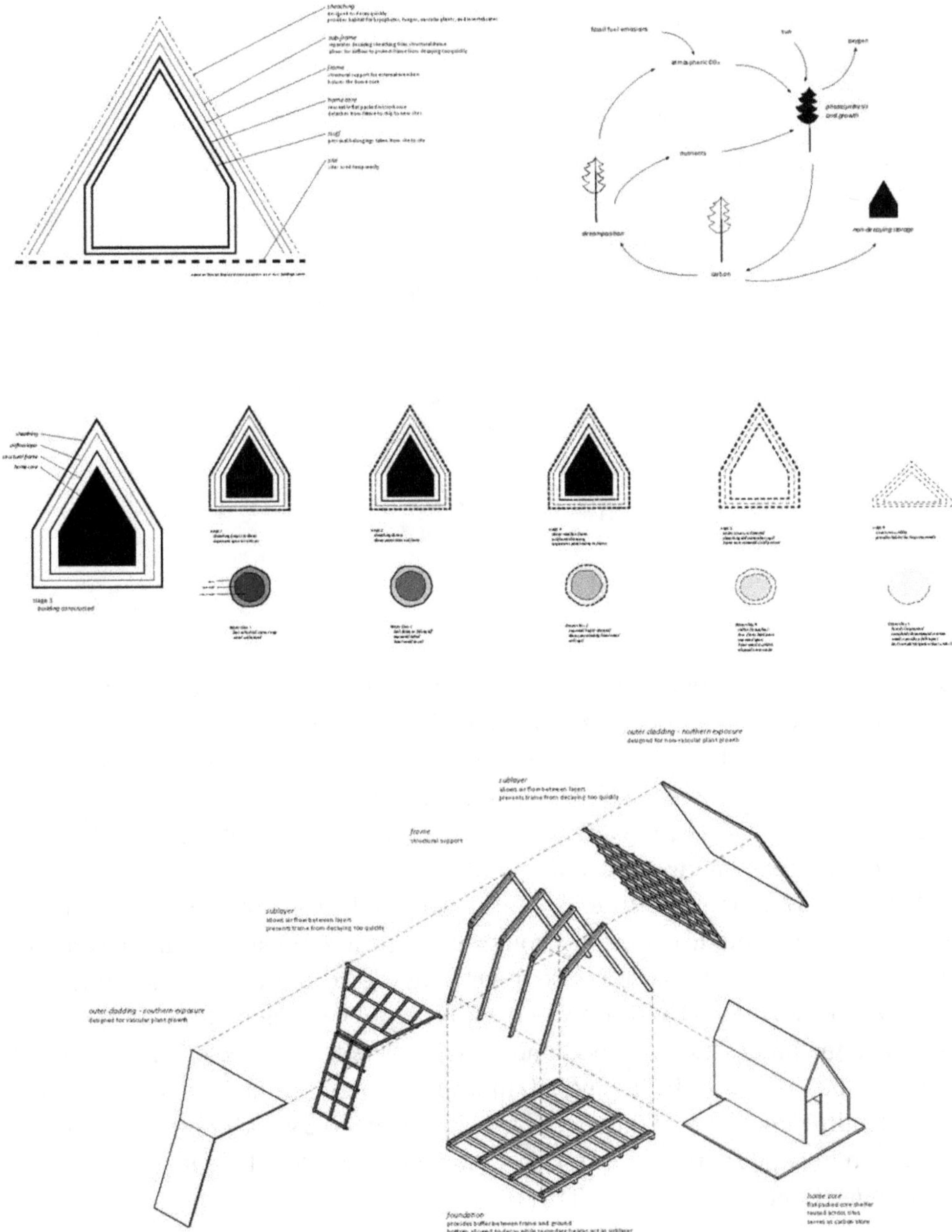

By moving beyond the initial question of "wood panels" and/or "mass timber" and toward an inquiry based on unpacking wood construction down to the scale of the unit, the studio was able to develop a new wood construction logic optimized

Figure 9. Anoka County learning Kiosk, October 2017

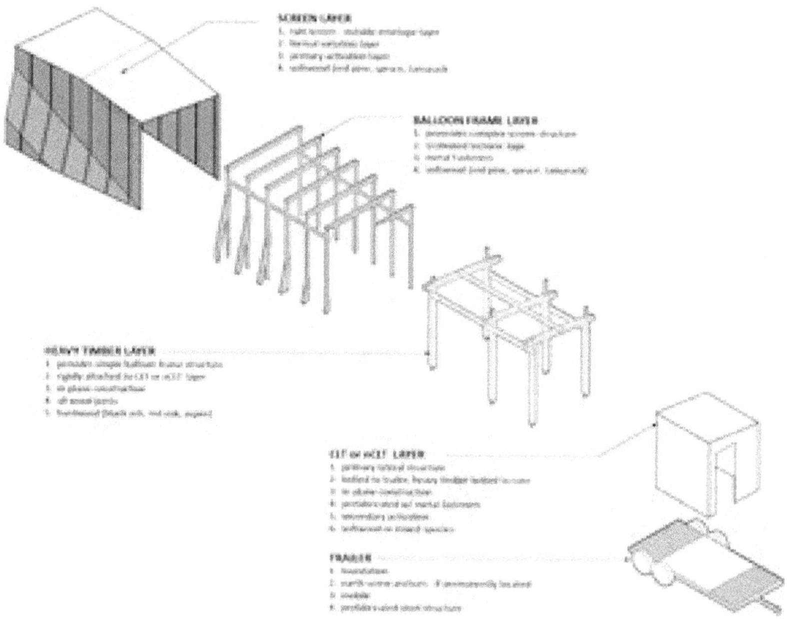

to increase wood utilization. Wood's performance was not optimized thermally or structurally but rather through the architecture's ability to utilize otherwise unusable forest resources. In the context of Minnesota, this is important since the local pulp industry is waning, and manufacturers are looking toward new markets to maintain local economies—which, as The Littleton Trials showed, is a critical aspect of local forest management and forest health.

What the studio's "construction logic" resolved itself into reflects what the studio learned through harvesting some of its own stumpage, milling some of its own timber, and constructing its own buildings (Figure 10). The building's optimization strategy—based on wood utilization—necessitated the design team to engage the processes that wood undergoes as it is transformed from a tree into a product on the way to becoming a building. Through this abductive inquiry, the team identified moments of intervention where normative wood construction logics could be rethought. This process yielded a unique extended relational network with wood industry partners.

A unique collaboration emerged from this network that eventually received a USFS Wood Innovation Grant entitled "*Minnesota Made Transitional Nail Cross Laminated Timber Panels*." The funding secured by this grant allowed the University of Minnesota to procure a robot with one of the network's industrial partners, Savanna Pallets, a manufacturer and sawyer from north-central Minnesota who

Figure 10. Author chainsaw milling wood harvested for the Anoka County learning kiosks, May 2017

Figure 11. Robot procured through the "Minnesota Made Transitional Nail Cross Laminated Timber Panels" grant, June 2019

Figure 12. Savanna Pallets sawmill McGregor, MN, May 2017

Figure 13. Pallet house system diagram, November 2015

specializes in the utilization of a range of wood species and wood qualities. The robot hosts a range of tool attachments for nailing, lagging, and drilling that allow Savanna Pallets to partially automate the production of large format pallets and for the design team to automate the partial production of a hybrid construction system that consists of lightweight wood-framed and mass timber panels. The grant was written to both secure and enhance the manufacturing capacity of Savanna Pallets, extending its reach into the well-established pallet market that already utilizes low-quality wood. As their robot-based manufacturing capacities improve, the scale of the costume pallets they can produce will increase until they are eventually pallets the scale of "walls." At this point, the production regime for the robot will transition toward the manufacture of architectural panels. The project seeks to utilize advanced manufacturing techniques to create an architectural product out of low-values material that can sell at a price higher than what the materials used to make the product would sell for if they were configured into pallets, but less than that other existing prefabricated wall panel systems (SIP/CLT) cost.

While the Anoka County Kiosk and the Minnesota Made Transitional Panel projects differ greatly in physical size, the scale of their relational networks align. Both projects rethink construction logics in order to utilize low-quality material in order to generate local architectures that co-produce economic development opportunities. Similar to the Littleton Trials, the project leveraged the Post-Normal framework to shift the scope of the project from panel design toward the design of a construction logic. The initial panel inquiry still exists; only it has been redirected into a project that operates within the manufacturing sector.

One House Many Nations and The Muskrat Hut

Location: *The Pas, Manitoba*
Project Partners: *Opaskwayak Cree Nation, Idle No More, University of Minnesota, Crosby Hill Plumbing and Heating, University of Saskatchewan, University of Manitoba,*
Project Year: OHMN 2015 to present; Muskrat Hut 2018-2019

Our involvement in the One House Many Nations (OHMN) project began in 2015 when Alex Wilson of the Opaskwayak Cree Nation (OCN) asked if we could design and construct a house made of repurposed wood pallets that were accumulating at the local dump on her reserve in Northern Manitoba. The OHMN campaign was established earlier that year to raise awareness around, and provide solutions for, the growing housing crisis on First Nations reservations in Canada (Assembly of First Nations, 2013). The campaign was established by Idle No More, a woman-led Indigenous land defense and protest movement that was established in Canada in

2012. The "pallet house" that was initially designed, but never built, consisted of a nail-laminated panel system that used repurposed pallet stock. Constructing a one-off house out of repurposed pallets was not economically feasible. However, the design concept clarified an underlying irony attached to the project that would later structure an extended relational network and co-productive agenda. Developing a housing system made of wood scraps extracted from contested lands that had been processed into a product used to ship commodities, established a forced consumer dependency on the very people whose housing shortage our design sought to remedy. This situation clarified that the housing crisis was a symptom of much larger systemic issues (Figure 13). Material performances attached to production, procurement, labor, and land-use are all tied to long-running history of colonial injustices attached to housing (Perry, 2003). In this respect, material performance has to be optimized toward recognizing treaty rights, education, economic development, land ownership, and public health/wellness. The project requires a material agenda architecturally optimized for these broader systemic issues—an architecture of decolonization.

For this to happen, we determined that a house needed to do several things. First, its design needed to be based on traditional Indigenous ideas that enhance cultural values and is adaptive of resilient to change; the material expression of its architecture needed to break with the imagery of colonial housing. Second, the house's performance needed to be conceptually optimized at the scale of the community to create trans-generational housing opportunities that can enhance traditional art, language, and knowledge transfer. Providing standalone houses would not solve the crisis, the housing needed to create communities that could push against the systems actively undermining them. Third, the housing needed to be constructed of a system that can be sourced, constructed, and maintained by the communities that will use them to minimize environmental impact and support local economic development; a system that broke down the initial ironies identified in the "pallet house" concept.

Nearly two years later, the team completed our first OHMN house prototype (Figure 14) and installed it at EDIT 2017 in Toronto. The house now lives in The Pas, Manitoba, where two members of the OCN community live in it. The design of this prototype consisted of two types of modular units. The first type was called a modular "shelter" unit. These units enclosed the main living spaces and contained no additional services. The second type was called a modular "service" unit. These connected the shelter units to each other and provided all of the services the house needed to function. The service unit's triangular geometry (Figure 15) was developed to allow houses to sit nimbly on under-valued land parcels; to maneuver delicately through sites with irregular footprints or those with landscape features that needed to remain undisturbed (Wilson & Mans, 2019). The shelter modules were constructed from CLT panels. This decision was based in part on geography, in part in on local economic development, and in part on durability. OCN is located

Figure 14. One House Many Nations prototype house at EDIT Toronto, October 2017

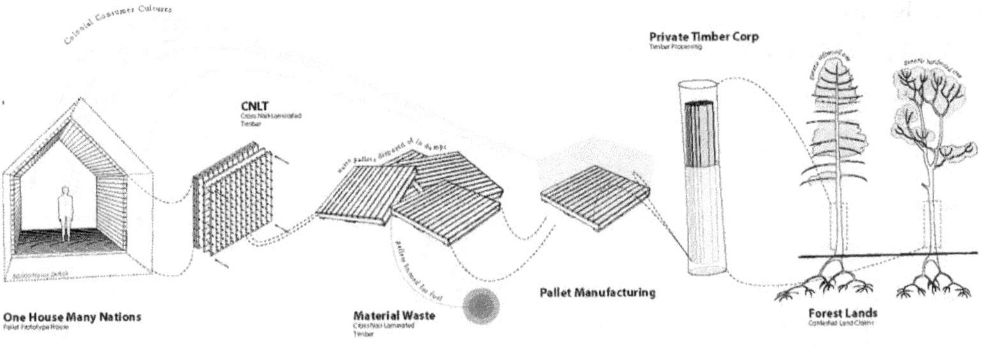

adjacent to The Pas, approximately 630 KM north of Winnipeg and accessible by road but logistically remote. This area falls within the Boreal forest region, which informed our decision to use monolithic wood construction as an import replacement strategy to replace imported lightweight systems such as SIPs and/or wood-framed construction (Jacobs, 1985).

Building the prototype created a unique relational network that revealed several issues with our initial design. One of the main challenges of our initial concept was raising the capital necessary to scale from prototype to production. Labor challenges also shifted our strategy from training local labor in new construction techniques toward utilizing existing systems that were well established. Over time, the unique circumstances of the project forced us to rethink our co-production strategy and the scale of the system for which our architecture needed to be optimized.

As a result, based on both refined material costs and labor considerations, the design for the current housing system has changed dramatically since the initial prototype. The project still looks to leverage prefabrication; however, we are shifting focus away from locally prefabricating everything and toward remotely prefabricating utility cores (Figure 16) in a location that has more skilled labor to help bring down labor costs and improve quality (Richard, 2005), (Kieran, & Timberlake, 2004). Rather than basing our local economic development strategy on the production of CLT panels, we are pursuing a more typical wood-framed/site-built envelope that can better utilize OCN's existing labor force and couples it with off-site prefabricated utility cores to minimize overall project costs. The actual design of the enclosure system is flexible and can be adjusted to meet the material and cultural context of different communities that use the system.

The changes we have made to the prototype design are geared toward minimizing project costs. The project's post-normal framework is also focused on defining a co-productive agenda where the construction of houses can impact the colonial systems

Figure 15. One House Many Nations EDIT house construction logic diagram

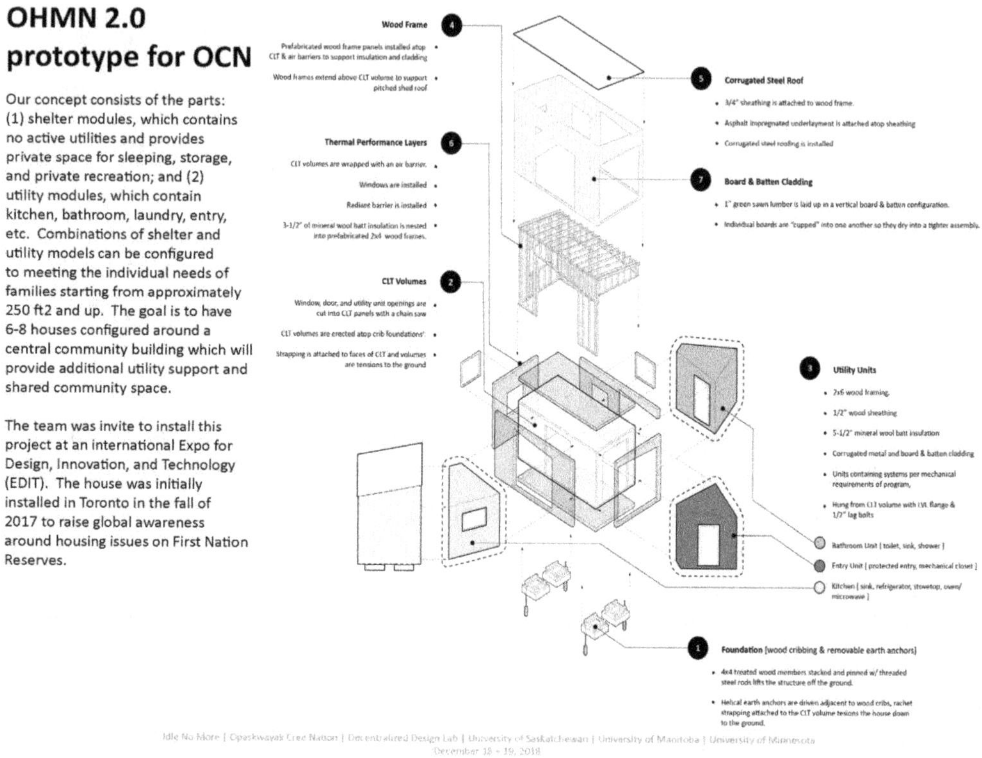

that established the housing crisis. To this end, we are developing construction logics that incorporate educational and service opportunities for those who are building houses. These processed connect back to the project's initial goals of utilizing Indigenous knowledge, creating trans-generation opportunities, and utilizing local materials. The project is extending to work with four different Indigenous communities to develop a sharable youth build program to incorporate the OHM utility core into the design of their own local housing solutions. Utilizing the same core design will further bring down construction and design costs and will allow the OHMN's team to work more effectively with individual communities to develop their own culturally and geographically specific envelope systems.

As we develop the next round of OHMN houses, the team is also looking at how other architectures can engage colonial systems attached to housing. The Wachusko Weesti (Muskrat Hut) (Figure 17) project is a community-led and interdisciplinary collaboration between Indigenous and non-Indigenous academics, professionals, students, natural resource experts, and community members. The project addresses two critical and well-documented issues that affect First Nation communities: 1)

Figure 16. One House Many Nations updated "Utility Core" house construction logic diagram

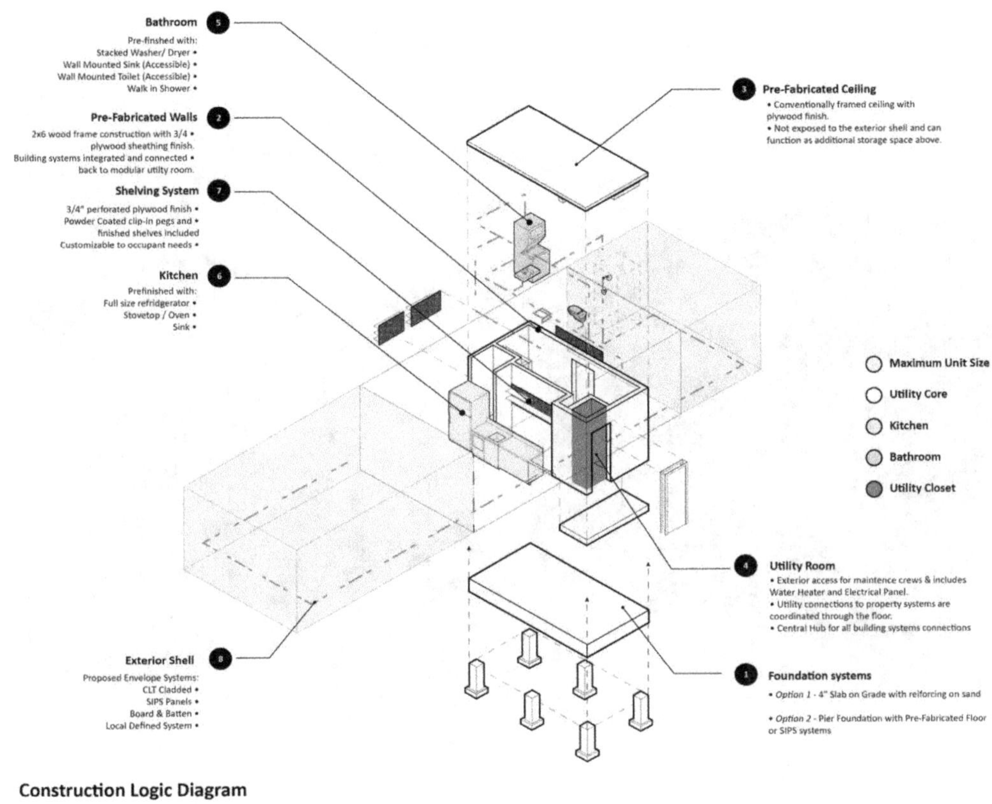

inadequate access to safe water, and 2) a shortage of adequate housing. Community design that ensures the availability of water, wastewater management, and housing infrastructure is foundational to the structure and maintenance of a healthy society (Hennessy, 2008). The project utilized many of the design and construction workshop strategies that the OHMN project will deploy to co-produce a small building (a simplified bathroom on wheels) that simultaneously acts as a piece of systemic infrastructure to enable land-based education, community engagement, and land defense (Figure 17).

DIRECTIONS FOR FUTURE RESEARCH

It is not unusual to talk about building as a vocation or as a creative practice; it is less common to talk about it as a kind of research. On the occasion it is discussed,

Figure 17. Muskrat Hut portable bathroom, kitchen, and sauna in OCN, Manitoba, August 2019

it is usually done in deductively or inductively and rarely in discursive terms that could generate *new* questions. The divisions that exist between architecture's academic and practice communities result from the boundaries we draw to isolate our inquiries from another and to restrict the movement of knowledge from one discipline to another. A logical step toward a collective discourse is to reestablish practices that allow more boundary-spanning across individual actors, sectors, and disciplines. Materials are boundary spanning. When considered openly, they are immediately trans-scalar. There is an opportunity through a post-normal material practice to optimize architectural performance to more meaningfully impact wicked problems through its co-production.

The end goal of this essay is not to solve, but to ask a better and more impactful question. Once the appropriate scale of optimization is defined, future research will be required to understand architectural co-production impacts more thoroughly. While this essay does not offer a particular metric or method to characterize these impacts, it introduces the concept of Post-Normal Science as a framework for establishing material practices that engage extended communities of practice. These communities are critical to defining research plans and collecting data needed to inform decisions surrounding wicked problems.

For the projects outlined above, we are working with our extended relational networks to establish methods for collecting data on forest productivity, manufacturing efficiency, and community health. The methods for collecting this data extend beyond architectural expertise, and the data collection period extends beyond schedule for building individual projects. This is another reason why creating long-term relationships is so important as we begin to require longer-termed data requirements.

All this said, while there are increased uncertainties attached to the impact of this work, we believe the work is asking better questions even if we are not getting explicitly measurable results on the co-productive aspects of the projects. Moving forward, there is a great deal of future research to be done on both identifying opportunities for architecture to be optimized toward specific co-production as well as research needed to evaluate the impact of this co-production on the larger than building scale systems we seek to improve.

CONCLUSION

The physical buildings described above have a minuscule impact on the systems they engage. They are a tiny, tiny part of nature. A tiny, tiny part of the relational networks to which they ascribed, networks that do connect to systems of planetary significance. Having built them, I confess that I think about design agency, material performance, system optimization, and building (co)production differently. I openly prescribe a higher value to installing "junk" wood in buildings when I know its utilization improves the economic viability of New England forests and increases their management. Small wood buildings are more interesting to me than tall wood buildings if utilizing advanced wood technology to can help utilize materials that maintain local economies. Moreover, and perhaps not surprisingly, when you read the acknowledgments to this chapter, I am thinking more and more that redesigning the bathroom might just save us all.

Much of the research completed to date is qualitative—otherwise, it is anecdotal. That said, research that strays beyond the conventional architectural boundaries is needed to transform who architects serve, how architecture is practiced, and the impact architecture is designed to have. More research is needed to understand how we assess the impacts of what we are co-producing. The buildings discussed in this chapter are, for me, the beginnings of this research. As a colleague once said, and that I feel so wonderfully and counterintuitively sums this up, "R-values are not *our* values" (Moe, 2019). It is part thermal word-craft and part perfect explanation.

Performance is relative. This research suggests we ask that something new. I believe that inquiry of building helps us ask new questions.

ACKNOWLEDGMENT

I would l like to thank David Kennedy, Thomas Fisher, Cecilio Ortiz Garcia, and Marla Perez Lugo for their contributions to different ideas developed in this manuscript. The overarching practice concept presented here has developed in line with the framework outlined above, crisscrossing disciplines and shaped between both intellectual and practice centered relationships. I have, in effect, tried to organize and compile these ideas into a consistent argument.

Funding to support these projects was acquired through the United States Forest Service's Wood Innovation Grant, the University of Minnesota's Office of the Vice Provost of Research, Social Sciences and Humanities Research Council of Canada, the Softwood Lumber Board, and Anoka County

Most importantly, I would like to thank my perpetual project partner Wayne Mans from Crosby Hill Plumbing and Heating. Most of the project work shown above would be partially complete without him. Acknowledging and working with divergent views is critical to this kind of thinking. Few vocations in my mind have contributed more to civilization than plumbing. It is, to me, one of the noblest trades ever generally overlooked by those who typically make claims about such things. I suspect that the general discourse on architecture and materials would have progressed much further if those in engaged in these conversations would have only consulted a good plumber.

REFERENCES

Ackoff, R. (2003). What's wrong with what's wrong with. *Interfaces*, *33*(5), 78–82. doi:10.1287/inte.33.5.78.19242

Assembly of First Nations. (2013). *Fact sheer first nation housing on-reserve*. Retrieved from https://www.afn.ca/uploads/files/housing/factsheet-housing.pdf

Fisher, T. (2016). *Designing our way to a better world*. Minneapolis, MN: University of Minnesota Press. doi:10.5749/minnesota/9780816698875.001.0001

Forest Products Laboratory (FPL). (2010). *Wood handbook wood as an engineering material* (Centennial ed., General technical report FPL; 190). Madison, WI: U.S. Dept. of Agriculture, Forest Service, Forest Products Laboratory.

Funtowitz, S., & Ravetz, J. (1993). *Science for the post-normal age. Futures*. Pergamon.

Hameury, S., & Lundström, T. (2004). Contribution of indoor exposed massive wood to a good indoor climate. *Energy and Building*, *36*(3), 281–292. doi:10.1016/j.enbuild.2003.12.003

Hennessy, T. W., Ritter, T., Holman, R. C., Bruden, D. L., Yorita, K. L., Bulkow, L., ... Smith, J. (2008). The relationship between in-home water service and the risk of respiratory tract, skin, and gastrointestinal tract infections among rural Alaska natives. *American Journal of Public Health*, *98*(11), 2072–2078. doi:10.2105/AJPH.2007.115618 PMID:18382002

Jacobs, J. (1985). *Cities and the wealth of nations*. New York, NY: Vintage Books.

Kao, J. (2007). Innovation nation: How America is losing its innovation edge, why it matters, and what we can do to get it back. New York: Free Press.

Kay, J., Regier, H., Boyle, M., & Francis, G. (1999). An Ecosystem Approach for Sustainability: Addressing the Challenge of Complexity. *Futures*, *31*(7), 721–742. doi:10.1016/S0016-3287(99)00029-4

Kieran, S., & Timberlake, J. (2004). *Refabricating architecture: How manufacturing methodologies are poised to transform building construction*. New York: McGraw-Hill.

Kuhn, T. (1962). *The Structure of Scientific Revolutions*. Chicago: University of Chicago Press.

Mans, J., Kennedy, D., & Peek, B. (2018). The Little Trials: An Abductive Thermal Inquiry. In K. Moe, J. Hutton, & D. Ibañez (Eds.), *Wood Urbanism: From the Molecular to the Territorial* (pp. 194–207). Barcelona: Aktar.

Moe, K. (2014). *Insulating modernism: Isolated and non-isolated thermodynamics in architecture*. Basel: Birkhäuser Verlag GmbH. doi:10.1515/9783038213215

Moe, K. (2018). Intro: R-Values Are Not Our Values. In K. Moe, J. Hutton, & D. Ibañez (Eds.), *Wood Urbanism: From the Molecular to the Territorial* (pp. 156–163). Barcelona: Aktar.

Neumeier, M. (2009). *The designful company: How to build a culture of nonstop innovation: A whiteboard overview*. Berkeley, CA: New Riders.

Peirce, C., Houser, N., & Kloesel, C. (1992). *The essential Peirce: Selected philosophical writings*. Bloomington, IN: Indiana University Press.

Peirce, C. S., & Buchler, J. (1955). *Philosophical writings of Peirce*. New York: Dover Publications.

Perry, A. (2003). From "the hot-bed of vice" to the "good and well-ordered Christian home": First Nations Housing and Reform in Nineteenth-Century British Columbia. *Ethnohistory (Columbus, Ohio)*, *50*(4), 587–610. doi:10.1215/00141801-50-4-587

Peterson, A. (2008). Ignorance and Ethics. In W. Vitek & W. Jackson (Eds.), The virtues of ignorance: Complexity, sustainability, and the limits of knowledge (Culture of the land) (pp. 109-134). Lexington, KY: University Press of Kentucky.

Radmanovic, K., Dukic, I., & Pervan, S. (2014). Specific heat capacity of wood. *Specificni toplinski kapacitet drva. Drvna Industrija*, *65*(2), 151–157. doi:10.5552/drind.2014.1333

Ravetz, J. (1987). Usable Knowledge, Usable Ignorance: Incomplete Science with Policy Implications. *Science Communication*, *9*(1), 87–116.

Richard, R. B. (2005). Industrialised building systems: Reproduction before automation and robotics. *Automation in Construction*, *14*(4), 442–451. doi:10.1016/j.autcon.2004.09.009

Rittel, H., & Webber, W. (1973). Dilemmas in a general theory of planning. *Policy Sciences*, *4*(2), 155–169. doi:10.1007/BF01405730

Roy, D., & Zeckhauser, R. (2015). Grappling with Ignorance: Frameworks from Decision Theory, Lessons from Literature. *Journal of Benefit-Cost Analysis*, *6*(1), 33–65. doi:10.1017/bca.2015.8

Shank, G. (1998). The Extraordinary Ordinary Powers of Abductive Reasoning. *Theory & Psychology*, *8*(6), 841–860. doi:10.1177/0959354398086007

Vitek, W., & Jackson, W. (2008). *The virtues of ignorance: Complexity, sustainability, and the limits of knowledge (Culture of the land)*. Lexington, KY: University Press of Kentucky.

Wilson, A., & Mans, J. (2019). One House Many Nations: Indigenous Project-Based Collaboration. In M. Henderson (Ed.), *Catching Fire: Fuelling Inquiry and Passion Through Project-Based Learning* (pp. 171–184). Winnipeg, MB: Portage & Main Press.

Compilation of References

a4ws.org. (2019). *Alliance for Water Stewardship*. Retrieved from https://a4ws.org/

Abanda, F. H., Tah, J. H. M., & Cheung, F. K. T. (2013). Mathematical modelling of embodied energy, greenhouse gases, waste, time–cost parameters of building projects: A review. *Building and Environment*, *59*, 23–37. doi:10.1016/j.buildenv.2012.07.014

Abd El-Hameed, A. K., Mansour, Y. M., & Faggal, A. A. (2017). Benchmarking water efficiency of architectural finishing materials based on a "cradle-to-gate" approach. *Journal of Building Engineering*, *14*, 73–80. doi:10.1016/j.jobe.2017.10.001

Ackoff, R. (2003). What's wrong with what's wrong with. *Interfaces*, *33*(5), 78–82. doi:10.1287/inte.33.5.78.19242

Acquaye, A. (2010). *A stochastic hybrid embodied energy and CO2_eq intensity analysis of building and construction processes in Ireland*. Academic Press.

Acquaye, A. A., Duffy, A. P., & Basu, B. (2011). Stochastic hybrid embodied CO2-eq analysis: An application to the Irish apartment building sector. *Energy and Building*, *43*(6), 1295–1303. doi:10.1016/j.enbuild.2011.01.006

Acuqaye, A. (2010). *A Stochastic Hybrid Embodied Energy and CO2-eq Intensity Anal-ysis of Building and Construction Processes in Ireland* (PhD thesis). Dublin Institute of Technology.

Adalberth, K. (1997). Energy use during the life cycle of single-unit dwellings: Examples. *Building and Environment*, *32*(4), 321–329. doi:10.1016/S0360-1323(96)00069-8

Adalberth, K., Almgren, A., & Petersen, E. H. (2001). Life cycle assessment of four multi-family buildings. *International Journal of Low Energy and Sustainable Buildings*, *2*(1), 1–21.

Addington, M., & Schodek, D. (2005). Smart Materials and New Technologies for the architecture and design profession. Oxford, UK: Architectural Press.

Addington, D. M. (2001). The history and future of ventilation. In J. D. Spengler, J. M. Samet, & J. F. McCarthy (Eds.), *Indoor air quality handbook* (pp. 2.1–2.16). New York, NY: McGraw-Hill.

Addis, B. (2007). *Building: 3000 years of design, engineering and construction*. New York: Phaidon Press.

Akadiri, P. O., Olomolaiye, P. O., & Chinyio, E. A. (2013). Multi-criteria evaluation model for the selection of sustainable materials for building projects. *Automation in Construction, 30*, 113–125. doi:10.1016/j.autcon.2012.10.004

Alanne, K., & Saari, A. (2004). Sustainable small-scale CHP technologies for buildings: The basis for multi-perspective decision-making. *Renewable & Sustainable Energy Reviews*, *8*(5), 401–431. doi:10.1016/j.rser.2003.12.005

Allen, S., Allen, D., Phoenix, V., Le Roux, G., Durántez Jiménez, P., Simonneau, A., ... Galop, D. (2019, May). Atmospheric transport and deposition of microplastics in a remote mountain catchment. *Nature Geoscience*, *12*(5), 339–344. doi:10.103841561-019-0335-5

Allplan Inc. (2019). *The Allplan solution portfolio*. Retrieved October 25, 2019, from https://www.allplan.com/us_en/products/solution-portfolio/

Almeida, M., Ferreira, M., & Barbosa, R. (2018). Relevance of Embodied Energy and Carbon Emissions on Assessing Cost Effectiveness in Building Renovation—Contribution from the Analysis of Case Studies in Six European Countries. *Buildings*, *8*(8), 103. doi:10.3390/buildings8080103

American Chemistry Council. (2009). *Plastics, Major Markets: Building and Construction.* Retrieved from https://plastics.americanchemistry.com/B-C/

American Institute of Architects (AIA), & Arup. (2018). *Prescription for Healthier Building Materials: A Design and Implementation Protocol.* Retrieved from http://content.aia.org/sites/default/files/2018-06/2018-06-07-SAI18_Materials_Protocol_Handbook_v04_FINAL.pdf

American Institute of Architects (AIA). (2018). *Ethics from the Office of General Counsel - 2018 Code of Ethics and Professional Conduct.* Retrieved from https://aianova.org/pdf/codeofethics.pdf

Amziane, S., & Arnaud, L. (2013). *Bio-aggregate-based building materials applications to hemp concretes*. Hoboken, NJ: Wiley. doi:10.1002/9781118576809

An, X., & Fan, H. (2016, September 15). Hybrid design and energy absorption of luffa-sponge-like hierarchical cellular structures. *Materials & Design*, *106*, 247–257. doi:10.1016/j.matdes.2016.05.110

Aparcana, S., & Salhofer, S. (2013). Development of a social impact assessment methodology for recycling systems in low-income countries. *The International Journal of Life Cycle Assessment*, *18*(5), 1106–1115. doi:10.100711367-013-0546-8

Arens, E., Humphreys, M. A., de Dear, R., & Zhang, H. (2010). Are 'class A' temperature requirements realistic or desirable? *Building and Environment*, *45*(1), 4–10. doi:10.1016/j.buildenv.2009.03.014

Arrigoni, A., & Pelosato, R. (2017). Life cycle assessment of natural building materials: the role of carbonation, mixture components and transport in the environmental impacts of hempcrete blocks. *Journal of Cleaner Production, 149.* doi:10.1016/j.jclepro.2017.02.161

Arroyo, P., Tommelein, I., & Ballard, G. (2012). Deciding a sustainable alternative by 'choosing by advantages' in the AEC industry. In *Proc. 20th Conf. of the International Group for Lean Construction (IGLC), San Diego, CA* (pp. 41-50). Academic Press.

Ashby, M. F. (1980). Engineering materials: an introduction to their properties and applications. New York, NY: Pergamon Press.

Ashby, M. F. (2017). *Materials selection in mechanical design*. Academic Press.

ASHRAE. (1960). *Heating, ventilating, air conditioning guide*. New York, NY: ASHRAE.

ASHVE. (1924). Celebrate thirty years of progress. *Journal of the American Society of Heating and Ventilating Engineers, 30*(3).

ASMI. (2019). *Athena Impact Estimator*. Athena Sustainable Materials Institute (ASMI). Retrieved August 4, 2019, from https://calculatelca.com/software/impactestimator/

Assefa, G. (2019). Life Cycle Insights for Creating Sustainable Cities. In Sustainable Cities-Authenticity, Ambition and Dream. IntechOpen. doi:10.5772/intechopen.81633

Assefa, G., & Ambler, C. (2017). To demolish or not to demolish: Life cycle consideration of repurposing buildings. *Sustainable Cities and Society, 28*, 146–153. doi:10.1016/j.scs.2016.09.011

Assembly of First Nations. (2013). *Fact sheer first nation housing on-reserve*. Retrieved from https://www.afn.ca/uploads/files/housing/factsheet-housing.pdf

Atkinson, C., Hobbs, S., West, J., & Edwards, S. (1996). Life cycle embodied energy and carbon dioxide emissions in buildings. *Industry and Environment, 19*(2), 29–31.

Autodesk Inc. (2019a). *AutoCAD*. Retrieved October 25, 2019, from https://www.autodesk.com/products/autocad/overview

Autodesk Inc. (2019b). *Revit®*. Retrieved October 25, 2019, from https://www.autodesk.com/products/revit/overview

Autodesk Inc. (2019c). *Dynamo Studio*. Retrieved October 25, 2019, from https://www.autodesk.com/products/dynamo-studio/overview

Azari, R., & Abbasabadi, N. (2018). Embodied energy of buildings: A review of data, methods, challenges, and research trends. *Energy and Building, 168*, 225–235. doi:10.1016/j.enbuild.2018.03.003

Baboian, R. (1985). *Materials degradation caused by acid rain*. Paper presented at the 20th State-of-the-Art Symposium of the American Chemical Society, Arlington, VA.

Badger, D. D., & Gayle, M. (1982). Badger's Illustrated Catalogue of Cast-Iron Architecture. New York: Dover Publications.

Baitz, M., Kreißig, J., Byrne, E., Makishi, C., & Kupfer, T. (2004). *Life-Cycle Assessment of PVC and of principal competing materials*. European Commission. Retrieved from https://ec.europa.eu/environment/waste/studies/pdf/pvc-final_report_lca.pdf

Baltic Cluster Forum. (2017, June 28) *Switching to the circular economy in India.* Retrieved from http://circularconstruction.eu/

Banham, R. (1969). *The architecture of the well-tempered environment.* Cambridge, MA: MIT Press.

Basbagill, J., Flager, F., Lepech, M., & Fischer, M. (2013). Application of Life-Cycle Assessment to Early Stage Building Design for Reduced Embodied Environmental Impacts. *Building and Environment, 60*(February), 81–92. doi:10.1016/j.buildenv.2012.11.009

Bath Inventory of Carbon and Energy (ICE) | Greenhouse Gas Protocol. (n.d.). Retrieved August 6, 2019, from https://ghgprotocol.org/Third-Party-Databases/Bath-ICE

Bayer, C., Gamble, M., Gentry, R., & Joshi, S. (2010). *AIA Guide to Building Life-Cycle Assessment in Practice*. Retrieved from https://www.brikbase.org/sites/default/files/aiab082942.pdf

BC Assessment & the Revenue Services Vancouver, and Community Services Vancouver. (2016). Property Tax Report Data. *City of Vancouver Open Data Catalogue*. Retrieved from https://data.vancouver.ca/datacatalogue/propertyTax.htm

BC Assessment, . (2015a). *Bulk Extract for KPI Sales Detail, A09*, 2005–2015. BC Assessment and University of British Columbia.

BC Assessment. (2015b). *BIR - A09 - 2014-2017. BC* Assessment and University of British Columbia.

BC Hydro. (2017). *About BC Hydro.* Accessed July 18. https://www.bchydro.com/about.html

Beck, U. (2006). Living in the world risk society. *Economy and Society, 35*(3), 329–345. doi:10.1080/03085140600844902

Bedane, A., Eić, M., Farmahini-Farahani, M., & Xiao, H. (2016). Theoretical modeling of water vapor transport in cellulose-based materials. *Cellulose (London, England), 23*(3), 1537–1552. doi:10.100710570-016-0917-y

Benini, L., Mancini, L., Sala, S., Manfredi, S., Schau, E. M., & Pant, R. (2014). Normalisation method and data for Environmental Footprints. Luxembourg: European Union. EUR, 26842.

Ben-Joseph, E. (2005). *The Code of the City: Standards and the Hidden Language of Place Making*. Cambridge, MA: The MIT Press.

Benoit, C., Andrews, E., Barthel, L., & Beck, T. (2009). *Guidelines for Social Life Cycle Assessment of Products*. United Nations Environment Programme.

Benyus, J. M. (1997). *Biomimicry: innovation inspired by nature*. New York: Quill.

Bernard, C. (1879). *Leçons sur les phénomènes de la vie communs aux animaux et aux végétaux*. Paris: J. B. Baillière. doi:10.5962/bhl.title.103133

Bilow, D. N., & Kamara, M. E. (2008). Fire and Concrete Structures. *Structures Congress 2008*, 1–10.

Bionova. (2018). *OneClickLCA*. Retrieved August 4, 2019, from https://www.oneclicklca.com/construction/

Bionova. (2018). *The Embodied Carbon Review*. Retrieved January 20, 2020, from www.embodiedcarbonreview.com

Birgisdottir, H., Moncaster, A., Wiberg, A. H., Chae, C., Yokoyama, K., Balouktsi, M., ... Malmqvist, T. (2017). IEA EBC annex 57 'evaluation of embodied energy and CO2eq for building construction'. *Energy and Building*, *154*, 72–80. doi:10.1016/j.enbuild.2017.08.030

Blengini, G. A. (2009). Life cycle of buildings, demolition and recycling potential: A case study in Turin, Italy. *Building and Environment*, *44*(2), 319–330. doi:10.1016/j.buildenv.2008.03.007

Bollmann, U. E., Fernández-Calviño, D., Brandt, K. K., Storgaard, M. S., Sanderson, H., & Bester, K. (2017). Biocide Runoff from Building Facades: Degradation Kinetics in Soil. *Environmental Science & Technology*, *51*(7), 3694–3702. doi:10.1021/acs.est.6b05512 PMID:28287716

Börjesson, P., & Gustavsson, L. (2000). Greenhouse gas balances in building construction: Wood versus concrete from life-cycle and forest land-use perspectives. *Energy Policy*, *28*(9), 575–588. doi:10.1016/S0301-4215(00)00049-5

Braubach, M., & Fairburn, J. (2010). Social inequities in environmental risks associated with housing and residential locations: A review of evidence. *European Journal of Public Health*, *20*(1), 36–42. doi:10.1093/eurpub/ckp221 PMID:20047933

Bribián, I., Usón, A. A., & Scarpellini, S. (2009). Life Cycle Assessment in Buildings: State-of-the-Art and Simplified LCA Methodology as a Complement for Building Certification. *Building and Environment*, *44*(12), 2510–2520. doi:10.1016/j.buildenv.2009.05.001

British Columbia, Ministry of Environment. (2016). *2016/17 B.C. Best Practices Methodology for Quantifying Greenhouse Gas Emissions: Including Guidance for Public Sector Organizations, Local Governments and Community Emissions*. Author.

British Columbia. Government of. (2012). *Vancouver Community Energy & Emissions Inventory*. Retrieved from https://www2.gov.bc.ca/gov/content/environment/climate-change/data/ceei

Broman, G. I., & Robèrt, K. H. (2017). A framework for strategic sustainable development. *Journal of Cleaner Production*, *140*, 17–31. doi:10.1016/j.jclepro.2015.10.121

Brown, N. W., Olsson, S., & Malmqvist, T. (2014). Embodied greenhouse gas emissions from refurbishment of residential building stock to achieve a 50% operational energy reduction. *Building and Environment*, *79*, 46–56. doi:10.1016/j.buildenv.2014.04.018

Bruce-Hyrkäs, T., Pasanen, P., & Castro, R. (2018). Overview of Whole Building Life Cycle Assessment for Green Building Certification and Ecodesign through Industry Surveys and Interviews. *Procedia CIRP*, *69*, 178–183. doi:10.1016/j.procir.2017.11.127

Brummer, G., & Pienaar, C. (2008). Embodied Carbon in Construction. *Green Design*. Retrieved from http://www.davislangdon.com/upload/Africa%20Research%20Publications/Article%20-%20Embodied%20Carbon%20Jnl%203%2008.pdf

Brunklaus, B., Thormark, C., & Baumann, H. (2010). Illustrating Limitations of Energy Studies of Buildings with LCA and Actor Analysis. *Building Research and Information*, *38*(3), 265–279. doi:10.1080/09613211003654871

BSRIA. (n.d.). *Embodied energy calculations design tool*. Accessed June 17, 2019. https://www.bsria.co.uk/news/article/embodied-calculations/

Buchanan, A. H., & Honey, B. G. (1994). Energy and carbon dioxide implications of building construction. *Energy and Building*, *20*(3), 205–217. doi:10.1016/0378-7788(94)90024-8

Building for Environmental and Economic Sustainability (BEES). Retrieved from https://www.nist.gov/services-resources/software/bees and https://ws680.nist.gov/bees/(A(s4j522g11QEkAAAAMTZhMjA2ZDgtYmZkYy00OWJjLTgwYjAtMGUyNjk1NzNkNGRl92eu0lpz1iKrFVoBsvKGUeA4fms1))/default.aspx

Building Green Inc. (n.d.). *The urgency of embodied carbon and what you can do about it*. Accessed June 19, 2019. https://www.buildinggreen.com/feature/urgency-embodied-carbon-and-what-you-can-do-about-it

Building Research Establishment. (2018). *BREEAM UK New Construction – Technical Manual*. Author.

Buyle, M., Braet, J., & Audenaert, A. (2013). Life Cycle Assessment in the Construction Sector: A Review. *Renewable & Sustainable Energy Reviews*, *26*(October), 379–388. doi:10.1016/j.rser.2013.05.001

C.C.R.O.T. (2019). *Polymer Databse*. Retrieved from http://polymerdatabase.com/

Cabanac, M. (1971). Physiological role of pleasure. *Science. New Series*, *173*(4002), 1103–1107. PMID:5098954

Cabeza, L. F., Rincón, L., Vilariño, V., Pérez, G., & Castell, A. (2014). Life Cycle Assessment (LCA) and Life Cycle Energy Analysis (LCEA) of Buildings and the Building Sector: A Review. *Renewable & Sustainable Energy Reviews*, *29*(January), 394–416. doi:10.1016/j.rser.2013.08.037

California Building Standards Commission. (2017). *California Building Standards Code (California Code of Regulations, Title 24)*. Retrieved from http://www.bsc.ca.gov/codes.aspx

Canadian Commission on Building and Fire Codes. (2015). *National Building Code of Canada, 2015*. Ottawa: National Research Council of Canada. Retrieved from http://ubc.summon.serialssolutions.com/ 2.0.0/link/0/ eLvHCXMwfV09C8Iw ED0UF8FBq2K1QiYnK2 nahmQWiz_AvbT5mKQu-v-91 KqlVCFDSOBIQpLHS-7 uAcTsQMPOnRBppsrUSqtYaiy ThaFUKa5TLbDUGn7_XX W6oXa9vzMIPVJw JPBDJB0fNQMEVbc1 I6cY2wqtd_i RTcH7atoSvNcfrjKD ganmsGtSU19J2UhU ExdnTm6W1KkDij1 B-EwXsM5Ol-M5RNt5 8-iSN0NhS5gUzl W9utchbXoFpHAp AHWMqMDiRAoth VGWcpNoHllhmQ9 ejyUfgnbre_Y5 MliKHOZXN3Jfjmd 23WtzA2M3gde LQgAjizvebOtFegJy03-a

Canadian Mortgage and Housing Corporation. (2016). *Rental Market Report: Vancouver CMA 2016*. Author.

Cannon, W. B. (1932). *The wisdom of the body*. New York, NY: W. W. Norton. doi:10.1097/00000441-193212000-00028

Carothers, W. H., Williams, I., Collins, A. M., & Kirby, J. E. (1931, November). Acetylene Polymers and Their Derivatives. II. A New Synthetic Rubber: Chloroprene and Its Polymer. *Journal of the American Chemical Society*, *53*(11), 4203–4225. doi:10.1021/ja01362a042

Cavalliere, C., Dell'Osso, G. R., Pierucci, A., & Iannone, F. (2018). Life cycle assessment data structure for building information modeling. *Journal of Cleaner Production*, *199*, 193–204. doi:10.1016/j.jclepro.2018.07.149

Cavalliere, C., Habert, G., Dell'Osso, G. R., & Hollberg, A. (2019). Continuous BIM-based assessment of embodied environmental impacts throughout the design process. *Journal of Cleaner Production*, *211*, 941–952. doi:10.1016/j.jclepro.2018.11.247

Cellura, M., Guarino, F., Longo, S., & Mistretta, M. (2014). Energy life-cycle approach in Net zero energy buildings balance: Operation and embodied energy of an Italian case study. *Energy and Building*, *72*, 371–381. doi:10.1016/j.enbuild.2013.12.046

CEN. (2010). *CEN/TR 15941:2010 Sustainability of construction works - Environmental product declarations - Methodology for selection and use of generic data*. Brussels: European Committee for Standardization.

CEN. (2011a). *EN 15942:2011 Sustainability of construction works - Environmental product declarations - Communication format business-to-business*. Brussels: European Committee for Standardization.

CEN. (2011b). *EN 15978: 2011 Sustainability of construction works—assessment of environmental performance of buildings—calculation method*. Brussels: European Committee for Standardization.

CEN. (2012). *EN 15804:2012 Sustainability of construction works—environmental product declarations—core rules for the product category of building products*. Brussels: European Committee for Standardization.

CEN. (2017a). *CEN/TR 16970:2017 Sustainability of construction works - Guidance for the implementation of EN 15804*. Brussels: European Committee for Standardization.

CEN. (2017b). *CEN/TR 17005:2017 Sustainability of construction works - Additional environmental impact categories and indicators - Background information and possibilities*. Brussels: European Committee for Standardization.

Census Bureau. (1940). *1940 Census of Housing: Volume 2. General Characteristics*. Retrieved from United States Department of Commerce website: https://www.census.gov/library/publications/1943/dec/housing-vol02.html

Census Bureau. (1950). *1950 Census of Housing: Volume 1. General Characteristics*. Retrieved from United States Department of Commerce website: https://www.census.gov/library/publications/1953/dec/housing-vol-01.html

Center for International Environmental Policy (CIEL). (2019). *Plastics & Health: The Hidden Costs of a Plastic Planet*. Retrieved from https://www.ciel.org/wp-content/uploads/2019/02/Plastic-and-Health-The-Hidden-Costs-of-a-Plastic-Planet-February-2019.pdf

Chang, J.-H. (2016). Thermal comfort and climatic design in the tropics: An historical critique. *Journal of Architecture (London)*, *21*(8), 1171–1202. doi:10.1080/13602365.2016.1255907

Chang, Y., Huang, Z., Ries, R. J., & Masanet, E. (2016). The embodied air pollutant emissions and water footprints of buildings in China: A quantification using disaggregated input–output life cycle inventory model. *Journal of Cleaner Production*, *113*, 274–284. doi:10.1016/j.jclepro.2015.11.014

Chang, Y., Ries, R. J., & Lei, S. (2012). The embodied energy and emissions of a high-rise education building: A quantification using process-based hybrid life cycle inventory model. *Energy and Building*, *55*, 790–798. doi:10.1016/j.enbuild.2012.10.019

Chang, Y., Ries, R. J., & Wang, Y. (2010). The embodied energy and environmental emissions of construction projects in China: An economic input–output LCA model. *Energy Policy*, *38*(11), 6597–6603. doi:10.1016/j.enpol.2010.06.030

Chastas, P., Theodosiou, T., & Bikas, D. (2016). Embodied energy in residential buildings-towards the nearly zero energy building: A literature review. *Building and Environment*, *105*, 267–282. doi:10.1016/j.buildenv.2016.05.040

Chen, Y., & Su, N. (2017). In-Depth Analysis of the Structure and Properties of Two Varieties of Natural Luffa Sponge Fibers. *Materials Basel*, *10*(5), 1-25. doi:10.3390/ma10050479

Chen, T. Y., Burnett, J., & Chau, C. K. (2001). Analysis of embodied energy use in the residential building of Hong Kong. *Energy*, *26*(4), 323–340. doi:10.1016/S0360-5442(01)00006-8

Chester, T. (1942). Comfort with summer air conditioning. *ASHVE Transactions*, *48*, 107–122.

Chow, R. (2002). *Suburban Space: The Fabric of Dwelling* (1st ed.). Berkeley, CA: University of California Press.

Chyka, P. A., Butler, A. Y., & Herman, M. I. (2001). Ferrous sulfate adsorption by activated charcoal. *Veterinary and Human Toxicology*, *43*(1), 11–13. PMID:11205069

Cinelli, M., Coles, S. R., Nadagouda, M. N., Błaszczyński, J., Słowiński, R., Varma, R. S., & Kirwan, K. (2015). A green chemistry-based classification model for the synthesis of silver nanoparticles. *Green Chemistry*, *17*(5), 2825–2839. doi:10.1039/C4GC02088J

Clarke, J. (2007). *Energy simulation in building design*. Routledge. doi:10.4324/9780080505640

Cole, R. J., & Kernan, P. C. (1996). Life-cycle energy use in office buildings. *Building and Environment*, *31*(4), 307–317. doi:10.1016/0360-1323(96)00017-0

Collier, Z. A., Wang, D., Vogel, J. T., Tatham, E. K., & Linkov, I. (2013). Sustainable roofing technology under multiple constraints: A decision-analytical approach. *Environment Systems & Decisions*, *33*(2), 261–271. doi:10.100710669-013-9446-5

Comstock, S. (1995). *Proclaiming the truth: An illustrated history of the American Society of Heating, Refrigerating and Air-Conditioning Engineers, Inc.* Atlanta, GA: ASHRAE.

Cook, G., & Ponssard, J.-P. (2011). A proposal for the renewal of sectoral approaches building on the Cement Sustainability Initiative. *Climate Policy*, *11*(5), 1246–1256. doi:10.1080/14693062.2011.602552

Cooper, G. (1998). *Air-conditioning America: Engineers and the controlled environment, 1900–1960*. Baltimore, MD: Johns Hopkins University Press.

Crampton, J. W., & Krygier, J. (2018). *An Introduction to Critical Cartography*. Retrieved from http://beu.extension.unicen.edu.ar/xmlui/handle/123456789/359

Crawford, R. H., & Stephan, A. (2013, January). The significance of embodied energy in certified passive houses. In Proceedings of World Academy of Science, Engineering and Technology (No. 78, p. 453). World Academy of Science, Engineering and Technology (WASET).

Crawford, R. H. (2008). Validation of a hybrid life-cycle inventory analysis method. *Journal of Environmental Management*, *88*(3), 496–506. doi:10.1016/j.jenvman.2007.03.024 PMID:17485159

Crenshaw, K. (1989). Demarginalizing the Intersection of Race and Sex: A Black Feminist Critique of Antidiscrimination Doctrine, Feminist Theory and Antiracist Politics. *University of Chicago Legal Forum*, *1989*, 139–168.

Cronon, W. (1992). *Nature's Metropolis: Chicago and the Great West (Reprint edition)*. New York: W. W. Norton & Company.

Cruse, A. (2019). The temperature of a SANAA. In B. Wilke (Ed.), *Ryue Nishizawa / SANAA: Grace Farms* (pp. 62–69). New York, NY: Applied Research and Design.

Dahlgren, J., Cecchini, M., Takhar, H., & Paepke, O. (2007). Persistent organic pollutants in 9/11 world trade center rescue workers: Reduction following detoxification. *Chemosphere*, *69*(8), 1320–1325. doi:10.1016/j.chemosphere.2006.05.127 PMID:17234251

Dahlstrom, O., Sornes, K., Eriksen, S. T., & Hertwich, E. G. (2012). Life Cycle Assessment of a Single-Family Residence Built to Either Conventional- or Passive House Standard. *Energy and Building*, *54*(November), 470–479. doi:10.1016/j.enbuild.2012.07.029

Dahmen, J., von Bergmann, J., & Das, M. (2018). Teardown Index: Impact of property values on carbon dioxide emissions of single-family housing in Vancouver. *Energy and Building*, *170*, 95–106. doi:10.1016/j.enbuild.2018.03.012

Dakwala, M., Mohanty, B., & Bhargava, R. (2014). Simultaneous water and energy conservation through graphical and mathematical programming: A case study for float glass industry. *Journal of Cleaner Production*, *78*, 78. doi:10.1016/j.jclepro.2014.04.059

Davidson, S. (2019). *Grasshopper®*. Retrieved October 25, 2019, from https://www.grasshopper3d.com/

de Dear, R. (2011). Revisiting an old hypothesis of human thermal perception: Alliesthesia. *Building Research and Information*, *39*(2), 108–117. doi:10.1080/09613218.2011.552269

de Dear, R. J., & Brager, G. S. (1998). Developing an adaptive model of thermal comfort and preference. *ASHRAE Transactions*, *104*(1), 145–167.

De Felice, F., Campagiorni, F., & Petrillo, A. (2013). Economic and environmental evaluation via an integrated method based on LCA and MCDA. *Procedia: Social and Behavioral Sciences*, *99*, 1–10. doi:10.1016/j.sbspro.2013.10.465

De Wolf, C. (n.d.). *deQo: Database of Embodied Quantity Outputs*. Retrieved August 5, 2019, from deQo: Database of Embodied Quantity Outputs website: https://www.carbondeqo.com/

De Wolf, C. C. E. L. (2014). *Material Quantities in Building Structures and Their Environmental Impact*. Massachusetts Institute of Technology. Retrieved from https://dspace.mit.edu/handle/1721.1/91298

De Wolf, C., Cerezo, C., Murtadhawi, Z., Hajiah, A., Al Mumin, A., Ochsendorf, J., & Reinhart, C. (2017). Life cycle building impact of a Middle Eastern residential neighborhood. *Energy*, *134*, 336–348. doi:10.1016/j.energy.2017.06.026

De Wolf, C., Pomponi, F., & Moncaster, A. (2017). Measuring Embodied Carbon Dioxide Equivalent of Buildings: A Review and Critique of Current Industry Practice. *Energy and Building*, *140*(April), 68–80. doi:10.1016/j.enbuild.2017.01.075

Deru, M. P., & Torcellini, P. A. (2007). *Source Energy and Emission Factors for Energy Use in Buildings*. National Renewable Energy Laboratory. Retrieved from https://www.nrel.gov/docs/fy07osti/38617.pdf

Di Cesare, S., Silveri, F., Sala, S., & Petti, L. (2018). Positive impacts in social life cycle assessment: State of the art and the way forward. *The International Journal of Life Cycle Assessment*, *23*(3), 406–421. doi:10.100711367-016-1169-7

Dixit, M. K., Fernández-Solís, J. L., Lavy, S., & Culp, C. H. (2010). Identification of parameters for embodied energy measurement: A literature review. *Energy and Buildings, 42*(8), 1238-1247.

Dixit, M. K. (2017). Life Cycle Embodied Energy Analysis of Residential Buildings: A Review of Literature to Investigate Embodied Energy Parameters. *Renewable & Sustainable Energy Reviews, 79*(November), 390–413. doi:10.1016/j.rser.2017.05.051

Dixit, M. K., Culp, C. H., & Fernandez-Solis, J. L. (2015). Embodied energy of construction materials: Integrating human and capital energy into an IO-based hybrid model. *Environmental Science & Technology, 49*(3), 1936–1945. doi:10.1021/es503896v PMID:25561008

Dixit, M. K., Fernandez-Solis, J. L., Lavy, S., & Culp, C. H. (2010). Identiðcation of parameters for embodied energy measurement—A literature review. *Journal of Energy and Buildings, 42*(8), 1238–1247. doi:10.1016/j.enbuild.2010.02.016

Dixit, M. K., Fernández-Solís, J. L., Lavy, S., & Culp, C. H. (2012). Need for an embodied energy measurement protocol for buildings: A review paper. *Renewable & Sustainable Energy Reviews, 16*(6), 3730–3743. doi:10.1016/j.rser.2012.03.021

Doh, J. H., & Panuwatwanich, K. (2014). Variations in embodied energy and carbon emission intensities of construction materials. *Environmental Impact Assessment Review, 49*, 31–48. doi:10.1016/j.eiar.2014.06.003

Dong, Y., & Ng, S. (2015). A social life cycle assessment model for building construction in Hong Kong. *The International Journal of Life Cycle Assessment, 20*(8), 1166–1180. doi:10.100711367-015-0908-5

Dove, C. A., Bradley, F. F., & Patwardhan, S. V. (2019). A material characterization and embodied energy study of novel clay-alginate composite aerogels. *Energy and Building, 184*, 88–98. doi:10.1016/j.enbuild.2018.10.045

Dow Chemical Company. (2019). *Vinyl Chloride Monomer, Product Safety Assessment.* Retrieved from http://msdssearch.dow.com/PublishedLiteratureDOWCOM/dh_08d7/0901b803808d77d0.pdf

Dowden, M. (2013). Climate change and sustainable development: Law, policy and practice. *Estates Gazette.*

Downing, A. S., Bhowmik, A., Collste, D., Cornell, S., Donges, J. F., Fetzer, I., ... Mooij, W. (2019). Matching scope, purpose and uses of planetary boundaries science. *Environmental Research Letters, 14*(7), 073005. doi:10.1088/1748-9326/ab22c9

Duthinh, D. (2014). Structural Design for Fire: A Survey of Building Codes and Standards. *Technical Note (NIST TN) - 1842.* Retrieved from https://www.nist.gov/publications/structural-design-fire-survey-building-codes-and-standards

Eaton, K. J., & Amato, A. (1998). A comparative life cycle assessment of steel and concrete framed office buildings. *Journal of Constructional Steel Research, 1*(46), 286–287. doi:10.1016/S0143-974X(98)00074-1

Eberhardt, L., Birgisdottir, H., & Birkved, M. (2019a). Comparing life cycle assessment modeling of linear vs. circular building components. In Iop Conference Series: Earth and Environmental Science (Vol. 225, No. 1, p. 012039). IOP Publishing.

Eberhardt, L. C. M., Birgisdóttir, H., & Birkved, M. (2019b). Life cycle assessment of a Danish office building designed for disassembly. *Building Research and Information*, *47*(6), 666–680. doi:10.1080/09613218.2018.1517458

Ellen MacArthur Foundation. (2013). The art of design for disassembly. *Engineering the circular economy a field manual for re-designing a regenerative economy.* Retrieved from https://www.ellenmacarthurfoundation.org/assets/downloads/news/EMF_Engineering-the-Circular-Economy_300913.pdf

Ellen MacArthur Foundation. (2016). The case for a circular economy in India. *Circular economy in India: rethinking growth for long-term prosperity, 11.* Retrieved from https://www.ellenmacarthurfoundation.org/

Elliott, M. R., Wang, Y., Lowe, R. A., & Kleindorfer, P. R. (2004). Environmental justice: Frequency and severity of US chemical industry accidents and the socioeconomic status of surrounding communities. *Journal of Epidemiology and Community Health*, *58*(1), 24–30. doi:10.1136/jech.58.1.24 PMID:14684723

Emami, N., Heinonen, J., Marteinsson, B., Säynäjoki, A., Junnonen, J. M., Laine, J., & Junnila, S. (2019). A Life Cycle Assessment of Two Residential Buildings Using Two Different LCA Database-Software Combinations: Recognizing Uniformities and Inconsistencies. *Buildings*, *9*(1), 20. doi:10.3390/buildings9010020

Energy4farms. (2009). *Hempcrete. American Lime Technology*. Retrieved from https://energy4farms.eu/hempcrete/

Environmental Protection Agency (EPA). (2010). *Toxicological Review of Chloroprene, (CAS No. 126-99-8) In Support of Summary Information on the Integrated Risk Information System (IRIS),* Retrieved from https://cfpub.epa.gov/ncea/iris/iris_documents/documents/toxreviews/1021tr.pdf

Ercin, E. (2012). *Governance of globalized water resources: the application of the water footprint to inform corporate strategy and government policy* (Unpublished doctoral dissertation). University of Twente, Ankara, Turkey.

Ercin, A. E., & Hoekstra, A. Y. (2014). Water footprint scenarios for 2050: A global analysis. *Environment International*, *64*, 71–82. doi:10.1016/j.envint.2013.11.019 PMID:24374780

Essington, M. E. (2003). Soil and Water Chemistry an Integrative Approach: An Integrative Approach. Baton Rouge, LA: CRC Press LLC.

European Commission. (2017). *PEFCR Guidance document, - Guidance for the 14 development of Product Environmental Footprint Category Rules (PEFCRs), version 6.3, December 15 2017.* Retrieved July 25, 2019, from https://ec.europa.eu/environment/eussd/smgp/pdf/PEFCR_guidance_v6.3.pdf

European Commission. (n.d.). *REACH*. Retrieved from https://ec.europa.eu/environment/chemicals/reach/reach_en.htm

European Committee for Standarization. (n.d.). *CEN/TC 350- Sustainability of construction works*. Accessed June 24, 2019. https://standards.cen.eu/dyn/www/f?p=204:7:0:FSP_ORG_ID:481830&cs=181BD0E0E925FA84EC4B8BCCC284577F8

European Parliament, Council of the European Union. (2010). *Directive 2010/31/Eu on the Energy Performance of Buildings*. Official Journal of the European Union.

European Parliament, Council of the European Union. (2012). *Directive 2012/27/Eu on Energy Efficiency*. Official Journal of the European Union.

Evans, R. (1997). Figures, doors and passages. In *Translations from drawing to building* (pp. 54–91). Cambridge, MA: MIT University Press.

Falk, R., & McKeever, D. (2004). *Recovering wood for reuse and recycling: a United States perspective*. Paper presented at the European COST E31 Conference.

Fanger, P. O. (1970). *Thermal comfort: Analysis and applications in environmental engineering*. Copenhagen, Denmark: Danish Technical Press.

Fernandez, J. (2006). Material Architecture, emergent materials for innovative buildings and ecological construction. London, UK: Routledge.

Fernández-Galiano, L. (2000). *Fire and memory: On architecture and energy* (G. Cariño, Trans.). Cambridge, MA: MIT Press.

Ferreira, F. A. F., Spahr, R. W., Sunderman, M. A., & Jalali, M. S. (2018). A prioritisation index for blight intervention strategies in residential real estate. *The Journal of the Operational Research Society*, *69*(8), 1269–1285. doi:10.1080/01605682.2017.1390535

Finkbeiner, M., Ackermann, R., & Bach, V. (2014). Challenges in Life Cycle Assessment: An Overview of Current Gaps and Research Needs. In *Background and Future Prospects in Life Cycle Assessment*. Springer Science+Business Media.

Fisher, T. (2016). *Designing our way to a better world*. Minneapolis, MN: University of Minnesota Press. doi:10.5749/minnesota/9780816698875.001.0001

Fisk, W. J., Lei-Gomez, Q., & Mendell, M. (2007). Meta-analyses of the associations of respiratory health effects with dampness and mold in homes. *Indoor Air*, *17*(4), 284–296. doi:10.1111/j.1600-0668.2007.00475.x PMID:17661925

Fitzgerald, F. E., Standfast, S. J., Youngblood, L. G., Melius, J. M., & Janerich, D. T. (1986, November/December). Assessing the Health Effects of Potential Exposure to PCBs, Dioxins, and Furans from Electrical Transformer Fires: The Binghamton State Office Building Medical Surveillance Program. *Archives of Environmental Health*, *41*(6), 368–376. doi:10.1080/00039896.1986.9935781 PMID:3113345

Foraboschi, P., Mercanzin, M., & Trabucco, D. (2014). Sustainable structural design of tall buildings based on embodied energy. *Energy and Building*, *68*, 254–269. doi:10.1016/j.enbuild.2013.09.003

Forest Products Laboratory (FPL). (2010). *Wood handbook wood as an engineering material* (Centennial ed., General technical report FPL; 190). Madison, WI: U.S. Dept. of Agriculture, Forest Service, Forest Products Laboratory.

Fouguet, M., Jusselme, T., & Vareilles, J. (2017). Establishing building environmental targets to implement a low carbon objective at the district level: methodology and case study. In *Proceedings of 33rd PLEA International Conference* (*Vol. 3*, No. CONF, pp. 49-56). Academic Press.

Fox, J. (2019, February 13). Why America's New Apartment Buildings All Look the Same. *Bloomberg.Com.* Retrieved from https://www.bloomberg.com/news/features/2019-02-13/why-america-s-new-apartment-buildings-all-look-the-same

Franklin Associates. (1991). *Comparative Energy Evaluation of Plastic Products and Their Alternatives for the Building and Construction and Transportation Industries.* Retrieved from https://www.greenbuildingsolutions.org/wp-content/uploads/2016/05/Comparative-Energy-Evaluation-of-Plastic-Products-1991.pdf

Franks, F. (1984). Water (Rev. 1st ed.). London: Royal Society of Chemistry.

Freiesleben Hansen, P. (2009). *The Science of Construction Materials.* Berlin: Springer. doi:10.1007/978-3-540-70898-8

Freinkel, S. (2011). *Plastic: A Toxic Love Story.* Boston: Houghton Mifflin Harcourt.

Funtowitz, S., & Ravetz, J. (1993). *Science for the post-normal age. Futures.* Pergamon.

Ganassali, S., Lavagna, M., & Campioli, A. (2016). LCA benchmarks in building's environmental certification systems. In *41st IAHS World Congress on Housing Sustainability and Innovation for the Future* (pp. 1-10). PRT.

Gayle, C., & Gayle, M. (1998). *Cast-Iron Architecture in America: The Significance of James Bogardus* (1st ed.). New York: W. W. Norton & Company.

Gerbens-Leenes, P. W., Hoekstra, A. Y., & Bosman, R. (2018). The blue and greywater footprint of construction materials: Steel, cement and glass. *Water Resources and Industry*, *19*, 1–12. doi:10.1016/j.wri.2017.11.002

Geronimus, A. T. (2000). To mitigate, resist, or undo: Addressing structural influences on the health of urban populations. *American Journal of Public Health*, *90*(6), 867–872. doi:10.2105/AJPH.90.6.867 PMID:10846503

Gertsev, V. I., & Gertseva, V. V. (2004). Classification of mathematical models in ecology. *Ecological Modelling*, *178*(3-4), 329–334. doi:10.1016/j.ecolmodel.2004.03.009

Getter, K. L., Rowe, D. B., Robertson, G. P., Cregg, B. M., & Andresen, J. A. (2009). Carbon sequestration potential of extensive green roofs. *Environmental Science & Technology, 43*(19), 7564–7570. doi:10.1021/es901539x PMID:19848177

Geyer, R., Jambeck, J. R., & Lavender Law, K. (2017). Production, Use, and Fate of All Plastics Ever Made. *Science Advances, 3*(7), 1–5. doi:10.1126ciadv.1700782 PMID:28776036

Gibson, L. J., & Ashby, M. F. (1999). *Cellular Solids: Structure and Properties*. Cambridge University Press.

Glass, S., & Zelinka, S. (2010). *Moisture Relations and Physical Properties of Wood*. Academic Press.

Goggins, J., Keane, T., & Kelly, A. (2010). The assessment of embodied energy in typical reinforced concrete building structures in Ireland. *Energy and Building, 42*(5), 735–744. doi:10.1016/j.enbuild.2009.11.013

Gold, K. (2017). Amid a Housing Crisis, Vancouver Buys Houses and Leaves Them Empty. *The Globe and Mail*. Retrieved from https://www.theglobeandmail.com/real-estate/vancouver/amid-a-housing-crisis-vancouver-buys-houses-and-leaves-them-empty/article35513761/

Gómez-Laserna, O., Olazabal, M. Á., Morillas, H., Prieto-Taboada, N., Martinez-Arkarazo, I., Arana, G., & Madariaga, J. M. (2013). In-situ spectroscopic assessment of the conservation state of building materials from a Palace house affected by infiltration water. *Journal of Raman Spectroscopy: JRS, 44*(9), 1277–1284. doi:10.1002/jrs.4359

Government, H. M. (2010). *Innovation and growth team, in: Low Carbon Construction-Final Report*. Retrieved from http://www.bis.gov.uk/assets/biscore/business-sectors/docs/l/10-1266-low-carbon-construction-igt-ðnal-report.pdf

Graphisoft. (2019). *ArchiCAD*. Retrieved October 25, 2019, from https://www.graphisoft.com/archicad/

Griffin, C. T., Bynum, L., Green, A., Marandyuk, S., Namgung, J., Burkhardt, A., & Hoffman, M. (2013, June). Comparing the embodied energy of structural systems in parking garages. In *2nd International Conference on Structures and Architecture, ICSA 2013* (pp. 945-952). 10.1201/b15267-133

Gupta, S., Kua, H. W., & Low, C. Y. (2018) Use of biochar as carbon sequestering additive in cement mortar. *Journal of Cement and Concrete Composites, 149*. doi:10.1016/j.cemconcomp.2017.12.009

Haapio, A., & Viitaniemi, P. (2008). A Critical Review of Building Environmental Assessment Tools. *Environmental Impact Assessment Review, 28*(7), 469–482. doi:10.1016/j.eiar.2008.01.002

Habibi, Y., Lucia, L. A., & Wiley, I. (2012). *Polysaccharide building blocks a sustainable approach to renewable materials*. Hoboken, NJ: John Wiley & Sons, Inc. doi:10.1002/9781118229484

Häkkinen, T., Antuña, C., Mäkeläinen, T., Lützkendorf, T., Balouktsi, M., Immendörfer, A., . . . Van Dessel, J. (2012). *Sustainability and Performance Assessment and Benchmarking of a Building – Final Report*. Retrieved August 4, 2019, from http://cic.vtt.fi/superbuildings/. Finland: VTT.

Häkkinen, T., Kuittinen, M., Ruuska, A., & Jung, N. (2015). Reducing Embodied Carbon during the Design Process of Buildings. *Journal of Building Engineering*, *4*(December), 1–13. doi:10.1016/j.jobe.2015.06.005

Hameury, S., & Lundström, T. (2004). Contribution of indoor exposed massive wood to a good indoor climate. *Energy and Building*, *36*(3), 281–292. doi:10.1016/j.enbuild.2003.12.003

Hamilton-MacLaren, F., Loveday, D. L., & Mourshed, M. (2009). *The calculation of embodied energy in new build UK housing*. Academic Press.

Han, H.-S. (2014). The Impact of Abandoned Properties on Nearby Property Values. *Housing Policy Debate*, *24*(2), 311–334. doi:10.1080/10511482.2013.832350

Harikae, A. (2016, July). Vo Trong Nghia Architects. *Architecture and Urbanism, 550.*

Harquitectes. (2016). Organizing matter. *2G: Harquitectes, 74.*

Häyhä, T., Lucas, P. L., van Vuuren, D. P., Cornell, S. E., & Hoff, H. (2016). From Planetary Boundaries to national fair shares of the global safe operating space - How can the scales be bridged? *Global Environmental Change*, *40*, 60–72. doi:10.1016/j.gloenvcha.2016.06.008

Hedge, A. (2019), *Global Neoprene Market size is set to achieve 2.3% CAGR over 2019 to 2025.* Retrieved from https://www.reuters.com/brandfeatures/venture-capital/article?id=101765

Hennessy, T. W., Ritter, T., Holman, R. C., Bruden, D. L., Yorita, K. L., Bulkow, L., ... Smith, J. (2008). The relationship between in-home water service and the risk of respiratory tract, skin, and gastrointestinal tract infections among rural Alaska natives. *American Journal of Public Health*, *98*(11), 2072–2078. doi:10.2105/AJPH.2007.115618 PMID:18382002

Heschong, L. (1979). *Thermal delight in architecture*. Cambridge, MA: MIT Press.

Hewlett, P. C., Liska, M., & Lea, F. M. (2019). Lea's Chemistry of Cement and Concrete (5th ed.). Elsevier Science.

Highley, T. L. (1995). Comparative durability of untreated wood in use above ground. *International Biodeterioration & Biodegradation*, *35*(4), 409–419. doi:10.1016/0964-8305(95)00063-1

Hinshaw, M. (2015, April 19). *The new Seattle, where everything looks the same*. Retrieved August 5, 2019, from https://crosscut.com/2015/04/the-new-seattle-where-everything-looks-the-same

Hirt, S. (2013). Home, Sweet Home: American Residential Zoning in Comparative Perspective. *Journal of Planning Education and Research*, *33*(3), 292–309. doi:10.1177/0739456X13494242

History of APA, Plywood, and Engineered Wood. (n.d.). Retrieved July 30, 2019, from APA – The Engineered Wood Association website: https://www.apawood.org/apas-history

Hochhäusl, S., Lange, T., Adams, R. E., Barber, D. A., Bierig, A., Cupers, K., ... Vronskaya, A. (2018). Architecture and the environment. *Architectural Histories*, *6*(1), 1–13. doi:10.5334/ah.259

Hoekstra, A. (2015). The water footprint of industry. In Assessing and Measuring Environmental Impact and Sustainability (pp. 221-254). Academic Press.

Hoekstra, A. (2017). Water Footprint Assessment: Evolvement of a New Research Field. *An International Journal - Published for the European Water Resources Association, 31*(10), 3061-3081.

Hoekstra, A. Y. (2016). A critique on the water-scarcity weighted water footprint in LCA. *Ecological Indicators*, *66*(C), 564–573. doi:10.1016/j.ecolind.2016.02.026

Hollberg, A., Lützkendorf, T., & Habert, G. (2019). Top-down or bottom-up? – How environmental benchmarks can support the design process. *Building and Environment*, *153*, 148–157. doi:10.1016/j.buildenv.2019.02.026

Holzhey, M., & Skoczek, M. (2016). *UBS Global Real Estate Bubble Index: For Housing Markets of Select Cities*. UBS.

Homi, K. (2017). The Unprecedented Expansion of Global Middle Class, An Update; Brookings Instituition. *Global Economy & Development.*

Hossain, M. U., & Ng, S. T. (2018). Critical consideration of buildings' environmental impact assessment towards adoption of circular economy: An analytical review. *Journal of Cleaner Production*, *205*, 763–780. doi:10.1016/j.jclepro.2018.09.120

Hossain, M., Poon, U., Dong, C., Lo, S., & Cheng, Y. (2018). Development of social sustainability assessment method and a comparative case study on assessing recycled construction materials. *The International Journal of Life Cycle Assessment*, *23*(8), 1654–1674. doi:10.100711367-017-1373-0

Hosseinijou, S., Mansour, A., & Shirazi, S. (2014). Social life cycle assessment for material selection: A case study of building materials. *The International Journal of Life Cycle Assessment*, *19*(3), 620–645. doi:10.100711367-013-0658-1

Houghten, F. C., & Yaglou, C. P. (1923). Psychrometric chart with equal comfort lines or effective temperatures for still air. *ASHVE Journal*, *29*, 165–176.

Hoxha, E., Jusselme, T., Brambilla, A., Cozza, S., Andersen, M., & Rey, E. (2016). *Impact targets as guidelines towards low carbon buildings: preliminary concept*. No. CONF.

Huang, Y., Niu, J., & Tse-ming, C. (2012). Energy and Carbon Emission Payback Analysis for Energy-Efficient Retrofitting in Buildings: Overhang Shading Option. *Energy and Building*, *44*(January), 94–103. doi:10.1016/j.enbuild.2011.10.027

Huijbregts, M., Heijungs, R., & Hellweg, S. (2004). Complexity and Integrated Resource Management: Uncertainty in LCA. *The International Journal of Life Cycle Assessment*, *9*(5), 341–342. doi:10.1007/BF02979428

Hu, M. (2017). Balance between energy conservation and environmental impact: Life-cycle energy analysis and life-cycle environmental impact analysis. *Energy and Building*, *140*, 131–139. doi:10.1016/j.enbuild.2017.01.076

Hung, M. K., Johnson, A. U., & Zamin, J. (2014). A Review on the Use of Agriculture Waste Material as Lightweight Aggregate for Reinforced Concrete Structural Members. *Advances in Materials Science and Engineering*. doi:10.1155/2014/365197

Hunt, R., Franklin, W., & Hunt, R. (1996). LCA— How it came about. *The International Journal of Life Cycle Assessment*, *1*(1), 4–7. doi:10.1007/BF02978624

Husain, D., & Prakash, R. (2019). Life Cycle Ecological Footprint Assessment of an Academic Building. *Civil, Architectural. Environmental and Agricultural Engineering*, *100*(1), 97–110.

I.S.O. (2019). ISO 14025:2006. In Environmental labels and declarations - Type III environmental declarations - Principles and procedures. ISO Standards.

Ibn-Mohammed, T., Greenough, R., Taylor, S., Ozawa-Meida, L., & Acquaye, A. (2013). Operational vs. embodied emissions in buildings—A review of current trends. *Energy and Building*, *66*, 232–245. doi:10.1016/j.enbuild.2013.07.026

Iddon, C. R., & Firth, S. K. (2013). Embodied and operational energy for new-build housing: A case study of construction methods in the UK. *Energy and Building*, *67*, 479–488. doi:10.1016/j.enbuild.2013.08.041

IEA & OECD. (2016). *Energy Technology Perspectives 2016*. Paris: International Energy Agency.

Indian Chamber of Commerce. (2017). Outlook. *The Indian coal sector: Challenges and future outlook. Impact of coal shortate. 14.* Retrieved from https://www.pwc.in/assets/pdfs/industries/power-mining/icc-coal-report.pdf

Indian Chamber of Commerce. (2017). Overview. *The Indian coal sector: Challenges and future outlook. Impact of coal shortate. 6.* Retrieved from https://www.pwc.in/assets/pdfs/industries/power-mining/icc-coal-report.pdf

Innovations, K. T. (2016). *Tally*. Retrieved August 4, 2019, from https://www.choosetally.com/

International Association of Certified Home Inspectors (InterNACHI). (2019). *InterNACHI's Standard Estimated Life Expectancy Chart for Homes, InterNACHI*. Retrieved from https://www.nachi.org/life-expectancy.htm

International Code Council (ICC). (2017). *2018 International Building Code*. ICC.

International Energy Agency. (n.d.). *The future of cooling*. Accessed March 18, 2019. https://webstore.iea.org/the-future-of-cooling

International Living Future Institute. (2018). *Living Product Challenge 2.0.* Author.

International Living Future Institute. (n.d.). Retrieved from https://living-future.org/education/Life-Cycle-assessment-environmental-product-declarations-need-know/

International Organization for Standardization (ISO). (2015). *ISO 14001:2015 "Environmental management systems: Requirements with guidance for use."* Retrieved from https://www.iso.org/cms/render/live/en/sites/isoorg/contents/data/standard/06/08/60857.html

International Organization for Standardization (ISO). (2017). *ISO 21930:2017, "Sustainability in buildings and civil engineering works — Core rules for environmental product declarations of construction products and services."* Retrieved from https://www.iso.org/obp/ui/#iso:std:iso:21930:ed-2:v1:en

IPCC. (2014a). Buildings. *In Climate Change 2014: Mitigation of Climate Change. Contribution of Working Group III to the Fifth Assessment Report of the Intergovernmental Panel on Climate Change*, (pp. 671–738). Cambridge, UK: Cambridge University Press. Retrieved from http://pure.iiasa.ac.at/11117

IPCC. (2014b). Climate Change 2014: Synthesis Report. Contribution of Working Groups I, II and III to the Fifth Assessment Report of the Intergovernmental Panel on Climate Change. *IPCC*. Retrieved from https://epic.awi.de/37530/

IPCC. (2014c). Summary for Policy Makers. In *Climate Change 2014: Impacts, Adaptations, and Vulnerability. Part A: Global and Sectoral Aspects. Contribution of Working Group II to the Fifth Assessment Report of the Intergovernmental Panel on Climate Change* (pp. 1–32). Cambridge, UK: Cambridge University Press.

IPCC. (2018) Summary for Policymakers. In Global Warming of 1.5°C. An IPCC Special Report on the impacts of global warming of 1.5°C above pre-industrial levels and related global greenhouse gas emission pathways, in the context of strengthening the global response to the threat of climate change, sustainable development, and efforts to eradicate poverty. In Press.

ISO. (2006). *ISO 14040:2006. Environmental Management—Life Cycle Assessment—Principles and Framework*. International Organization for Standardization. Retrieved from https://www.iso.org/obp/ui/#iso:std:iso:14040:ed-2:v1:en

ISO. (2006a). *14025: 2006, Environmental labels and declarations - Type III environmental declarations -Principles and procedures*. Geneva: International Organization for Standardization.

ISO. (2006b). *14040: 2006, Environmental management - life cycle assessment - principles and framework*. Geneva: International Organization for Standardization.

ISO. (2006c). *14044: 2010, Environmental management - life cycle assessment - requirements and guidelines*. Geneva: International Organization for Standardization.

ISO. (2010). *21931-1: 2010, Sustainability in building construction. Framework for methods of assessment of the environmental performance of construction works. Part 1: Buildings*. Geneva: International Organization for Standardization.

ISO. (2017). *21930: 2017, Sustainability in buildings and civil engineering works - Core rules for environmental product declarations of building products and services*. Geneva: International Organization for Standardization.

ISO. (2019a). *ISO/DIS 21678 Sustainability in buildings and civil engineering works - Indicators and benchmarks - Principles for the development and use of benchmarks*. Geneva: International Organization for Standardization.

ISO. (2019b). *ISO/WD 22057 Enabling use of Environmental Product Declarations (EPD) at construction works level using building information modeling (BIM)*. Geneva: International Organization for Standardization.

Jacobs, J. (1985). *Cities and the wealth of nations*. New York, NY: Vintage Books.

Jaffal, I., Ouldboukhitine, S. E., & Belarbi, R. (2012). A comprehensive study of the impact of green roofs on building energy performance. *Renewable Energy*, *43*, 157–164. doi:10.1016/j.renene.2011.12.004

Jalaei, F., Zoghi, M., & Khoshand, A. (2019). Life cycle environmental impact assessment to manage and optimize construction waste using Building Information Modeling (BIM). *International Journal of Construction Management*, 1-18.

Jensen, A. A., Hoffman, L., Møller, B. T., & Schmidt, A. (1998). Life-Cycle assessment (LCA): a guide to approaches, experiences and information sources. *Environmental Issues Series*. Retrieved from https://www.eea.europa.eu/publications/GH-07-97-595-EN-C/Issue-report-No-6.pdf/view

Jones, R. A. L. (2014). *Soft Condensed Matter*. Oxford, UK: Oxford University Press.

Jørgensen, A., Dreyer, L., & Wangel, C. (2012). Addressing the effect of social life cycle assessments. *The International Journal of Life Cycle Assessment*, *17*(6), 828–839. doi:10.100711367-012-0408-9

JRC European Commission & Institute for Environmental Sustainability. (2010). *Analysis of Existing Environmental Impact Assessment Methodologies for Use in Life Cycle Assessment*. Author.

Junnila, S., & Horvath, A. (2003). Life-Cycle Environmental Effects of an Office Building. *Journal of Infrastructure Systems*, *9*(4), 157–166. doi:10.1061/(ASCE)1076-0342(2003)9:4(157)

Jupp, J. R., & Singh, V. (2014). Similar concepts, distinct solutions, common problems: learning from PLM and BIM deployment. In *IFIP International Conference on Product Lifecycle Management* (pp. 31-40). Springer. 10.1007/978-3-662-45937-9_4

Kabir, M. R., Rooke, B., Dassanayake, G. M., & Fleck, B. A. (2012). Comparative life cycle energy, emission, and economic analysis of 100 kW nameplate wind power generation. *Renewable Energy*, *37*(1), 133–141. doi:10.1016/j.renene.2011.06.003

Kaltenbach, F. (2004). Plastics -- translucent semi-finished sheet products. In F. Kaltenbach (Ed.), Translucent Materials - Glass, Plastic, Metals. Birkhauser, Edition Detail, 41.

Kao, J. (2007). Innovation nation: How America is losing its innovation edge, why it matters, and what we can do to get it back. New York: Free Press.

Katami, T., Yasuhara, A., Okuda, T., & Shibamoto, T. (2002). Formation of PCDDs, PCDFs, and Coplanar PCBs from Polyvinyl Chloride during Combustion in an Incinerator. *Environmental Science & Technology, 36*(6), 1320–1324. doi:10.1021/es0109904 PMID:11944687

Kavgic, M., Mavrogianni, A., Mumovic, D., Summerfield, A., Stevanovic, Z., & Djurovic-Petrovic, M. (2010). A Review of Bottom-up Building Stock Models for Energy Consumption in the Residential Sector. *Building and Environment, 45*(7), 1683–1697. doi:10.1016/j.buildenv.2010.01.021

Kay, J., Regier, H., Boyle, M., & Francis, G. (1999). An Ecosystem Approach for Sustainability: Addressing the Challenge of Complexity. *Futures, 31*(7), 721–742. doi:10.1016/S0016-3287(99)00029-4

Kellett, R., Christen, A., Coops, N. C., van der Laan, M., Ben Crawford, T. R. T., & Olchovski, I. (2013). A Systems Approach to Carbon Cycling and Emissions Modeling at an Urban Neighborhood Scale. *Landscape and Urban Planning, 110*(February), 48–58. doi:10.1016/j.landurbplan.2012.10.002

Kennedy, C. A., Stewart, I., Facchini, A., Cersosimo, I., Mele, R., Chen, B., ... Dubeux, C. (2015). Energy and material flows of megacities. *Proceedings of the National Academy of Sciences of the United States of America, 112*(19), 5985–5990. doi:10.1073/pnas.1504315112 PMID:25918371

Kennedy, C., Steinberger, J., Gasson, B., Hansen, Y., Hillman, T., Havránek, M., ... Mendez, G. V. (2009). Greenhouse Gas Emissions from Global Cities. *Environmental Science & Technology, 43*(19), 7297–7302. doi:10.1021/es900213p PMID:19848137

Kermani, B., & Harrop, D. (2019). *Water Chemistry*. Chichester, UK: John Wiley & Sons, Ltd.

Kieran, S., & Timberlake, J. (2004). *Refabricating architecture: How manufacturing methodologies are poised to transform building construction*. New York: McGraw-Hill.

Kiss, B., & Szalay, Z. (2019). A Visual Method for Detailed Analysis of Building Life Cycle Assessment Results. *Applied Mechanics and Materials, 887*, 319–326. doi:10.4028/www.scientific.net/AMM.887.319

Knickerbocker, W. S. (1927). *Classics of modern science (Copernicus to Pasteur)*. New York, New York: A.A. Knopf.

Kohler, N., König, H., Kreissig, J., & Lützkendorf, T. (2013). *A Life-Cycle Approach to Buildings: Principles, Calculations, Design Tools*. DETAIL, Green Books.

Korvenmaa, P. (1990). The Finnish Wooden House Transformed: American prefabrication, war-time housing and Alvar Aalto. *Construction History, 6*, 47–61.

Krantz, J., Larsson, J., Lu, W., & Olofsson, T. (2015). Assessing embodied energy and greenhouse gas emissions in infrastructure projects. *Buildings, 5*(4), 1156–1170. doi:10.3390/buildings5041156

Krieg, E., & Rybtchinski, B. (2011). Noncovalent Water-Based Materials: Robust yet Adaptive. *Chemistry (Weinheim an der Bergstrasse, Germany)*, *17*(33), 9016–9026. doi:10.1002/chem.201100809 PMID:21726009

Kua, H. W., & Maghimai, M. (2017). Steel-versus-Concrete Debate Revisited: Global Warming Potential and Embodied Energy Analyses based on Attributional and Consequential Life Cycle Perspectives. *Journal of Industrial Ecology*, *21*(1), 82–100. doi:10.1111/jiec.12409

Kuhn, T. (1962). *The Structure of Scientific Revolutions*. Chicago: University of Chicago Press.

Kuramochi, T., Höhne, N., Schaeffer, M., Cantzler, J., Hare, B., Deng, Y., ... Mir, G. U. R. (2018). Ten key short-term sectoral benchmarks to limit warming to 1.5 C. *Climate Policy*, *18*(3), 287–305. doi:10.1080/14693062.2017.1397495

Kurdowski, W. (2014). *Cement and concrete chemistry*. Dordrecht: Springer. doi:10.1007/978-94-007-7945-7

Langer, W. (2011). *Aggregate Resource Availability in the Conterminous United States, Including Suggestions for Addressing Shortages, Quality, and Environmental Concerns* (United States Geological Survey No. Open-File Report 2011-1119). United States Department of the Interior.

Laughland, O., & Lartey, J. (2019a). Residents of US Cancer Town diagnosed at ‘highly unusual’ rates, study says. *The Guardian*. Retrieved from https://www.theguardian.com/us-news/2019/jul/24/reserve-louisiana-cancer-highly-unusual-rates-study

Laughland, O., & Lartey, J. (2019b). First slavery, then a chemical plant and cancer deaths: one town’s brutal history. *The Guardian*. Retrieved from https://www.theguardian.com/us-news/2019/may/06/cancertown-louisiana-reserve-history-slavery

Lee, B., Trcka, M., & Hensen, J. L. (2011). Embodied energy of building materials and green building rating systems—A case study for industrial halls. *Sustainable Cities and Society*, *1*(2), 67–71. doi:10.1016/j.scs.2011.02.002

Leeds, L. (1871). *Treatise on ventilation*. New York, NY: John Wiley & Sons.

Legget, R. F. (1958). *The Durability of Buildings*. doi:10.1520/STP39480S

Lenzen, M. (2000). Errors in conventional and Input-Output—based Life—Cycle inventories. *Journal of Industrial Ecology*, *4*(4), 127–148. doi:10.1162/10881980052541981

Lenzen, M., & Treloar, G. (2002). Embodied energy in buildings: Wood versus concrete—reply to Börjesson and Gustavsson. *Energy Policy*, *30*(3), 249–255. doi:10.1016/S0301-4215(01)00142-2

Lenzen, M., Wood, R., & Wiedmann, T. (2010). Uncertainty analysis for multi-region input–output models–a case study of the UK’s carbon footprint. *Economic Systems Research*, *22*(1), 43–63. doi:10.1080/09535311003661226

Leopold, C. S. (1947). Conditions for Comfort. *ASHVE Transactions*, *53*, 295–306.

Leoto, R., & Lizarraldea, G. (2019). Challenges in evaluating strategies for reducing a building's environmental impact through Integrated Design. *Building and Environment, 155*, 34–46. doi:10.1016/j.buildenv.2019.03.041

Leslie, T. (2013). *Chicago Skyscrapers, 1871-1934*. University of Illinois Press. doi:10.5406/illinois/9780252037542.001.0001

Lichtfouse, E., Schwarzbauer, J., & Robert, D. (2015). Pollutants in Buildings, Water and Living Organisms: Vol. 7. *Environmental Chemistry for a Sustainable World.* Cham: Springer International Publishing.

Li, J., & Colombier, M. (2009). Managing carbon emissions in China through building energy efficiency. *Journal of Environmental Management, 90*(8), 2436–2447. doi:10.1016/j.jenvman.2008.12.015 PMID:19344996

Listokin, D., & Hattis, D. B. (2005). Building Codes and Housing. *Cityscape (Washington, D.C.), 8*(1), 21–67.

Lithner, D., Nordensvan, I., & Göran, D. (2012). Comparative acute toxicity of leachates from plastic products made of polypropylene, polyethylene, PVC, acrylonitrile-butadiene-styrene, and epoxy to *Daphina manga. Environmental Science and Pollution Research International, 19*(5), 1763–1772. doi:10.100711356-011-0663-5 PMID:22183785

Lloyd, S. M., & Ries, R. (2007). Characterizing, Propagating, and Analyzing Uncertainty in Life-Cycle Assessment: A Survey of Quantitative Approaches. *Journal of Industrial Ecology, 11*(1), 161–179. http://onlinelibrary.wiley.com/doi/10.1162/jiec.2007.1136/full. doi:10.1162/jiec.2007.1136

Lotteau, M., Loubet, P., & Sonnemann, G. (2017). An analysis to understand how the shape of a concrete residential building influences its embodied energy and embodied carbon. *Energy and Building, 154*, 1–11. doi:10.1016/j.enbuild.2017.08.048

Lovell, J. (2010). *Building Envelopes: An Integrated Approach* (1st ed.). New York: Princeton Architectural Press.

Lstiburek, J. W. (2007b). The Perfect Wall. ASHRAE Journal, 49(5), 74,76,78.

Lstiburek, J. W. (2007a). The material view of mold. *ASHRAE Journal.*

Ludwig, N., Rosina, E., & Sansonetti, A. (2018). Evaluation and monitoring of water diffusion into stone porous materials by means of innovative IR thermography techniques. *Measurement, 118*, 348–353. doi:10.1016/j.measurement.2017.09.002

Lützkendorf, T., Kohler, N., & König, H. (2012). Integrated Life Cycle Assessment: Benchmarks and Uncertainty. In *Proceedings of the International Symposium on Life Cycle Assessment and Construction–Civil engineering and buildings* (pp. 10-12). Academic Press.

Lynch, K. (1960). *The Image of the City*. Cambridge, MA: MIT Press.

Ma, Z., Cooper, P., Daly, D., & Ledo, L. (2012). Existing building retrofits: Methodology and state-of-the-art. *Energy and Buildings, 55*, 889–902.

Make in India. (2016). Retrieved from http://www.makeinindia.com/article/-/v/make-in-india-reason-vision-for-the-initiative

Mans, J., Kennedy, D., & Peek, B. (2018). The Little Trials: An Abductive Thermal Inquiry. In K. Moe, J. Hutton, & D. Ibañez (Eds.), *Wood Urbanism: From the Molecular to the Territorial* (pp. 194–207). Barcelona: Aktar.

Marr, G. (2017). Vancouver Real Estate Market Heating up Again as Sales and Prices Recover from Buyers Tax. *Financial Post*. Retrieved from https://business.financialpost.com/personal-finance/mortgages-real-estate/vancouver-real-estate-market-heating-up-again-as-sales-and-prices-rise

Matthews, H. S., Hendrickson, C. T., & Weber, C. L. (2008). The Importance of Carbon Footprint Estimation Boundaries. *Environmental Science & Technology, 42*(16), 5839–5842. doi:10.1021/es703112w PMID:18767634

McDermott, S. (2013). *Sasol Heeds Community's Call for Environmental Justice in Mossville, LA*. Trillium Asset Management. Retrieved from https://trilliuminvest.com/sasol-heeds-communitys-call-for-environmental-justice-in-mossville-la/

Means, P., & Guggemos, A. (2015). Framework for life cycle assessment (LCA) based environmental decision making during the conceptual design phase for commercial buildings. *Procedia Engineering, 118*, 802–812. doi:10.1016/j.proeng.2015.08.517

Meex, E., Hollberg, A., Knapen, E., Hildebrand, L., & Verbeeck, G. (2018). Requirements for applying LCA-based environmental impact assessment tools in the early stages of building design. *Building and Environment, 133*, 228–236. doi:10.1016/j.buildenv.2018.02.016

Mendenhall, E., & Yun, L. (2018). *2018 Profile of Home Buyers and Sellers*. Retrieved from National Association of Realtors website: https://www.nar.realtor/sites/default/files/documents/2018%20HBS%20Highlights.pdf

Menzies, F., & Banfill, P. F. G. (2007). Lifecycle assessment and embodied energy: a review. *Proceedings of the Institution of Civil Engineers: Construction Materials,* (160), 135-143. doi:10.1680/coma2007.160.4.135

Menzies, G. F. (2011). *Embodied energy considerations for existing buildings*. Historic Scotland Technical Paper, 13.

Menzies, G. F., Turan, S., & Banfill, P. F. G. (2007). Life-Cycle Assessment and Embodied Energy: A Review. *Proceedings of the Institution of Civil Engineers - Construction Materials, 160*(4), 135–43. 10.1680/coma.2007.160.4.135

Mewis, R. (Ed.). (2018). Building Construction Costs With RSMeans Data 2019 (Annual edition). Gordian.

Milà i Canals, L., Chenoweth, J., Chapagain, A., Orr, S., Antón, A., & Clift, R. (2009). Assessing freshwater use impacts in LCA: Part I—inventory modelling and characterisation factors for the main impact pathways. *The International Journal of Life Cycle Assessment, 14*(1), 28–42. doi:10.100711367-008-0030-z

Miller, D., Doh, J. H., & Mulvey, M. (2015). Concrete slab comparison and embodied energy optimisation for alternate design and construction techniques. *Construction & Building Materials, 80*, 329–338. doi:10.1016/j.conbuildmat.2015.01.071

Milne, G. (2013). *Australian Government "Your Home, Australia's guide to environmentally sustainable homes*. Retrieved from https://www.yourhome.gov.au/materials/embodied-energy

Minnesota B3. (2018). *Buildings*. Benchmarks, & Beyond.

Mithraratne, N., & Vale, B. (2004). Life cycle analysis model for New Zealand houses. *Building and Environment, 39*(4), 483–492. doi:10.1016/j.buildenv.2003.09.008

Moe, K. (2013). *Convergence: An Architectural Agenda for Energy* (1st ed.). London: Routledge. doi:10.4324/9780203489444

Moe, K. (2014). *Insulating modernism*. Boston, MA: Birkhäuser. doi:10.1515/9783038213215

Moe, K. (2018). Intro: R-Values Are Not Our Values. In K. Moe, J. Hutton, & D. Ibañez (Eds.), *Wood Urbanism: From the Molecular to the Territorial* (pp. 156–163). Barcelona: Aktar.

Moffitt, L. (2016). *A Toxic Tour Through the Gulf South, Compass,* Retrieved from https://content.sierraclub.org/coal/posts/toxic-tour-through-gulf-south

Moghtadernejad, S., Chouinard, L. E., & Mirza, M. S. (2018). Multi-criteria decision-making methods for preliminary design of sustainable facades. *Journal of Building Engineering, 19*, 181–190. doi:10.1016/j.jobe.2018.05.006

Mohammadpourkarbasi, H., & Sharples, S. (2017). *Eco-Retrofitting Very Old Dwellings*. Accessed July 18. https://mediatum.ub.tum.de/doc/1169276/1169276.pdf

Mohanta, N., & Acharya, S. K. (2016). Investigation of mechanical properties of luffa cylindrica fiber reinforced epoxy hybrid composite. *International Journal of Engineering Science and Technology, 7*(1), 1. doi:10.4314/ijest.v7i1.1

Moncaster, A. M., & Symons, K. E. (2013). A method and tool for 'cradle to grave'embodied carbon and energy impacts of UK buildings in compliance with the new TC350 standards. *Energy and Building, 66*, 514–523. doi:10.1016/j.enbuild.2013.07.046

Moore, S. A., & Wilson, B. B. (2013). *Questioning Architectural Judgment* (1st ed.). London: Routledge. doi:10.4324/9780203796658

Mueller, L. (2015). Sanborn Fire Insurance Maps: History, Use, Availability. *The Primary Source, 26*(2).

Müezzinoğlu, A., & Toprak, H. (1995). *A Life Cycle Study for Common Construction Materials.* Paper presented at the Cleaner Technologies and Cleaner Products for Sustainable Development, Berlin, Germany.

Myllyviita, T., Holma, A., Antikainen, R., Lähtinen, K., & Leskinen, P. (2012). Assessing environmental impacts of biomass production chains–application of life cycle assessment (LCA) and multi-criteria decision analysis (MCDA). *Journal of Cleaner Production, 29*, 238–245. doi:10.1016/j.jclepro.2012.01.019

Nässén, J., Holmberg, J., Wadeskog, A., & Nyman, M. (2007). Direct and indirect energy use and carbon emissions in the production phase of buildings: An input–output analysis. *Energy, 32*(9), 1593–1602. doi:10.1016/j.energy.2007.01.002

National Center for Environmental Assessment (NCEA). (2006). *An Inventory of Sources and Environmental Releases of Dioxin-Like Compounds in the United States for the Years 1987, 1995, and 2000.* Retrieved from https://cfpub.epa.gov/ncea/risk/recordisplay.cfm?deid=159286

National Environmental Health Association. (2003). Efflorescence. (Technical Briefs). *Journal of Environmental Health, 65*(10), 35. PMID:12800819

National Institute of Environmental Health Science (NIEH). (2010). *Endocrine Disruptors.* Washington, DC: National Institutes of Health. Retrieved from https://www.niehs.nih.gov/health/materials/endocrine_disruptors_508.pdf

National Oceanic and Atmospheric Administration Office for Coastal Management (NOAA). (2019). *Sea Level Rise Viewer.* Retrieved August 6, 2019, from https://coast.noaa.gov/digitalcoast/tools/slr

Natural Resources Canada. (2014). *Comprehensive Energy Use Database - Residential Sector – British Columbia.* Retrieved from https://oee.nrcan.gc.ca/corporate/statistics/neud/dpa/menus/trends/comprehensive/trends_res_bc.cfm

Nault, É., Jusselme, T., & Andersen, M. (2018). Setting contextual life-cycle objectives in urban design: requirements for a decision-support method. In *International Conference for Sustainable Design of the Built Environment SDBE 2018.* Academic Press.

Naushad, M. (2018). *A New Generation Material Graphene: Applications in Water Technology.* Cham: Springer.

Nessy, M. (2015). *Island Buried by Nature: The Dark Colonial Past of Jungle Book Island.* Retrieved from https://www.messynessychic.com

Neumeier, M. (2009). *The designful company: How to build a culture of nonstop innovation: A whiteboard overview.* Berkeley, CA: New Riders.

Nevins, R. (1961). Psychrometrics and modern comfort. *ASHRAE Transactions, 67*, 609–621.

Nghia, V. T. (2016, December). *Lifestyles in Asia in the future—Ideas from history, culture and the environment* [Video file]. Retrieved from https://www.youtube.com/watch?v=PSdD16ox48s

Ngo, T., Mirza, A., Gammampila, R., Aye, L., Crawford, R., & Mendis, P. (2009, September). Life cycle energy of steel and concrete framed commercial buildings. In *Solar09, Proceedings of the 47th ANZSES Annual Conference* (*Vol. 29*, pp. 1-10). Academic Press.

Nicol, F., Humphreys, M., & Roaf, S. (2012). *Adaptive thermal comfort: Principles and practice*. New York, NY: Routledge. doi:10.4324/9780203123010

NIST. (2016). *BEES*. Retrieved August 4, 2019, from https://www.nist.gov/services-resources/software/bees

NOAA. (2019). *Global carbon dioxide growth in 2018 reached 4th highest on record*. Accessed on August 2, 2019 https://www.noaa.gov/news/global-carbon-dioxide-growth-in-2018-reached-4th-highest-on-record

Norman, J., MacLean, H. L., & Kennedy, C. A. (2006). Comparing High and Low Residential Density: Life-Cycle Analysis of Energy Use and Greenhouse Gas Emissions. *Journal of Urban Planning and Development*, *132*(1), 10–21. doi:10.1061/(ASCE)0733-9488(2006)132:1(10)

NYC Open Data. (2019). *311 Service Requests from 2010 to Present*. Retrieved from New York City Department of Information Technology & Telecommunications website: /Social-Services/311-Service-Requests-from-2010-to-Present/erm2-nwe9, /Social-Services/311-Service-Requests-from-2010-to-Present/erm2-nwe9

Nyrud, A. Q., & Bringslimark, T. (2010). Is interior wood use psychologically beneficial? A review of psychological responses toward wood. *Wood and Fiber Science*, *42*(2), 202–218.

O'Neil, D. W., Fanning, A. L., Lamb, W. F., & Steinberger, J. K. (2018). A good life for all within planetary boundaries. *Nature Sustainability*, *1*(2), 88–95. doi:10.103841893-018-0021-4

Occupational Safety and Health Administration (OSHA). (2018). *1910.1017 - Vinyl chloride*. Washington, DC. United States Department of Labor. Retrieved from https://www.osha.gov/laws-regs/regulations/standardnumber/1910/1910.1017

Opher, T., Shapira, A., & Friedler, E. (2018). A comparative social life cycle assessment of urban domestic water reuse alternatives. *The International Journal of Life Cycle Assessment*, *23*(6), 1315–1330. doi:10.100711367-017-1356-1

Organization, I. (2006a). *ISO 14040:2006-Environmental Management - Life Cycle Assessment - Principles and Framework, 2006*. Author.

Organization, I. (2006b). *ISO 14044:2006-Environmental Management - Life Cycle Assessment e Requirements and Guidelines, 2006*. Academic Press.

Ozoemena, M., Cheung, W. M., Hasan, R., & Hackney, P. (2014). *A hybrid Data Quality Indicator and statistical method for improving uncertainty analysis in LCA of a small off-grid wind turbine*. Academic Press.

Öztaş, S. K., & Ipekci, C. A. (2013). Evaluation of the embodied energy for building materials in Turkey. *Advanced Materials Research*, *689*, 273–277. doi:10.4028/www.scientific.net/AMR.689.273

Palko, M. (2014). Advanced Building Construction and Materials II. Durnten: Trans Tech Publications, Limited.

Pander, S. (2019). How Vancouver is pushing to build a city of zero emissions buildings. *Fast Company*. Accessed Dec. 3, 2019. https://www.fastcompany.com/90421205/how-vancouver-is-pushing-for-all-its-new-buildings-to-be-zero-emissions

Pan, J., Jeffrey, P., Marc, A. E., & Amy, P. (2015). Impact of Water Chemistry, Pipe Material and Stagnation on the Building Plumbing Microbiome. *PLoS One*, *10*(10), e0141087. doi:10.1371/journal.pone.0141087 PMID:26495985

Papadopoulos, A., & Giama, E. (2007). Environmental performance evaluation of thermal insulation materials and its impact on the building. *Building and Environment*, *42*(5), 2178–2187. doi:10.1016/j.buildenv.2006.04.012

Parkinson, T., de Dear, R., & Candido, C. (2016). Thermal pleasure in built environments: Alliesthesia in different thermoregulatory zones. *Building Research and Information*, *44*(1), 20–33. doi:10.1080/09613218.2015.1059653

Pavlík, Z., Žumár, J., Medved, I., & Černý, R. (2012). Water Vapor Adsorption in Porous Building Materials: Experimental Measurement and Theoretical Analysis. *Transport in Porous Media*, *91*(3), 939–954. doi:10.100711242-011-9884-9

Peirce, C. S., & Buchler, J. (1955). *Philosophical writings of Peirce*. New York: Dover Publications.

Peirce, C., Houser, N., & Kloesel, C. (1992). *The essential Peirce: Selected philosophical writings*. Bloomington, IN: Indiana University Press.

Pérez-Lombard, L., Ortiz, J., & Pout, C. (2008). A Review on Buildings Energy Consumption Information. *Energy and Building*, *40*(3), 394–398. doi:10.1016/j.enbuild.2007.03.007

Perkins + Will. (2015). *Healthy Environments: What's New (and What's Not) With PVC.* Retrieved from https://perkinswill.com/site/default/filesPerkinsWll_PVC_2015_Whitepaper_2.pdf

Perry, A. (2003). From "the hot-bed of vice" to the "good and well-ordered Christian home": First Nations Housing and Reform in Nineteenth-Century British Columbia. *Ethnohistory (Columbus, Ohio)*, *50*(4), 587–610. doi:10.1215/00141801-50-4-587

Peterson, A. (2008). Ignorance and Ethics. In W. Vitek & W. Jackson (Eds.), The virtues of ignorance: Complexity, sustainability, and the limits of knowledge (Culture of the land) (pp. 109-134). Lexington, KY: University Press of Kentucky.

Petti, L., Serreli, M., & Di Cesare, S. (2018). Systematic literature review in social life cycle assessment. *The International Journal of Life Cycle Assessment*, *23*(3), 422–431. doi:10.100711367-016-1135-4

Peuportier, B. L. P. (2001). Life cycle assessment applied to the comparative evaluation of single family houses in the French context. *Energy and Buildings, 33*(5), 443-450.

Pezzini, M. (2012). *An emerging middle class.* Retrieved from http://oecdobserver.org/news/fullstory.php/aid/3681/An_emerging_middle_class.html

Pharos Project. (n.d.). Retrieved August 5, 2019, from https://www.pharosproject.net/dashboard/

Plastics Market Watch. (2016). *Plastics – Building for a more sustainable future.* The Plastics Industry Trade Association. Retrieved from www.plasticsmarketwatch.org

PlasticsEurope. (2017). *Plastics - The Facts 2017: An analysis of European plastics production, demand and waste data.* Retrieved from https://www.plasticseurope.org/application/files/5715/1717/4180/Plastics_the_facts_2017_FINAL_for_website_one_page.pdf

Plinke, E., Wenk, N., Wolff, G., Castiglione, D., & Palmark, M. (2000). *Mechanical Recycling of PVC Wastes, Study for DG XI of the European Commission.* Retrieved from https://ec.europa.eu/environment/waste/studies/pvc/mech_recylce.pdf

Podesto, L. (2015). Maximizing Value with Mid-Rise Construction. Wood Products Council.

Pomponi, F., & Moncaster, A. (2016). Embodied carbon mitigation and reduction in the built environment–What does the evidence say? *Journal of Environmental Management, 181*, 687–700. doi:10.1016/j.jenvman.2016.08.036 PMID:27558830

Porteous, J. (2007). *Structural timber design to Eurocode 5.* Malden, MA: Blackwell Pub. doi:10.1002/9780470697818

Portland State University. (n.d.). *Embodied energy + Carbon Calculators.* Accessed June 17, 2019. http://web.pdx.edu/~cgriffin/eecc/

Pré Consultants, B. V. (2019). *SimaPro.* Retrieved August 3, 2019, from https://simapro.com/

Pullen, S. (1996). *Data quality of embodied energy methods* (Doctoral dissertation). Deakin University.

Pullen, S. (2000). Estimating the embodied energy of timber building products. Journal of the institute of wood. *Science, 15*(3), 147–151.

Quartz: open data for a healthier, more sustainable future. (2015). Retrieved August 5, 2019, from Quartz website: http://quartzproject.org/

Radford, T. (2016, August 15). All hail the humble moss, bringer of oxygen and life to Earth. *The Guardian.* Retrieved from https://www.theguardian.com/international

Radmanovic, K., Dukic, I., & Pervan, S. (2014). Specific heat capacity of wood. *Specificni toplinski kapacitet drva. Drvna Industrija, 65*(2), 151–157. doi:10.5552/drind.2014.1333

Raleigh, E., & Galster, G. (2015). Neighborhood Disinvestment, Abandonment, and Crime Dynamics. *Journal of Urban Affairs, 37*(4), 367–396. doi:10.1111/juaf.12102

Ramesh, T., Prakash, R., & Shukla, K. K. (2010). Life cycle energy analysis of buildings: An overview. *Energy and Buildings, 42*(10), 1592-1600.

Ramesh, T., Prakash, R., & Shukla, K. K. (2010). Life Cycle Energy Analysis of Buildings: An Overview. *Energy and Building*, *42*(10), 1592–1600. doi:10.1016/j.enbuild.2010.05.007

Ravetz, J. (1987). Usable Knowledge, Usable Ignorance: Incomplete Science with Policy Implications. *Science Communication*, *9*(1), 87–116.

Raworth, K. (2014). Defining a safe and just space for humanity. In State of the World 2013: Is Sustainability Still Possible? (pp. 28-38). Island Press-Center for Resource Economics.

Real Estate Board of Vancouver. (2017). *June 2017 MLS® Home Price Index*. Accessed July 9. https://www.rebgv.org/home-price-index?region=Greater+Vancouver&type=all&date=2017-06-01

Reddy, B. V., & Jagadish, K. S. (2003). Embodied energy of common and alternative building materials and technologies. *Energy and Buildings, 35*(2), 129-137.

Reddy, Venkatarama, & Jagadish. (2003). Embodied Energy of Common and Alternative Building Materials and Technologies. *Energy and Buildings, 35*(2), 129–137. Retrieved from https://www.sciencedirect.com/science/article/pii/S0378778801001414

Reimerink, W. & Kleut., D. (1999). Air pollution control by adsorption. *Air Pollution Control by Adsorption*, *120*(Part B), 807-819. doi:10.1016/S0167-2991(99)80380-2

Reinhart, C. F., & Cerezo Davila, C. (2016). Urban building energy modeling – A review of a nascent field. *Building and Environment*, *97*, 196–202. doi:10.1016/j.buildenv.2015.12.001

Reports and Data. (2019). *Construction Plastics Market To Reach USD 128.12 Billion By 2026*. Retrieved from https://www.globenewswire.com/news-release/2019/06/03/1863469/0/en/Construction-Plastics-Market-To-Reach-USD-128-12-Billion-By-2026-Reports-And-Data.html

Reyna, J. L., & Chester, M. V. (2015). The Growth of Urban Building Stock: Unintended Lock-in and Embedded Environmental Effects. *Journal of Industrial Ecology*, *19*(4), 524–537. doi:10.1111/jiec.12211

Rezaei, F., Bulle, C., & Lesage, P. (2019). Integrating building information modeling and life cycle assessment in the early and detailed building design stages. *Building and Environment*, *153*, 158–167. doi:10.1016/j.buildenv.2019.01.034

Richard, R. B. (2005). Industrialised building systems: Reproduction before automation and robotics. *Automation in Construction*, *14*(4), 442–451. doi:10.1016/j.autcon.2004.09.009

Riera Perez, M. G., Rey, E., Liman, U., Roulet, Y., & Favris-Donzel, A. (2014). *SméO, a sustainability assessment tool targeting the 2000 Watts society*. No. CONF.

Rietschel, H. (1927). *Heating and ventilating: A handbook for architects and engineers* (C. W. Brabbee, Trans.). New York, NY: McGraw-Hill.

Ristow, W. W. (1968). United States Fire Insurance and Underwriters Maps · 1852—1968. *Quarterly Journal. Library of Congress*, *25*(3), 194–218.

Rittel, H., & Webber, W. (1973). Dilemmas in a general theory of planning. *Policy Sciences*, *4*(2), 155–169. doi:10.1007/BF01405730

Robati, M., Daly, D., & Kokogiannakis, G. (2019). A method of uncertainty analysis for whole-life embodied carbon emissions (CO2-e) of building materials of a net-zero energy building in Australia. *Journal of Cleaner Production*, *225*, 541–553. doi:10.1016/j.jclepro.2019.03.339

Robert McNeel & Associates. (2019). Retrieved October 25, 2019, from https://www.rhino3d.com

Robèrt, K. H. (2000). Tools and concepts for sustainable development, how do they relate to a general framework for sustainable development, and to each other? *Journal of Cleaner Production*, *8*(3), 243–254. doi:10.1016/S0959-6526(00)00011-1

Robèrt, K. H. (2002). *The Natural Step Story, Seeding a Quiet Revolution*. Gabriola Island, BC, Canada: New Catalyst Books.

Robins, C. (2018). *How technology is opening up the gates of bigger opportunities for India*. Retrieved from https://economictimes.indiatimes.com/

Rockström, J., Steffen, W., Noone, K., Persson, Å., Chapin III, F. S., Lambin, E. F., ... Nykvist, B. (2009b). A safe operating space for humanity. *Nature*, *461*(7263), 472.

Rockström, J., Steffen, W. L., Noone, K., Persson, Å., Chapin, F. S. III, Lambin, E., ... Nykvist, B. (2009a). Planetary boundaries: Exploring the safe operating space for humanity. *Ecology and Society*, *14*(2), art32. doi:10.5751/ES-03180-140232

Rodger, L. (2014, February 9). *The History of Activated Carbon*. Retrieved from https://www.jurassiccarbon.com/blogs/news/12186281-the-history-of-activated-carbon

Roh, S., Tae, S., Kim, R., & Park, S. (2019). Probabilistic Analysis of Major Construction Materials in the Life Cycle Embodied Environmental Cost of Korean Apartment Buildings. *Sustainability*, *11*(3), 846. doi:10.3390u11030846

Rosenthal, S. S., & Helsley, R. W. (1994). Redevelopment and the Urban Land Price Gradient. *Journal of Urban Economics*, *35*(2), 182–200. doi:10.1006/juec.1994.1012

Rose, W. B. (2005). *Water in Buildings: An Architect's Guide to Moisture and Mold* (1st ed.). Hoboken, NJ: Wiley.

Rose, W. B., Kaskel, B. S., Kudder, R. J., Mitchell, M. R., & Link, R. E. (2011). Insulation Draws Water. *Journal of Testing and Evaluation*, *39*(1), 102972. doi:10.1520/JTE102972

Rott, U. (2003). Multiple Use of Water in Industry—The Textile Industry Case. *Journal of Environmental Science and Health. Part A, Toxic/Hazardous Substances & Environmental Engineering*, *38*(8), 1629–1639. doi:10.1081/ESE-120021485 PMID:12929813

Rowley, H. V., Peters, G. M., Lundie, S., & Moore, S. J. (2012). Aggregating sustainability indicators: Beyond the weighted sum. *Journal of Environmental Management, 111*, 24–33. doi:10.1016/j.jenvman.2012.05.004 PMID:22813856

Roy, D., & Zeckhauser, R. (2015). Grappling with Ignorance: Frameworks from Decision Theory, Lessons from Literature. *Journal of Benefit-Cost Analysis, 6*(1), 33–65. doi:10.1017/bca.2015.8

Ryberg, M. W., Owsianiak, M., Richardson, K., & Hauschild, M. Z. (2016). Challenges in implementing a planetary boundaries based life-cycle impact assessment methodology. *Journal of Cleaner Production, 139*, 450–459. doi:10.1016/j.jclepro.2016.08.074

Rysanek, A., & Choudhary, R. (2013). Optimum building energy retrofits under technical and economic uncertainty. *Energy and Building, 57*, 324–337.

Ryu, H.-Y., Yoon, S. H., Han, D.-H., Hafeez, H., Paluvai, N. R., Lee, C. S., & Park, J.-G. (2016). Fabrication of hydrophobic/hydrophilic switchable aluminum surface using poly(N-isopropylacrylamide). *Progress in Organic Coatings, 99*, 295–301. doi:10.1016/j.porgcoat.2016.06.008

Saadatian, O., Sopian, K., Salleh, E., Lim, C. H., Riffat, S., Saadatian, E., ... Sulaiman, M. Y. (2013). A review of energy aspects of green roofs. *Renewable & Sustainable Energy Reviews, 23*, 155–168. doi:10.1016/j.rser.2013.02.022

Sala, S., Cerutti, A. K., & Pant, R. (2018). *Development of a weighting approach for the environmental footprint*. Luxembourg: Publications Office of the European Union.

Salazar, J., & Meil, J. (2009). Prospects for Carbon-Neutral Housing: The Influence of Greater Wood Use on the Carbon Footprint of a Single-Family Residence. *Journal of Cleaner Production, 17*(17), 1563–1571. doi:10.1016/j.jclepro.2009.06.006

Sandy Inundation Zone | NYC Open Data. (n.d.). Retrieved August 4, 2019, from https://data.cityofnewyork.us/Environment/Sandy-Inundation-Zone/uyj8-7rv5

Sanjurjo-Sánchez, J., & Alves, C. (2012). *Pollutant-Induced Decay of Building Materials*. Dordrecht: Springer Netherlands. doi:10.1007/978-94-007-2439-6_2

Sartori, I., & Hestnes, A. G. (2007). Energy Use in the Life Cycle of Conventional and Low-Energy Buildings: A Review Article. *Energy and Building, 39*(3), 249–257. doi:10.1016/j.enbuild.2006.07.001

Sathre, R., & Gustavsson, L. (2009). Using Wood Products to Mitigate Climate Change: External Costs and Structural Change. *Applied Energy, 86*(2), 251–257. doi:10.1016/j.apenergy.2008.04.007

Sawislak, K. (1996). *Smoldering City: Chicagoans and the Great Fire, 1871-1874*. University of Chicago Press.

Sazedj, S., Morais, A. J., & Jalali, S. (2014). *Comparison of embodied energy and carbon dioxide emissions of brick and concrete based on functional units*. Academic Press.

Scheuer, C., Keoleian, G. A., & Reppe, P. (2003). Life cycle energy and environmental performance of a new university building: modeling challenges and design implications. *Energy and Buildings, 35*(10), 1049-1064.

Schlegl, F., Gantner, J., Traunspurger, R., Albrecht, S., & Leistner, P. (2019). LCA of buildings in Germany: Proposal for a future benchmark based on existing databases. *Energy and Building, 194*, 342–350. doi:10.1016/j.enbuild.2019.04.038

Schyns, J. F., Booij, M. J., & Hoekstra, A. Y. (2017). The water footprint of wood for lumber, pulp, paper, fuel and firewood. *Advances in Water Resources, 107*, 490–501. doi:10.1016/j.advwatres.2017.05.013

Sellers, C. (1977). *Hazards of the job: From industrial disease to environmental health science*. Chapel Hill, NC: University of North Carolina Press.

Seto, K. C., Dhakal, S., Bigio, A., Blanco, H., Delgado, G. C., Dewar, D., ... Ramaswami, A. (2014). Human Settlements, Infrastructure and Spatial Planning. In O. Edenhofer, R. Pichs-Madruga, Y. Sokona, E. Farahani, S. Kadner, K. Seyboth, ... J. C. Minx (Eds.), *Climate Change 2014: Mitigation of Climate Change. Contribution of Working Group III to the Fifth Assessment Report of the Intergovernmental Panel on Climate Change*. Cambridge, UK: Cambridge University Press.

Shadram, F., Johansson, T. D., Lu, W., Schade, J., & Olofsson, T. (2016). An integrated BIM-based framework for minimizing embodied energy during building design. *Energy and Building, 128*, 592–604. doi:10.1016/j.enbuild.2016.07.007

Shank, G. (1998). The Extraordinary Ordinary Powers of Abductive Reasoning. *Theory & Psychology, 8*(6), 841–860. doi:10.1177/0959354398086007

Shmulsky, R., & Jones, P. D. (2011). *Wood and Water*. Academic Press.

Shoseyov, O., & Levy, I. (2008). *NanoBioTechnology bioinspired devices and materials of the future*. Totowa, NJ: Humana Press.

Shove, E. (2003). *Comfort, cleanliness and convenience: The social organization of normality*. New York, NY: Berg.

Siemens, A. G. (2013). *Production of Synthetic Rubber, Process Gas Chromatograph MAXUM edition II controls and optimizes Butyl Rubber production*. Retrieved from https://w3.siemens.com/mcms/sensor-systems/CaseStudies/CS_Butyl_Rubber_2013-01_en_Web.pdf, 2.

Silberglitt, R. (2019). *New and Critical Materials: Identifying Potential Dual-Use Areas*. doi:10.7249/CT513

Simonen, K., Droguett, B. R., Strain, L., & McDade, E. (2017). *Embodied Carbon Benchmark Study: LCA for Low Carbon Construction*. University of Washington. Retrieved from https://digital.lib.washington.edu/researchworks/handle/1773/38017

Simonen, K., Rodriguez, B. X., & Wolf, C. D. (2017). Benchmarking the Embodied Carbon of Buildings. *Technology|Architecture + Design, 1*(2), 208–218.

Simonen, K. (2014). *Life Cycle Assessment* (1st ed.). London: Routledge. doi:10.4324/9781315778730

Simpson, W., & TenWolde, A. (1999). Physical Properties and Moisture Relations of Wood. In *Wood handbook—Wood as an engineering material* (p. 463). Forest Products Laboratory.

Solecki, W., Rosenzweig, C., Dhakal, S., Roberts, D., Barau, A. S., Schultz, S., & Ürge-Vorsatz, D. (2018). City transformations in a 1.5 C warmer world. *Nature Climate Change*, *8*(3), 177–181. doi:10.103841558-018-0101-5

Solis-Guzman, J. (2015). *Ecological footprint assessment of building construction*. Sharjah: Bentham Science Publishers, Limited. doi:10.2174/97816810809871150101

Sousa, G., Jones, B. M., Mirzaei, P. A., & Robinson, D. (2017). A Review and Critique of UK Housing Stock Energy Models, Modelling Approaches and Data Sources. *Energy and Building*, *151*(September), 66–80. doi:10.1016/j.enbuild.2017.06.043

Soust-Verdaguer, B., Llatas, C., & García-Martínez, A. (2017). Critical review of BIM-based LCA method to buildings. *Energy and Building*, *136*, 110–120. doi:10.1016/j.enbuild.2016.12.009

Spirinckx, C., Thuring, M., Damen, L., Allacker, K., Mirabella, N., Ramon, D., ... Röck, M. (2018). Study on the Application of the PEF Method and related guidance documents to a newly office building. *European Union*.

Star, E. (2013). *Source Energy*. Portfolio Manager - Technical Reference.

Steffen, W., Richardson, K., Rockström, J., Cornell, S. E., Fetzer, I., Bennett, E. M., ... Folke, C. (2015). Planetary boundaries: Guiding human development on a changing planet. *Science*, *347*(6223), 1259855. doi:10.1126cience.1259855 PMID:25592418

Strang, V. (2015). *Water: Nature and Culture*. London: Reaktion Books, Limited.

Subramanian, C., Chau, C. K., & Yung, W. K. C. (2018). Relevance and feasibility of the existing social LCA methods and case studies from a decision-making perspective. *Journal of Cleaner Production*, *171*, 690–703. doi:10.1016/j.jclepro.2017.10.006

Swan, L. G., & Ismet Ugursal, V. (2009). Modeling of End-Use Energy Consumption in the Residential Sector: A Review of Modeling Techniques. *Renewable & Sustainable Energy Reviews*, *13*(8), 1819–1835. doi:10.1016/j.rser.2008.09.033

Swilling, M., Hajer, M., Baynes, T., Bergesen, J., Labbé, F., Musango, J. K., & Suh, S. (2018). *The Weight of Cities Resource Requirements of Future Urbanization*. International Resource Panel. United Nations Environment Programme.

Szabadvary, F. (1966). *The Period of Berzelius*. Elsevier Ltd. doi:10.1016/B978-0-08-010980-0.50009-2

Szalay, Z., & Kiss, B. G. (2019). Modular methodology for building life cycle assessment for a building stock model. In R. Caspeele, L. Taerwe, & D. M. Frangopol (Eds.), *Life Cycle Analysis and Assessment in Civil Engineering: Towards an Integrated Vision: Proceedings of the Sixth International Symposium on Life-Cycle Civil Engineering (IALCCE 2018)* (*Vol. 5*, pp. 855-861). London: CRC Press.

Tasoff, H. (2019). Plastic's Carbon Footprint, Researchers conduct first global assessment of the lifecycle of greenhouse gas emissions from plastics. *The CURRENT*. Retrieved from https://www.news.ucsb.edu/2019/019424/plastic-s-carbon-footprint?fbclid=IwAR3JqmOqIWj5EePTCYOmaLeMMfJSI0qYhSx__hR911b90r0emO4ICIxw3gk

Tecchio, P., Gregory, J., Ghattas, R., & Kirchain, R. (2019a). Structured under-specification of life cycle impact assessment data for building assemblies. *Journal of Industrial Ecology*, *23*(2), 319–334. doi:10.1111/jiec.12746

Tecchio, P., Gregory, J., Olivetti, E., Ghattas, R., & Kirchain, R. (2019b). Streamlining the Life Cycle Assessment of Buildings by Structured Under-Specification and Probabilistic Triage. *Journal of Industrial Ecology*, *23*(1), 268–279. doi:10.1111/jiec.12731

Tewari, M., & Godfrey, N. (2016). *Better Cities, Better Growth: India's Urban Opportunity.* Retrieved from http://newclimateeconomy.report/workingpapers

The Ozone Hole Inc. (2019). *The Ozone Hole 2018*. Retrieved August 4, 2019, from http://www.theozonehole.com/2018.htm

Thibodeau, C., Bataille, A., & Sié, M. (2019). Building rehabilitation life cycle assessment methodology–state of the art. *Renewable & Sustainable Energy Reviews*, *103*, 408–422. doi:10.1016/j.rser.2018.12.037

Thinkstep. (2019). *GaBi Software - to drive product sustainability!* Retrieved August 4, 2019, from http://www.gabi-software.com/software/gabi-software/

Thormark, C. (2006). The effect of material choice on the total energy need and recycling potential of a building. *Building and Environment*, *41*(8), 1019–1026. doi:10.1016/j.buildenv.2005.04.026

Togkalidou, T., Karoglou, M., Bakolas, A., Giakoumaki, A., & Moropoulou, A. (2013). Correlation of Water Vapor Permeability with Microstructure Characteristics of Building Materials Using Robust Chemometrics. *Transport in Porous Media*, *99*(2), 273–295. doi:10.100711242-013-0184-4

Tooke, T. R., Coops, N. C., & Webster, J. (2014a). Predicting Building Ages from LiDAR Data with Random Forests for Building Energy Modeling. *Energy and Building*, *68*(January), 603–610. doi:10.1016/j.enbuild.2013.10.004

Tooke, T. R., van der Laan, M., & Coops, N. C. (2014b). Mapping Demand for Residential Building Thermal Energy Services Using Airborne LiDAR. *Applied Energy*, *127*(August), 125–134. doi:10.1016/j.apenergy.2014.03.035

Torcellini, P., Long, N., & Judkoff, R. (2004). Consumptive Water Use for U.S. Power Production. *ASHRAE Transactions*, *110*, 96–100.

Tredgold, T. (1836). *Principles of warming and ventilation*. London, UK: M. Taylor.

Treloar, G. J., Gupta, H., Love, P. E. D., & Nguyen, B. (2003). An Analysis of Factors Influencing Waste Minimisation and Use of Recycled Materials for the Construction of Residential Buildings. *Management of Environmental Quality*, *14*(1), 134–145. doi:10.1108/14777830310460432

Treloar, G., Fay, R., Love, P. E. D., & Iyer-Raniga, U. (2000). Analysing the Life-Cycle Energy of an Australian Residential Building and Its Householders. *Building Research and Information*, *28*(3), 184–195. doi:10.1080/096132100368957

Trimble Inc. (2019). *SketchUp*. Retrieved October 25, 2019, from https://www.sketchup.com/

Tuomi, T., Reijula, K., Johnsson, T., Hemminki, K., Hintikka, E.-L., Lindroos, O., ... Haahtela, T. (2000). Mycotoxins in Crude Building Materials from Water-Damaged Buildings. *Applied and Environmental Microbiology*, *66*(5), 1899–1904. doi:10.1128/AEM.66.5.1899-1904.2000 PMID:10788357

UN Global Impact. (2019). *CEO Water Mandate*. Retrieved from https://ceowatermandate.org/

UNEP. (2013). *Energy efficiency for buildings*. Paris: United Nations Environment Programme.

United States Department of Commerce. (2015). *2015 United States Census via Social Explorer*. Retrieved from https://www.socialexplorer.com/explore-maps

United States Department of Housing and Urban Development. (2001). *Review of Structural Materials and Methods for Home Building in the United States: 1900 to 2000*. Retrieved from https://www.huduser.gov/portal/publications/destech/revstruc.html

University Network for Human Rights (UNHR). (2019). *Waiting to Die, Toxic Emissions and Disease Near the Louisiana Denka / DuPont Plant*. Retrieved from https://www.humanrightsnetwork.org/waiting-to-die

Upton, B., Miner, R., Spinney, M., & Heath, L. S. (2008). The Greenhouse Gas and Energy Impacts of Using Wood instead of Alternatives in Residential Construction in the United States. *Biomass and Bioenergy*, *32*(1), 1–10. doi:10.1016/j.biombioe.2007.07.001

Ürge-Vorsatz, D., Koeppel, S., & Mirasgedis, S. (2007). Appraisal of Policy Instruments for Reducing Buildings' CO_2 Emissions. *Building Research and Information*, *35*(4), 458–477. doi:10.1080/09613210701327384

US Green Building Council (USGBC). (2019). *LEED v4 for Building Design and Construction, Point Breakdown Spreadsheet.* Retrieved from https://www.usgbc.org/resources/leed-v4-building-design-and-construction-checklist

US Green Building Council (USGBC). (n.d.). *Buildings and Climate Change.* Retrieved from https://www.eesi.org/files/climate.pdf

USGBC. (2014). *LEED* (Vol. 4). Reference for Building Design and Construction.

Vancouver, City of. (2016). *Zero Emissions Building Plan*. Author.

Venezia, D., Joshi, S., & Szurgot, A. & Van Brunt, M. (2010). Energy-from-waste and dioxin emission control: is there a role for PVC separation. *Proceedings of the 18th Annual North American Waste-to-Energy Conference*, 1-8.

Verbeeck, G., & Hens, H. (2010a). Life Cycle Inventory of Buildings: A Calculation Method. *Building and Environment*, *45*(4), 1037–1041. doi:10.1016/j.buildenv.2009.10.012

Verbeeck, G., & Hens, H. (2010b). Life Cycle Inventory of Buildings: A Contribution Analysis. *Building and Environment*, *45*(4), 964–967. doi:10.1016/j.buildenv.2009.10.003

Verma, D., Gope, P., & Maheshwari, M. K. (2012). Coir Fiber Reinforcement and Application in Polymer Composites: A Review. *Journal of Materials and Environmental Science.*, *4*, 263–276.

Vesterberg, A., Hendermark, M., & Vass, A. M. (2005). *PVC in Medical Devices: An inventory of PVC and phthalates containing devices used in health care*. Retrieved from https://noharm.org/sites/default/files/lib/downloads/pvc/PVC_Med_Devices_Karolinska.pdf

Vickers, A. (2001). Handbook of water use and conservation. Choice Reviews Online, 39(3), 39-1594-1539-1594.

Vilches, A., Garcia-Martinez, A., & Sanchez-Montañes, B. (2017). Life Cycle Assessment (LCA) of Building Refurbishment: A Literature Review. *Energy and Building*, *135*(January), 286–301. doi:10.1016/j.enbuild.2016.11.042

Vitek, W., & Jackson, W. (2008). *The virtues of ignorance: Complexity, sustainability, and the limits of knowledge (Culture of the land)*. Lexington, KY: University Press of Kentucky.

Vollpracht, A., & Brameshuber, W. (2010). Investigations on the leaching behaviour of irrigated construction elements. (Report). *Environmental Science and Pollution Research International*, *17*(5), 1177–1182. doi:10.100711356-009-0264-8 PMID:19937278

von Bergmann, J., & Dahmen, J. (2016). *Teardown Index: Vancouver SFH Teardowns*. Retrieved from https://mountainmath.ca/teardowns

W.F.N. (2019). *Water Footprint Network*. Retrieved from https://waterfootprint.org/en/

Wackernagel, M. (1996). *Our ecological footprint: reducing human impact on the earth*. Philadelphia, PA: New Society Publishers.

Wallace, R., & Wallace, D. (1990). Origins of public health collapse in New York City: The dynamics of planned shrinkage, contagious urban decay and social disintegration. *Bulletin of the New York Academy of Medicine*, *66*(5), 391–434. PMID:2257377

Wang, H., & Zhai, Z. J. (2016). Advances in Building Simulation and Computational Techniques: A Review between 1987 and 2014. *Energy and Building*, *128*(September), 319–335. doi:10.1016/j.enbuild.2016.06.080

Wang, L., Ding, X., & Wu, X. (2014). Careful considerations when reporting and evaluating the greywater footprint of products. *Ecological Indicators*, *41*, 131–132. doi:10.1016/j.ecolind.2014.01.039

Wang, Y., Kuckelkorn, J., & Liu, Y. (2017). A State of Art Review on Methodologies for Control Strategies in Low Energy Buildings in the Period from 2006 to 2016. *Energy and Building*, *147*(July), 27–40. doi:10.1016/j.enbuild.2017.04.066

Wastiels, L., Schifferstein, H. N. J., Heylighen, A., & Wouters, I. (2012). Relating material experience to technical parameters: A case study on visual and tactile warmth perception of indoor wall materials. *Building and Environment*, *49*(1), 359–367. doi:10.1016/j.buildenv.2011.08.009

Wermiel, S. E. (2000). *The Fireproof Building: Technology and Public Safety in the Nineteenth-Century American City*. Baltimore, MD: The Johns Hopkins University Press.

Whiteman, G., Walker, B., & Perego, P. (2013). Planetary boundaries: Ecological foundations for corporate sustainability. *Journal of Management Studies*, *50*(2), 307–336. doi:10.1111/j.1467-6486.2012.01073.x

Wilson, A., & Mans, J. (2019). One House Many Nations: Indigenous Project-Based Collaboration. In M. Henderson (Ed.), *Catching Fire: Fuelling Inquiry and Passion Through Project-Based Learning* (pp. 171–184). Winnipeg, MB: Portage & Main Press.

Wilson, S., Hutson, M., & Mujahid, M. (2008). How Planning and Zoning Contribute to Inequitable Development, Neighborhood Health, and Environmental Injustice. *Environmental Justice*, *1*(4), 211–216. doi:10.1089/env.2008.0506

WOODPRODUCTSfi. (n.d.). *Adhesives in wood construction*. Retrieved from https://www.woodproducts.fi/content/adhesives-wood-construction

World Economic Forum. (2019). *Fourth Industrial Revolution: Beacons of Technology and Innovation in Manufacturing*. Retrieved from http://www3.weforum.org/docs/WEF_4IR_Beacons_of_Technology_and_Innovation_in_Manufacturing_report_2019.pdf

Xing, S., Xu, Z., & Jun, G. (2008). Inventory Analysis of LCA on Steel- and Concrete-Construction Office Buildings. *Energy and Building*, *40*(7), 1188–1193. doi:10.1016/j.enbuild.2007.10.016

Yaglou, C. P. (1932). How to use the effective temperature index and comfort charts. *ASHVE Transactions*, *38*, 411–423.

Yanowitz, B. (2013). *A Stronger, More Resilient New York*. Retrieved from New York City Economic Development Corporation website: https://www.nycedc.com/resource/stronger-more-resilient-new-york

Zemero, B. R., de Lima Tostes, M. E., Bezerra, U. H., dos Santos Batista, V., & Carvalho, C. C. M. (2019). Methodology for Preliminary Design of Buildings Using Multi-Objective Optimization Based on Performance Simulation. *Journal of Solar Energy Engineering*, *141*(4), 040801. doi:10.1115/1.4042244 PMID:30837779

Zhang, C., Yu, L., Ferdosian, F., Vijayaraghavan, S., Mesnager, J., Jollet, V., & Zhao, B. (2018). Behavior of Water/pMDI Emulsion Adhesive on Bonding Wood Substrates with Varied Surface Properties. *Industrial & Engineering Chemistry Research*, *57*(48), 16318–16326. doi:10.1021/acs.iecr.8b03485

Zhang, G. P., Hoekstra, A. Y., & Mathews, R. E. (2013). Water Footprint Assessment (WFA) for better water governance and sustainable development. *Water Resources and Industry*, *1-2*(C), 1–6. doi:10.1016/j.wri.2013.06.004

Zhang, W., Tan, S., Lei, Y., & Wang, S. (2014). Life Cycle Assessment of a Single-Family Residential Building in Canada: A Case Study. *Building Simulation*, *7*(4), 429–438. doi:10.100712273-013-0159-y

Zhang, X., Pint, C. L., Lee, M. H., Schubert, B. E., Jamshidi, A., Takei, K., & Javey, A. (2011). Optically- and thermally-responsive programmable materials based on carbon nanotube-hydrogel polymer composites. *Nano Letters*, *11*(8), 3239–3244. doi:10.1021/nl201503e PMID:21736337

Zhao, N. (2017, January 5). *The Aging Housing Stock*. Retrieved August 5, 2019, from Eye On Housing website: http://eyeonhousing.org/2017/01/the-aging-housing-stock-3/

Zheng, J., & Suh, S. (2019). Strategies to reduce the global carbon footprint of plastics. *Nature Climate Change*, *9*(5), 374–378. doi:10.103841558-019-0459-z

Zimmermann, M., Althaus, H. J., & Haas, A. (2005). Benchmarks for sustainable construction: A contribution to develop a standard. *Energy and Building*, *37*(11), 1147–1157. doi:10.1016/j.enbuild.2005.06.017

About the Contributors

Blaine Brownell is an architect and former Fulbright scholar to Japan with a focus on emergent materials and applications. He is a professor and interim head of the University of Minnesota School of Architecture. Brownell has authored seven books on advanced materials for architecture and design including the four-volume Transmaterial series, Matter in the Floating World, Material Strategies, and Hypernatural (co-authored with Marc Swackhamer). He has written the Mind & Matter column for Architect magazine since 2009, and his work has been published in over 70 architecture, design, science, and news journals.

* * *

Aletheia Aida is an Assistant Professor in Architecture at the University of Arizona with a research emphasis on critical material practice. She has more than twenty publications including Metropolis "Source Materials," Materials Research Society "Programmable Matter," Cambridge University Press MRS Advances, and IGI Global Press "Energetic Forms of Matter." Aletheia has presented at Yale's Center for Ecosystems and Architecture, and Princeton's Andlinger Center for Energy and the Environment. Her work is recognized by the United Nations Office for Disaster Risk Reduction, the National Science Foundation, and the 2018 AIA Arizona Research Design Award.

Getachew Assefa is one of the world's leading experts in life cycle assessment (LCA). Born in Ethiopia, he worked for many years as a senior researcher at the Royal Institute of Technology in Stockholm, Sweden. He has a BSc in chemical engineering, an MSc in environmental engineering and sustainable infrastructure, and a PhD in industrial ecology. Dr. Assefa was the Athena Chair in Life Cycle Assessment between 2010 and 2015.

Andrew Cruse is an associate professor of architecture at The Ohio State University and a registered architect. His academic work focuses on the evolution of human comfort, and its impact on architectural design and energy use. His design practice, Good Form Studio, recently completed a project for Novartis Pharma in Basel, Switzerland. Prior to joining the faculty at Ohio State, Cruse taught at Washington University in St. Louis and was an Associate at Machado and Silvetti LLC in Boston. He holds a BA in Art History from Columbia University and an M.Arch from Rice University.

Joseph Dahmen is an associate professor at the University of British Columbia School of Architecture and Landscape Architecture.

Ming Hu is an Assistant Professor at the School of Architecture, Planning and Preservation, University of Maryland, USA. She teaches technology courses which focus on the integration of architectural design with structural, materials, and building performance assessment. Her research seeks to tie a building's materials, system, and design technologies together with the goal of creating a carbon-neutral building. To make buildings greener and achieve the carbon-neutral goal, she has relied on three research approaches: zero impact building (ZIB), dynamic life cycle assessment (LCA), and smart technology and environmental health (STEH). Her first book, titled Net Zero Building: Predicated and Unintended Consequences, was released in April 2019 through Routledge. In this book, she presents a holistic overview of different net-zero building approaches and their related consequences, both positive and unintended.

Liz Kutschke is a Research Fellow at the Center for Sustainable Building Research. She received a Bachelor of Fine Arts in Interior Architecture and Design from the Milwaukee Institute of Art and Design in 2010, and a Masters of Architecture and a Master of Science in Architecture in Sustainable Design from the University of Minnesota in 2016 and 2017, respectively. Liz has a passion for research-driven architectural and urban design, and has focused on building energy use, learning from natural systems, and considerations of equity and social justice in the built environment. Liz has worked extensively on resilient design, building energy modeling and prototyping, and the Minnesota Sustainable Building Guidelines.

Jacob Mans is an architect and educator focused on understanding the feedback loops between building-scaled technical systems and large-scaled social and ecological systems. As architects and architectural researchers, we often describe these systems, and study them independently from one another. The reality is that architecture collects, channels, and distributes energy and materials across these immense, powerful and interconnected socio-technical systems. He believes architecture should ask questions and engage in critical inquiries that affect immense change rather than prioritizing research on the incremental improvement of preexisting architectural questions. Within a socio-technical research framework, a building can no longer be the sole scale of response to the question of building.

Shreyas More is the Co-founder and Co-director at the New Materials Research Centre [NMRC] at the ISDI School of Design and Innovation Mumbai. NMRC provides a platform to discuss the issues of sustainability through the lens of new materials and material systems. His work is centred at the intersection of performative materials, computation design, biological systems and digital fabrication. In 2018, his research initiative The Green Charcoal was exhibited at the Future of Architecture and Building Biennale, Mumbai to actuate conversations on the need to material innovation in the built industry to address the local and global crisis of resource availability, waste management, climate change and degrading urban environments. He has been the former interim Program Director - Interior Design and Rapid Prototyping Lab developer in the academic year 2016-17 at ISDI. He has attained Master in Advanced Architecture MAA01 from Institute for Advanced Architecture of Catalonia, Spain in 2015 with specialization in Digital Matter - Intelligent Constructions where he was investigating in the use of shape memory alloys to create passive self-shading deployable systems in architecture.

Jeana Ripple is an architect, founding principal of MIR Collective, Director of the Master of Architecture Program and Associate Professor at the University of Virginia, and a founding editor of TAD, Journal of Technology I Architecture + Design. Her research focuses on the impact of building codes on socio-economic vulnerabilities, innovative approaches to simulation-based design processes, and the invention of site-specific material systems. Ripple's work is recognized both nationally and internationally and draws on significant experience in practice and a previous career as a computer science engineer.

Meenal Sutaria is the Program Director for the Interior Design Program at the ISDI School of Design and Innovation, Parsons Mumbai. She is also the Co Founder and Co Director of the New Materials Research Center (NMRC) at ISDI. NMRC provides a platform to discuss the issues of sustainability through the lens of new material design and material systems which are practiced in the local, national and global societies, the central ethos being Circular Design and Circular Economy. It promotes Practice based research or Research by making at the intersection of design, science and Ecology. Meenal has a Bachelor in Architecture from Mumbai University and a Masters in Climate Responsive Design from University of Arizona,

Tucson. Her professional experience spans over two decades during which she has gathered strong private sector and semi government sector experience in the area of sustainable buildings in India and USA. She is the Founder and Principal Architect of Green Angle, a firm whose completed and ongoing works include 10 million sft of Residential works, 1 million sft of commercial and interior works, 40 million sft of studies on sustainability and high performance Buildings. She was also awarded as a key contributor to the 3.2 billion green footprint in India. Meenal has training and teaching experience for more than 5 years in Masters programs for Environmental Architecture and 10 years overall teaching experience as visiting faculty to various architecture colleges in Mumbai, Rajkot. She has trained overall 15 professionals to conduct charrettes for designs for energy efficiency, as a part of the Building Energy Efficiency Program funded by Swiss Development Corporation and Bureau of Energy Efficiency. She has also worked on donor funded programs towards Grid Efficiency for the energy sector as well as contributed to regulatory mechanisms.

Franca Trubiano is associate professor in Architecture at the University of Pennsylvania and a registered architect with l'Ordre des Architectes du Québec. Her research on "Fossil Fuels, the Building Industry, and Human Health" is sponsored by the Kleinman Energy Center. Her co-edited book Women Rebuild: Stories, Polemics and Futures is being published by ORO AR+D Publishing and her Design and Construction of High-Performance Homes: Building Envelopes, Renewable Energies and Integrated Practice (Routledge Press, 2012), was translated into Korean and winner of the 2015 Sejong Outstanding Scholarly Book Award. She is presently completing her manuscript Building Theories for Routledge, which challenges late 20th-century definitions and practices of architectural theory; and was co-organizer of the spring 2019 symposium "Architectural Theory Now?" held at the Weitzman School of Design. Franca was president of the Building Technology Educators Society (BTES) (2015); a founding member of the editorial board of the journal Technology, Architecture and Design (TAD); and a member of the Journal of Architectural Education (JAE) (2013-2016).

Jens von Bergmann is the founder and principal of MountainMath Software and Analytics, a Vancouver-based data analytics company.

Index

A

B

C

M

N

P

R

S

T

U

V

W

www.ingramcontent.com/pod-product-compliance
Lightning Source LLC
LaVergne TN
LVHW080602300926
849162LV00005B/1021